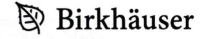

Einstein Studies

Editors: Don Howard   John Stachel

Published under the sponsorship
of the Center for Einstein Studies,
Boston University

Einstein and the Changing Worldviews of Physics

Edited by

Christoph Lehner, Jürgen Renn, and Matthias Schemmel

In cooperation with John Beckman and Eric Stengler

Managing Editor

Lindy Divarci

 Birkhäuser

Editors

Christoph Lehner
Max Planck Institute
for the History of Science
Boltzmannstraße 22
14195 Berlin
Germany
lehner@mpiwg-berlin.mpg.de

Jürgen Renn
Max Planck Institute
for the History of Science
and TOPOI Excellence Cluster
Boltzmannstraße 22
14195 Berlin
Germany
renn@mpiwg-berlin.mpg.de

Matthias Schemmel
Max Planck Institute
for the History of Science
and TOPOI Excellence Cluster
Boltzmannstraße 22
14195 Berlin
Germany
schemmel@mpiwg-berlin.mpg.de

ISBN 978-0-8176-4939-5   e-ISBN 978-0-8176-4940-1
DOI 10.1007/978-0-8176-4940-1
Springer New York Dordrecht Heidelberg London

Library of Congress Control Number: 2011943090

Mathematics Subject Classification (2010): 01-06, 83-06, 85-06

Printed on acid-free paper

Springer is part of Springer Science+Business Media
(www.birkhauser-science.com)

Dedicated to the memory of Jürgen Ehlers

Contents

Introduction

Does the universe have a beginning and does it have an end? What are the basic constituents of the universe? What determines the geometry of space and time? Classical and relativistic physics provide different answers to questions of this kind, thus constituting different scientific worldviews. How did the relativistic worldview emerge from the classical one? Which personal, cultural, and societal contexts played a role in this transition? How was this transition perceived by different communities? And what developments indicate that the worldview of physics keeps changing? These are some of the issues that are dealt with in the present volume.

The volume presents a collection of contributions to the seventh in a series of interdisciplinary conferences dedicated to the history and foundations of general relativity, which have been held since 1986 in locations alternating between the United States and Europe. One of the remarkable strengths of this series of conferences has been the dialogue it has fostered among historians, philosophers, and physicists, looking at the development and the foundations of general relativity from different perspectives. The seventh conference, jointly organized in La Orotava, Tenerife by the Fundación Canaria Orotava de Historia de la Ciencia, the Instituto Astrofísico de Canarias, and the Max Planck Institute for the History of Science, had a special character since it took place in 2005, marking both the centenary of Einstein's *annus mirabilis* and the fiftieth anniversary of his death.

The volume reviews conceptual conflicts at the foundations of physics now and in the past century. The focus is on the conditions and consequences of Einstein's pathbreaking achievements that sealed the decline of the classical notions of space, time, radiation, and matter. Particular attention is paid to the implications of conceptual conflicts for scientific views of the world at large, thus providing the basis for a comparison of the demise of the mechanical worldview around 1900 with the challenges presented by cosmology around 2000. In this regard, the present volume complements the four-volume series, *The Genesis of General Relativity* (Springer, 2007), which focuses on the emergence of Einstein's theory of gravitation from the knowledge of classical physics. As in the *Genesis* volumes, Einstein's contributions are not seen in isolation but rather are set into the wider intellectual context of dealing with the problem of gravitation in the twilight of classical physics. In the spirit of these volumes, the investigation of the historical development is pursued with a number of epistemological questions in mind, concerning in particular the transformation process of knowledge associated with the changing worldviews of physics.

At the Limits of the Classical Worldview

While general relativity constitutes a break with fundamental notions of classical physics, such as the assumption that space and time form a rigid framework serving as a stage for physical interactions, it is deeply rooted in knowledge of classical physics that has accumulated since the time of Newton. Even ideas such as the deflection of light by gravitational attraction, the relativity of inertia and its connection with gravitation, as well as the role of non-Euclidean geometry for the large-scale structure of the universe, have been discussed within the context of classical physics. The contribution by Renn and Schemmel discusses the broad array of theories of gravitation that were being proposed prior to the advent of general relativity. These theories reveal the potential of classical physics to respond to the challenges of rethinking gravitation in light of the advances by around 1900 of field theory, observational astronomy, mathematics, mechanics, and philosophy. The paper by Eisenstaedt looks further back at the way in which the relation of light and gravitation was dealt with in early Newtonian physics, anticipating insights of the later relativistic treatment. In his contribution, Gereon Wolters discusses Mach's reaction to Einstein's relativity theory, which traditionally has been understood as an outright rejection of the theory by one of its most important intellectual predecessors. Wolters thus sheds light on how the relation between heuristic principles taken from classical physics and the unexpected outcomes of their elaboration was reflected in the personal realm.

Contexts of the Relativity Revolution

Was the relativity revolution the outcome of a general cultural turn away from absolutism in societal, artistic, and scientific values to their relativization? In view of the role of the long-term development of knowledge for the emergence of both special and general relativity, such a naive view seems hardly tenable. Nevertheless, the cultural and political contexts of Einstein and his contemporaries did play a role in shaping the formulation of their scientific work, as well as its interpretation and reception. Einstein's political views, for instance, became part of his self-image as a freethinking, independent intellectual unbound by societal commitments, an image that must have fostered his intellectual independence in science, too. Schulmann's contribution traces Einstein's political views back to their roots in the period that also was formative for his science: the Swiss years prior to his *annus mirabilis*. Sánchez-Ron's analysis offers insights into the philosophical reception of relativity in Britain. It reveals a surprising agility of the proponents of different philosophical viewpoints in turning Einstein's scientific theories into support for their views.

In South America, the early reception of relativity also took place in the context of different philosophical frameworks and was triggered, in particular, by Einstein's visit to South America in 1925, as Tolmasquim points out in his reconstruction of the trip. But the relativity revolution had an impact not only on science and philosophy but also on literature and the arts, a theme that is taken up with a skeptical tone in the contribution of Hubert Goenner. As Goenner makes clear, it was ultimately Einstein's fascinating personality rather than the actual content of his scientific work that appealed to artists and writers.

The Emergence of the Relativistic Worldview

The emergence of a relativistic worldview was not a sudden event and was far from being completed when Einstein published his general theory of relativity in 1915. Important conceptual implications of the theory for the understanding of physical reality continue to be discussed even today. The contributions in this section present foundational issues, their contexts, and some of the protagonists in the profound conceptual development of general relativity since 1915. Brading and Ryckman return once more to the much-debated question of Hilbert's impact on this process, claiming that Hilbert's work on general relativity in the year 1915 was motivated by the concern to resolve the alleged tension between the requirement of general covariance and that of causality, which Einstein thought he had identified with the hole argument. Kennefick's contribution challenges the historical myth that the first observational confirmation of general relativity was due to an intellectual bias of Eddington and offers a new analysis of the question of the reliability of the results of the 1919 solar eclipse expedition. In their respective contributions, Goenner and Salisbury review the lives and works of two of the most influential scientists to shape the development of general relativity in the generation after Einstein: Peter Havas and Peter Bergmann. Their technical achievements in general relativity were closely associated with a deep philosophical, historical, and political awareness of Einstein's intellectual legacy. The paper by Schutz offers a broad survey of the conceptual revolution created by the development of general relativity, culminating in the equally surprising and profound claim that general relativity did not become a theory of physics until the 1970s, when a set of heuristic concepts emerged that was suited to communicate the results of the theory, connecting the results with insights achieved in other parts of physics, in particular astrophysics.

A New Worldview in the Making

The last section of the volume deals with a number of current issues. Recent developments confirm the impression that we are witnessing the emergence of a new worldview, triggered by new empirical findings as well as by theoretical achievements transcending the framework of general relativity. The section begins with a survey by John Beckman of the history and present state of observational tests of general relativity showing the solidity of Einstein's theory even in view of the most refined observational techniques currently available. Yet, the richness and complexity of recent observational data, as well as the expected yield from ongoing or planned observational projects and space-bound missions, indicate that the established understanding of the early universe and its development may be challenged. An example, discussed in the contribution by Battaner and Florido, is provided by primordial magnetic fields and their impact on the properties of the cosmic microwave background that may be detected by the forthcoming Planck space mission. But the established picture is also being challenged on the theoretical side. It is, for instance, remarkable to consider the extent to which even the mathematical elaboration of general relativity and the physical interpretation of its solutions, in particular of their properties such as singularities, are still under discussion, as the contribution by Senovilla reminds us. It is equally astonishing to see that, in spite of the many open issues still hampering the synthesis of general relativity and quantum theory, it is nevertheless possible to attain rather firm insights with regard to certain conditions that a future synthesis must satisfy. As in other cases in the history of physics, such progress could be achieved by exploiting an intermediate territory between the two theoretical frameworks to be integrated. The contribution by Wald offers an impressive review of the theoretical treatment of quantum fields in a curved spacetime, arguing that this treatment offers important hints for a future unification. The contribution by Dray addresses the borderline problems between relativity and quantum theory from a different perspective, focusing on the role of spinors in several approaches to the unification of quantum field theory and general relativity. In the final contribution to this volume, Ashtekar draws our attention to the fact that a unification of quantum field theory and general relativity might also force us to fundamentally revise our ideas about the beginning of the universe, conventionally understood according to the big bang model.

Taken together, the contributions to this volume make it evident that the dynamics of knowledge development driving the emergence of relativity theory, and involving both the integration of different knowledge resources as well as their conceptual transformation, is still at work in current research with the potential to again fundamentally change our physical worldview.

Max Planck Institute   *Christoph Lehner*
for the History of Science   *Jürgen Renn*
Berlin, August 2011   *Matthias Schemmel*

Part I

At the Limits of the Classical Worldview

1

Theories of Gravitation in the Twilight of Classical Physics

Jürgen Renn and Matthias Schemmel

Max Planck Institute for the History of Science, Germany

1.1 The Unfolding of Alternative Theories of Gravitation

More than is the case for any other theory of modern physics, general relativity is usually seen as the work of one man, Albert Einstein. In taking this point of view, however, one tends to overlook the fact that gravitation has been the subject of controversial discussion since the time of Newton. That Newton's theory of gravitation assumes action at a distance, i.e., action without an intervening mechanism or medium, was perceived from its earliest days as being problematical. Around the turn of the last century, in the twilight of classical physics, the problems of Newtonian gravitation theory had become more acute, also due to the rise of field theory suggesting alternative perspectives. Consequently, there was a proliferation of alternative theories of gravitation which were quickly forgotten after the triumph of general relativity. Yet in order to understand this triumph, it is necessary to compare general relativity to its contemporary competitors. General relativity owes much to this competition. The proliferation of theories of gravitation provides an exemplary case for studying the role of alternative pathways in the history of science. Thus, from this perspective, the emergence of general relativity constitutes an ideal topic for addressing longstanding questions in the philosophy of science on the basis of detailed historical evidence.

Different subdisciplines of classical physics generated different ways of approaching the problem of gravitation. The emergence of special relativity further increased the number of possible approaches and created new requirements that all approaches had to come to terms with. In this paper we will survey various alternative approaches to the problem of gravitation pursued around the turn of the last century and try to assess their potential for integrating the contemporary knowledge of gravitation.[1]

---

[1] This paper is closely based on our introduction to Renn 2007a, Vols. 3 and 4. There are numerous historical studies of the development of gravitation theories. In particular, we would like to mention (Roseveare 1982; Edward 2002; Whitrow and Morduch 1965; Lunteren 1991) and numerous contributions within the *Einstein Studies* series 1989–.

From the perspective of an epistemologically oriented history of science, the unfolding of alternative theories of gravitation in the twilight of classical physics can be interpreted as the realization of the potential embodied in the *knowledge system* of classical physics to address the problem of gravitation, this *knowledge system* eventually being transformed by the first relativistic revolution. The dynamics of this unfolding was largely governed by internal tensions of the *knowledge system* rather than by new empirical knowledge, which at best played only a minor role. A central problem of the Newtonian theory of gravitation was, as already mentioned, that it assumes the interaction between two attracting bodies to be instantaneous and that it does not provide any explanation for the instantaneous propagation of such interactions through arbitrary distances. This characteristic feature of the Newtonian gravitational force, called action at a distance, became even more dubious after the mid-nineteenth century when it was recognized that electromagnetic forces do not comply with the idea of action at a distance. This internal tension of the *knowledge system* of classical physics was intensified, but not created, by the advent of the theory of special relativity, according to which the notion of an instantaneous interaction between two bodies as it appears in Newton's force law can no longer be accepted.

The attempts to resolve these kinds of tensions typically crystallized around *mental models* representing the gravitational interaction on the basis of other familiar physical processes and phenomena. A mental model is conceived here as an internal *knowledge representation structure* serving to simulate or anticipate the behavior of objects or processes, like imagining electricity as a fluid or a stream of particles. Mental models are flexible structures of thinking that are suitable for grasping situations about which no complete information is available. They do so by relying on default assumptions that result from prior experiences and can be changed if additional knowledge becomes available without having to give up the model itself.[2]

Thus, in what may be called the *gas model*, gravitation could be conceived as resulting from pressure differences in a gaseous aether. Or, in what may be called the *umbrella model*, the attraction of two bodies could be imagined to result from the mutual shielding of the two bodies immersed in an aether whose particles rush in random directions and, in collisions with matter atoms, push them in the direction of the particles' motions. Or one could think of gravitation in analogy to the successful description of electromagnetism by the *Lorentz model*, accepting a dichotomy of gravitational field on the one hand and charged particles—masses—that act as sources of the field on the other. The elaboration of these approaches, with the help of mathematical formalism, led typically to a further proliferation of alternative approaches and, at the same time, provided the tools to explore these alternatives to a depth that made it possible to reveal new tensions. The history of these alternative approaches can thus be read, in a way similar to the dynamics inherent in Einstein's

---

[2] The notions of mental model and default assumption are taken from cognitive science; see, e.g., (Gentner and Stevens 1983, Minsky 1988). They are here combined and adapted to interpret historical developments; for more extensive discussions, see (Renn and Sauer 2007, 127; Renn and Damerow 2007).

own work,[3] as an interaction between the physical meaning embodied in various models and the mathematical formalism used to articulate them.

1.2 The Potential of Classical Physics

The history of treatments of gravitation in the nineteenth century reflects the transition from an era in which mechanics constituted the undisputed fundamental discipline of physics to an era in which mechanics became a subdiscipline alongside electrodynamics and thermodynamics.[4]

From the time of its inception, the action-at-a-distance conception of Newtonian gravitation theory was alien to the rest of mechanics, according to which interaction always involved contact. This explains the early occurrence of attempts to interpret the gravitational force by means of collisions, for instance, by invoking the umbrella model described above. During these early days the comparison of the gravitational force to electric and magnetic forces had already been suggested as well. However, the analogy with electricity and magnetism became viable only after theories on these subjects had been sufficiently elaborated. There were even attempts at thermal theories of gravitation after thermodynamics had developed into an independent subdiscipline of physics. Besides providing new foundational resources for approaching the problem of gravitation, the establishment of independent subdisciplines and the questioning of the primacy of mechanics that resulted from it affected the development of the theoretical treatment of gravitation in yet another way, namely, through the emergence of revisionist formulations of mechanics. This *heretical mechanics*, as we shall call it, consisted in attempts to revise the traditional formulation given to mechanics by Newton, Euler, and others, and often amounted to questioning its very foundations. Further stimuli for rethinking gravitation came, as we shall see in the following, from the development of astronomy and mathematics.

1.2.1 The Mechanization of Gravitation

Before the advent of the special theory of relativity, the validity of Newton's law of gravitation was essentially undisputed in mainstream physics. Alternative laws of gravitation were, of course, conceivable but Newton's law proved to be valid to a high degree of precision. While the minute discrepancies between the observed celestial motions and those predicted by Newtonian theory, most prominently the advance of Mercury's perihelion, could be resolved by one of these alternatives, they could also be resolved by adjusting lower-level hypotheses such as those regarding the distribution of matter in the solar system. In any case, the empirical knowledge at that time did not force a revision of Newtonian gravitation theory. The more pressing problem of this theory was that it did not provide a convincing model for the propagation of the gravitational force.[5]

---

[3] See (Renn 2007a, vols. 1 and 2).

[4] For a contemporary assessment, see (Zenneck 1903, translated in Renn 2007a, vol. 4).

[5] See (Zenneck 1903).

The most elaborate theories to address this problem made use of the umbrella model. These theories start from the idea of an impact of aether particles on matter, as formulated by Le Sage in the late eighteenth century (Le Sage 1784). The gravitational aether is imagined to consist of particles that move randomly in all directions. Whenever such an aether atom hits a material body it pushes the body in the direction of its movement. A single body remains at rest since the net impact of aether particles from all sides adds up to zero. However, if two bodies are present, they partly shield each other from the stream of aether particles. As a result, the impact of aether particles on their far sides outweighs that on their near sides and the two bodies are driven toward each other.

Caspar Isenkrahe, Sir William Thomson (Lord Kelvin), and others developed different theories based on this idea in the late nineteenth century (Thomson 1873; Isenkrahe 1879). But regardless of the details, this approach suffers from a fundamental problem related to the empirical knowledge about the proportionality of the force of gravity with mass. In order to take this into account one needs to allow the aether particles to penetrate a material body in such a way that they can interact equally with all of its parts. This requirement is better fulfilled the more transparent matter is to the aether particles. But, the more transparent matter is, the less shielding it provides from the aether particles on which the very mechanism for explaining gravity is based. Hence, without shielding there is no gravitational effect; without penetration there is no proportionality of the gravitational effect to the total mass. Furthermore, in theories explaining gravitation by the mechanical action of a medium, the problem of heat exchange between the medium and ordinary matter arises (in analogy to electromagnetic heat radiation), in most approaches leading to an extreme heating of matter.

From a broader perspective, such attempts at providing a mechanical explanation of gravity had lost their appeal by the end of the nineteenth century after the successful establishment of branches of physics that could not be reduced to mechanics, such as Maxwell's electrodynamics and Clausius' thermodynamics. Nevertheless, this development led indirectly to a contribution of the mechanical tradition to solving the problem of gravitation by provoking the emergence of revised formulations of mechanics, referred to here as heretical mechanics.

1.2.2 Heretical Mechanics

A critical revision of mechanics, pursued in different ways by Carl Neumann, Ludwig Lange, and Ernst Mach among others, had raised the question of the definition and origin of inertial systems and inertial forces, as well as their possible relation to the distribution of masses in the universe (Neumann 1870; Lange 1886; Mach 1883). Through the latter issue, this revision of mechanics was also important for the problem of gravitation. It also gave rise to attempts at formulating mechanics in purely relational terms, that is, exclusively in terms of the mutual distances of the particles and derivatives of these distances. Such attempts are documented, for instance, in texts by Immanuel and Benedict Friedlaender and of August Föppl.[6] As becomes clear from these texts, heretical mechanics contributed to understanding the relation between gravitational and inertial forces as both are due to the interaction of masses. According to Föppl there must be velocity-dependent forces between masses although he did not think of these forces as being gravitational. The Friedlaender brothers also conceived of inertia as resulting from an interaction between masses and did speculate on its possible relation to gravitation. In spite of such promising hints, heretical mechanics remained marginal within classical physics, in part because it lacked a framework with which one could explore the relation between gravitation and inertia. This relation was established by Einstein within the framework of field theory, first in 1907 through his principle of equivalence (Einstein 1907), and more fully with the formulation of general relativity.

Einstein's successful heuristic use of Machian ideas in his relativistic theory of gravitation encouraged the mechanical tradition to continue working toward a purely relational mechanics in the spirit of Mach. Attempts in this direction were made by Hans Reißner, Erwin Schrödinger, and, more recently, Julian Barbour and Bruno Bertotti.[7] The success of general relativity provided a touchstone for the viability of these endeavors. At the same time, the question of the extent to which the issues raised by heretical mechanics, such as a relational understanding of inertia, have been settled by general relativity is still being discussed today.[8]

1.2.3 From Peripheral Mathematics to a New Theory of Gravitation

The success or failure of a physical idea hinges to a large extent on the mathematical tools available for expressing it. In view of the crucial role of the mathematical concept of affine connection at a later state in the development of the general theory of relativity, it is interesting to consider the impact this tool might have had on the formulation of physical theories had it been part of mathematics by the latter half of the nineteenth century. That this counter-factual assumption is actually not that far-fetched can be seen from the work of Hermann Grassmann, Heinrich Hertz, Tullio Levi-Civita, and Elie Cartan.[9] Such a fictive development might have given rise to a kind of heretical gravitation theory driven by peripheral mathematics and formulated by some "Newstein" long before the advent of special relativity.[10] Perhaps the search for a different conceptualization of mechanics in which gravitation and inertia are treated alike, as is the case according to Einstein's equivalence principle, could have provided a physical motivation for such an alternative formulation of classical mechanics with the help of affine connections. Perhaps Heinrich Hertz's attempt to exclude forces from mechanics, replacing them by geometrical constraints, might

---

[6] (Friedlaender 1896; Föppl 1904); both texts are translated in (Renn 2007a, vol. 3). See further (Föppl 1905). See also (Hofmann 1904).

[7] See (Reißner 1914 and 1915; Schrödinger 1925; Barbour and Bertotti 1977, 1982).

[8] See, e.g., (Barbour and Pfister 1995).

[9] (Grassmann 1844), for a translation, see (Grassmann 1995; Levi-Civita 1916; Cartan 1986); these texts are partly reproduced in (Renn 2007a, vol. 4). See further (Hertz 1894).

[10] This idea has been developed in detail by John Stachel, see (Stachel 2007b).

have served as a starting point for such a development, triggering a geometrization of physics, had it not been so marginal to the mainstream of late nineteenth-century physics.

As with ordinary classical mechanics, Newstein's theory would have eventually conflicted with the tradition of electrodynamics and its implication of a finite propagation speed for physical interactions, which ultimately leads to the metrical structure of special relativity with its constraints on physical interactions. Then the problem that arose from this conflict could be—in contrast to the actual course of history—formulated directly in terms of the compatibility of two well-defined mathematical structures, the affine connection expressing the equality in essence of gravitation and inertia, and the metric tensor expressing the causal structure of spacetime. This formulation of the problem would have smoothed the pathway to general relativity considerably since the heretical aspect of Einstein's work—the incorporation of the equality in essence of gravitation and inertia—would have already been implemented in Newstein's predecessor theory. General relativity might thus have been the outcome of mainstream research.

1.2.4 The Potential of Astronomy

Another field of classical science that might have contributed more than it actually did to the emergence of general relativity is astronomy. This is made evident by the sporadic interventions by astronomers such as Hugo von Seeliger, who questioned the seemingly self-evident foundations of the understanding of the universe in classical science.[11] Their work was stimulated by new mathematical developments such as the emergence of non-Euclidean geometries and by heretical mechanics insofar as it raised questions relevant to astronomy, for instance, concerning the definition of inertial systems. It was further stimulated by the recognition of astronomical deviations from the predictions of Newton's law (such as the perihelion advance of Mercury), or by the paradoxes resulting from applying classical physics to the universe-at-large when this is assumed to be infinite (such as the lack of definiteness in the expression of the gravitational force, or Olbers' paradox of the failure of the night sky to be as bright as the Sun).

Although the full extent to which these problems were connected became clear only after the establishment of general relativity, the astronomer Karl Schwarzschild, who was exceptional in his interdisciplinary outlook, addressed many of them and was even able to relate them to one another.[12] He explored, for instance, the cosmological implications of non-Euclidean geometry and considered the possibility of an anisotropic large-scale structure of the universe in which inertial frames can only be defined locally. With less entrenched disciplinary boundaries of late nineteenth-century classical science, such considerations could have had wider repercussions

---

[11] See, e.g., (Seeliger 1895 and 1909).

[12] See, for instance, (Schwarzschild 1897), (translated in Renn 2007a, vol. 3) and (Schwarzschild 1900). On Schwarzschild's prerelativistic work on foundational questions and its relation to his contribution to general relativity, see (Schemmel 2005), reproduced in (Renn 2007a, vol. 3).

on the foundations of physics, perhaps giving rise to the emergence of a nonclassical cosmology.

1.2.5 A Thermodynamic Analogy

In rejecting the assumption of an instantaneous propagation of gravitational interactions, it makes sense to modify classical gravitation theory by drawing upon analogies with other physical processes that have a finite propagation speed, such as the propagation of electromagnetic effects or the transport of heat in matter. Such analogies obviously come with additional conceptual baggage. A gravitational theory built according to the model of electrodynamic field theory, for instance, was confronted with the question of whether the gravitational analogue of electromagnetic waves really exists, or the question of why there is only one kind of charge (gravitational mass) in gravitation theory as opposed to two in electromagnetism (positive and negative charge). To avoid such complications, one could also consider amending Newtonian theory by extending the classical Poisson equation for the gravitational potential into a diffusion equation by adding a term with a first-order time derivative, exploiting the analogy with heat transport in thermodynamics. In 1911, such a theory was proposed by Gustav Jaumann without, however, taking into account the spacetime framework of special relativity (Jaumann 1911 and 1912). As a consequence, it had little impact.

1.2.6 Electromagnetism as a Paradigm for Gravitation

Since early modern times magnetism has served as a model for action at a distance as it apparently occurs between the constituents of the solar system. However, as long as there was no mathematical formulation describing magnetic forces, no quantitative description of gravitation could be obtained from this analogy. After Newton had established a quantitative description of gravitation, this could now conversely be used as a model for describing magnetic and electric forces, as realized in the laws of Coulomb, Ampère, and Biot-Savart. With its further development as represented by velocity-dependent force laws and Maxwellian field theory, electromagnetic theory regained its paradigmatic potential for understanding gravitation. After the striking success of Einstein's field theory of gravitation, which describes the gravitational force in terms of the geometry of spacetime, gravitation took the lead again as attempts were made that aimed at a geometrical description of electromagnetism and the other fundamental interactions with a view toward the unification of all natural forces. The successful development of a quantum theory of the electromagnetic field made electrodynamics a model in a number of attempts at a quantization of gravitation. It seems, however, that the successful geometrical description of gravitation on one hand and the successful quantum field theoretic description of electrodynamics on the other have driven gravitation and electromagnetism conceptually further apart than ever. It is an open and controversial issue today, how elements from the two traditions have to be combined in order to achieve a quantum theory of gravitation or, even more ambitiously, a unified theory of all fundamental interactions.

The motive of unification also underlay nineteenth-century attempts to reduce gravitation to electricity, such as those of Ottaviano Fabrizio Mossotti and Karl Friedrich Zöllner, who interpreted gravity as a residual effect of electric forces (Zöllner et al. 1882). They assumed that the attractive electric force slightly outweighs the repulsive one, resulting in a universal attraction of all masses built up from charged particles. Ultimately, however, this interpretation amounts to little more than the statement that there is a close analogy between the fundamental force laws of electrostatics and Newtonian gravitation.

The paradigmatic role of electromagnetism for gravitation theory was boosted dramatically when electrodynamics emerged as the first field theory of physics. A field-theoretic reformulation of Newtonian gravity modeled on electrostatics was provided by the Poisson equation for the Newtonian gravitational potential. Even though the Poisson equation was merely a mathematical reformulation of Newton's law, it had profound implications for the physical interpretation of gravitation and introduced new possibilities for the modification of Newtonian gravitation theory. The analogy with electromagnetism raised the question of whether gravitational effects propagate with a finite speed like electromagnetic effects. A finite speed of propagation further suggested the existence of velocity-dependent forces among gravitating bodies, amounting to a gravitational analogue to magnetic forces. It also suggested the possibility of gravitational waves. In short, a field theory of gravitation opened up a whole new world of phenomena that might or might not be realized in nature.

The uncertainty of the existence of such phenomena was in any case not the most severe problem that a field theory of gravitation was confronted with. If gravitation is conceived of as a field with energy content, the fact that like "charges" always attract has a number of problematic consequences. First and foremost, ascribing energy to the gravitational field itself leads to a dilemma that does not occur in the electromagnetic case. In the latter case, the work performed by two attracting charges equal in magnitude as they approach each other can be understood to be extracted from the field, and the field energy disappears when the charges meet at one point. In contrast, while work can also be performed by two approaching gravitating masses, the field energy is enhanced, rather than diminished, as they come together at one point. (Accordingly no equivalent of a black hole is known in electrodynamics.) As Gustav Mie explains in a paper on the gravitational potential (Mie 1915, translated in Renn 2007a, vol. 4), the gravitational field is peculiar in that it becomes stronger when work is released. While a similar effect occurs with the magnetic field of two current-bearing conductors, the source of the energy is obvious in this case. The energy comes from an external energy supply such as a battery. Such an external supply is missing in the case of gravitation. A plausible escape strategy was to assume that the energy of the gravitational field is negative so that, when the field becomes stronger, positive energy is released, which can be exploited as work. For the plausible option of formulating a theory of gravitation in strict analogy to electrodynamics by simply postulating Maxwell's equations with appropriately changed signs for the gravitational field, this negative energy assumption has dramatic consequences when considering dynamic gravitational fields. A minute deviation of a gravitating system from equilibrium will cause the field to release more and more energy, while the system deviates further and further from its original state of equilibrium. In fact, due to the reversed sign, gravitational induction, if conceived in analogy to electromagnetic induction, becomes a self-accelerating process. This will be referred to here as the *negative energy problem*.

Despite this problem, Hendrik Antoon Lorentz took up the thread of Mossotti and others and proposed to treat gravitation as a residual force resulting from electromagnetism.[13] While the electromagnetic approach to gravitation offered, in principle, the possibility to account for observed deviations from Newtonian gravitation theory, the field theories actually elaborated by Lorentz and others failed to yield the correct value for the perihelion advance of Mercury, a commonly used touchstone.

All in all, the analogy of gravitation with electromagnetism, promising as it must have appeared, could not be as complete as advocated by its proponents. The considerable potential of the tradition of field theory for formulating a new theory of gravitation still needed to be explored and the key to disclosing its riches had yet to be discovered.

The attempts to subsume gravitation under the familiar framework of electromagnetism were later followed by approaches that aimed at a unification of physics on a more fundamental level, still focusing, however, on gravitation and electromagnetism. The most prominent attempts along these lines, contemporary with the genesis of general relativity, were those of Gustav Mie and David Hilbert.[14] These attempts, however, only led to a formal integration of the two forces without offering any new insights into the nature of gravity.

The key to successfully exploiting the resources of field theory for a new theory of gravitation was only found when the challenge of formulating a gravitational field theory was combined with insights from heretical mechanics. Instead of attempting a formal unification of two physical laws, Einstein combined the field theoretic approach with the idea of an equality in essence of gravitation and inertia, and eventually achieved an integration of two knowledge traditions hitherto separated due to the high degree of specialization of nineteenth-century physics.

1.3 The Challenge of Special Relativity for Gravitation

The advent of special relativity in 1905 made the need for a revision of Newtonian gravitation theory more urgent since an instantaneous propagation of gravitation was incompatible with the new spacetime framework in which no physical effect can propagate faster than the speed of light. A revision of this kind could be achieved in various ways. One could formulate an action-at-a-distance law involving a finite time of propagation as had been developed in electromagnetism, e.g., by Wilhelm Weber. Or one could formulate a genuine field theory of gravitation. The four-dimensional

---

[13] Lorentz 1900, reproduced in Renn 2007a, vol. 3. See also Gans 1905 and 1912.

[14] Mie 1912, 1914, 1915; Hilbert 1916, 1917; all these sources are translated (Mie 1912 only in part) in Renn 2007a, vol. 4.

formulation of special relativity emerging from the work of Henri Poincaré, Hermann Minkowski, and Arnold Sommerfeld brought about a set of clearly distinguished alternative approaches for realizing such a field theory of gravitation. Eventually, however, due to the implications of special relativity not only for the kinematic concepts of space and time but also for the dynamic concept of mass, gravitation was bursting out of the framework of special relativity.

1.3.1 A New Law of Gravitation Enforced by Special Relativity

The simplest way to make gravitation theory consistent with special relativity was to formulate a new direct particle interaction law of gravitation in accordance with the conditions imposed by special relativity, e.g., that the speed of propagation of the gravitational force be limited by the speed of light. This kind of approach, which was pursued by Poincaré in 1906 and by Minkowski in 1908,[15] could rely on the earlier attempts to introduce laws of gravitation with a finite speed of propagation. However, the stricter condition of Lorentz invariance now had to be satisfied.

While the formulation of a relativistic law of gravitation could solve the particular problem of consolidating gravitation theory with the new theory of special relativity, it disregarded older concerns about Newtonian gravitation, such as those relating to action at a distance. Furthermore, questions concerning the fulfillment of fundamental principles of physics, such as the equality of action and reaction, emerged in these formulations. In any case, the extent to which the modified laws of gravitation could be integrated into the larger body of physical knowledge remained unclear.

1.3.2 Toward a Field Theory of Gravitation

More important and more ambitious than the attempts at a new direct-particle interaction law of gravitation was the program of formulating a new field theory of gravitation. As pointed out above, if gravitation—in analogy to electromagnetism—is transmitted by a field with energy content, the fact that in the gravitational case like "charges" (masses) attract has problematic consequences, such as the negative energy problem. A promising approach to the negative energy problem was the assumption that masses also have energy content defined in such a way that the energy content of two attracting masses decreases when the masses approach each other. This effect can in turn be ascribed to a direct contribution of the gravitational potential to the energy content of the masses. Hence, there is a way to infer a relation between mass and energy content by considering the negative energy problem of a gravitational field theory.

The above considerations on the negative energy problem suggest that the potential plays a greater role in a gravitational field theory than it does in classical electromagnetic field theory. How to represent the gravitational potential is further directly connected with the question of how to represent the gravitational mass, or, more

---

[15] See (Poincaré 1906) and the Appendix to (Minkowski 1908); see also (Lorentz 1910). All three texts are (partly) translated in (Renn 2007a, vol. 3).

generally, the source of the gravitational field, since both are related through the field equation. The following three mathematical types of potentials were considered before the establishment of general relativity with the corresponding implications for the field strengths and the sources.

- *Scalar theories.* Potential and source are Lorentz scalars and the field strength is a (Lorentz) four-vector.
- *Vector theories.* Potential and source are four-vectors and the field is what was then called a "six-vector" (an antisymmetric second-rank tensor).
- *Tensor theories.* Potential and source are symmetric second-rank tensors and the field is represented by some combination of derivatives of the potential.

From what has been said above about a theory of gravitation construed in analogy with electrodynamics, the problems of a vector theory become apparent. In contrast to the electromagnetic case, where the charge density is one component of the four-current, the gravitational mass density is not one component of a four-vector. From this it follows in particular that no expression involving the mass is available to solve the negative energy problem by forming a scalar product of source and potential in order to adjust the energy expression.

Having thus ruled out vector theories, only scalar theories and tensor theories remain. Einstein's theories, in particular the *Entwurf* theory and his final theory of general relativity, belong to the latter class. Further alternative tensor theories of gravitation were proposed, but only after the success of general relativity, which is why they are not discussed here. As concerns scalar theories, a further branching of alternatives occurs as shall be explained in the following.

Every attempt to embed the classical theory of gravitation into the framework of special relativity had to cope not only with its kinematic implications, that is, the new spacetime structure which required physical laws be formulated in a Lorentz covariant manner, but also with its dynamical implications, in particular, the equivalence of energy and mass expressed by the formula $E = mc^2$. Since, in a gravitational field, the energy of a particle depends on the value of the gravitational potential at the position of the particle, the equivalence of energy and mass suggests that either the particle's mass or the speed of light (or both) must also be a function of the potential. Choosing the speed of light as a function of the potential immediately exits the framework of special relativity, which demands a constant speed of light. It thus may seem that choosing the inertial mass to vary with the gravitational potential is preferable since it allows one to stay within that framework.

According to contemporary evidence and later recollections,[16] Einstein in 1907 explored both possibilities, a variable speed of light and a variable mass. He quickly came to the conclusion that the attempt to treat gravitation within the framework of special relativity leads to the violation of a fundamental tenet of classical physics, which may be called *Galileo's principle*. It states that in a gravitational field all bodies fall with the same acceleration and that hence two bodies dropped from the same height with the same initial vertical velocity reach the ground simultaneously.

---

[16] For references to the historical sources, see (Renn 2007b and 2007c), and (Stachel 2007a).

The latter formulation generalizes easily to special relativity. If the inertial mass increases with the energy content of a physical system, as is implied by special relativity, a body with a horizontal component of motion will have a greater inertial mass than the same body without such a motion, and hence fall more slowly than the latter.

The same conclusion can be drawn by purely kinematic reasoning within the framework of special relativity. Consider two observers, one at rest, the other in uniform horizontal motion. When the two observers meet, they both drop identical bodies and watch them fall to the ground. From the viewpoint of the stationary observer, the body he has dropped will fall vertically, while the body the moving observer has dropped will fall along a parabolic trajectory. From the viewpoint of the moving observer, the roles of the two bodies are interchanged: the first body will fall along a parabolic trajectory while the second will fall vertically.

If one now assumes that, in the reference frame of the stationary observer, the bodies will touch the ground simultaneously, as is required by Galileo's principle in the above formulation, the same cannot hold true in the moving system due to the relativity of simultaneity. In other words, Galileo's principle cannot hold for both observers. Thus, the assumption of Galileo's principle leads to a violation of the principle of relativity. On the other hand, if one assumes, in accordance with the principle of relativity, that the two observers both measure the same time of fall for the body falling vertically in their respective frame of reference, the time needed for the body to fall along a parabolic path can be determined from this time by taking time dilation into account. It thus follows that the time needed for the fall along a parabolic path is longer than the time needed for the vertical fall, in accordance with the conclusion drawn from the dynamical assumption of a growth of inertial mass with energy content.

Each of the possibilities considered by Einstein, a dependence on the gravitational potential either of the speed of light or of the inertial mass, was later explored by Max Abraham and Gunnar Nordström, respectively. These theories represented the main competitors of Einstein's theories of gravitation.

1.3.3 The Problem of Gravitation as a Challenge for the Minkowski Formalism

The assumption of a dependence of the speed of light on the gravitational potential made it necessary to generalize the Minkowski formalism, although the full consequences of this generalization became clear only gradually. It was Max Abraham who took the first steps in this direction within this formalism by implementing Einstein's 1907 suggestion of a variable speed of light related to the gravitational potential.[17] Questioned by Einstein about the consistency of the modified formalism with Minkowski's framework, he introduced the variable line element of a nonflat four-dimensional geometry.[18]

Abraham's theory stimulated Einstein in 1912 to resume work on a theory of gravitation. Apart from developing his own theory, Abraham also made perceptive

---

[17] (Abraham 1912a, 1912b, 1912c, 1912d, 1912e); Abraham 1912a and 1912c are translated in (Renn 2007a, vol. 3).

[18] See the "Correction" to (Abraham 1912a), (*Physikalische Zeitschrift* 13: 176).

observations on alternative options for developing a relativistic theory of gravity, and on internal difficulties as well as on physical and astronomical consequences such as energy conservation in radioactive decay or the stability of the solar system.[19]

1.3.4 A Field Theory of Gravitation in the Framework of Special Relativity?

While Abraham explored the implications of a variable speed of light, Nordström pursued the alternative option of a variable mass.[20] Nordström thus remained within the kinematic framework of special relativity. As in all such approaches, however, he did so at the price of violating to some extent Galileo's principle.

More importantly, Nordström also faced the problem that in a special relativistic theory of gravitation the dynamical implications of special relativity need to be taken into account as well. These dynamical consequences suggested, for example, ascribing to energy not only an inertial but also a gravitational mass, which immediately implies that light rays are curved in a gravitational field. This conclusion, however, is incompatible with special relativistic electrodynamics in which the speed of light is constant.

Another implication of the dynamic aspects of special relativity concerns the source of the gravitational field. If any quantity other than the energy-momentum tensor of matter is chosen as a source-term in the gravitational field equation, as is the case in all scalar theories including Nordström's, gravitational mass cannot be fully equivalent to inertial mass, whose role has been taken in special relativistic physics by the energy-momentum tensor. However, while such conceptual considerations cast doubt on the viability of special relativistic theories of gravitation, they were not insurmountable hurdles for such theories. In fact, Nordström's final version of his theory remained physically viable as long as no counter-evidence was known. Einstein's successful calculation of Mercury's perihelion advance on the basis of general relativity in late 1915 undermined Nordström's theory, which did not yield the correct value.[21] This, however, did not constitute a fatal blow as long as other astrophysical explanations of Mercury's anomalous motion remained conceivable. The fatal blow only came when the bending of light in a gravitational field was observed in 1919. Nordström's theory did not predict such an effect. For the final version of his theory this can easily be seen by observing that it can be reformulated in a conformally flat space-time. Indeed, Einstein and Adriaan Fokker showed that Nordström's theory can be viewed as a special case of a metric theory of gravitation with the additional condition that the speed of light is a constant, thus excluding a dispersion of light waves that gives rise to the bending of light rays (Einstein and Fokker 1914).

Before Nordström's theory matured to its final version, which constitutes a fairly satisfactory special relativistic theory of gravitation, several steps were necessary in which the original idea was elaborated, in particular regarding the choice of an

---

[19] See, in particular, (Abraham 1913 and 1915), both translated in (Renn 2007a, vol. 3).

[20] (Nordström 1912, 1913a, 1913b); for translations of these texts, see (Renn 2007a, vol. 3).

[21] Planetary motion according to Nordström's theory was discussed in (Behacker 1913).

appropriate source expression. The most obvious choice and the first considered by Nordström is the rest mass density. The problem with this quantity, however, is that it is not a Lorentz scalar. Nordström's second choice was the Lagrangian of a particle. This, however, leads to a violation of the equality of gravitational and inertial mass. While according to special relativity, kinetic energy (e.g., the thermal motion of the particles composing a body) adds to the body's inertial mass, it is subtracted from the potential energy in the Lagrangian. If that Lagrangian hence describes the gravitational mass, the difference between the two masses increases as more kinetic energy is involved. In his final theory Nordström chose, at Einstein's suggestion, the trace of the energy-momentum tensor, the Laue scalar, thus extending the validity of the equivalence principle from mass points at rest to "complete static systems." A complete static system is a system for which there exists a reference frame in which it is in static equilibrium. In such a frame, the mechanical behavior of the system is essentially determined by a single scalar quantity. In fact, since in special relativity the inertial behavior of matter is determined by the energy-momentum tensor, the requirement of equality of inertial and gravitational mass implies that a scalar responsible for the coupling of matter to the gravitational field must be derived from the energy-momentum tensor.

The problem in choosing the Laue scalar as a source expression is how to deal with the transport of stresses in a gravitational field while maintaining energy conservation. Einstein argued that—unless appropriate provisions are taken—such stresses may be used to construct a *perpetuum mobile*, since it seems that one is able to switch gravitational mass on and off, so to speak, by creating or removing stresses. In other words, while the work required for creating a stress can simply be recovered by removing it, the gravitational mass created by the stress can meanwhile be used to perform work in the presence of a gravitational field. Given that stresses depend on the geometry of the falling object under consideration, a solution can be found by appropriately adjusting the geometry, as Nordström showed. Thus, the assumption that gravitational mass can be generated by stresses led, in conjunction with the requirement of energy-momentum conservation, to the conclusion that the geometry has to vary with the gravitational potential.

According to Einstein's assessment of Nordström's final theory in his Vienna lecture, the theory satisfies all one can require from a theory of gravitation based on contemporary knowledge, which did not yet include the observation of light deflection in a gravitational field.[22] At that time no known gravitation theory was able to explain Mercury's perihelion advance. Einstein's only remaining objection concerned the fact that what he considered to be Mach's principle—the assumption that inertia is caused by the interaction of masses—appears not to be satisfied in Nordström's theory.

But as we have seen, because of the role of stresses for gravitational mass, Nordström had to assume that the behavior of rods and clocks also depends on the gravitational potential. Indeed, as becomes clear with the hindsight provided by general relativity, it is arguable whether his theory really fits the special relativistic

---

[22] See also the discussion in (Laue 1917 and Giulini 2007).

framework, corresponding as it does to a spacetime theory that is only conformally flat, i.e., based on a metric that is flat up to a scalar factor. The relation in which Nordström's theory stands to general relativity in that it attributes transformations to material bodies, which in the later theory are understood as transformations of spacetime, is reminiscent of the way that Lorentz's theory of the aether stands to special relativity.

Appendix: Is Special Relativistic Gravitation a Theoretically Viable Option?

The following is part of a correspondence between John Norton and Domenico Giulini with whom we discussed the status of Nordström's final theory and the question of the possibility of consistent special relativistic theories of gravitation. The two continued the discussion in an e-mail exchange which we think, in its dialogical form, clarifies some of the points raised in the last subsection of our paper.[23]

John Norton You are worried that Einstein asserts a violation of conservation of energy in a theory that demonstrably conforms to energy conservation [Nordström's final theory]. But such a theory can be completely messed up if you "add" an extra assumption incompatible with the theory. Let us say, for example, that the theory requires bodies in mechanical equilibrium to change their length with gravitational potential (just as Lorentz' theory requires bodies to change their length with motion, as a mechanical effect). If you now add the assumption that such a body does not change its length as the gravitational potential changes, then you have an inconsistent set of assumptions. That inconsistency could be manifested in many different ways, including a violation of energy conservation.

Domenico Giulini Sure, if you add an assumption about the dynamical behavior of clocks and rods that simply is inconsistent with their dynamical laws, that's the end of our discussion. But do you have indications that Einstein actually did this—tacitly perhaps? What I know from his writings is that he always urged us to regard clocks and rods as "solutions to differential equations," which I read as "obeying consistent dynamical laws."

Clearly, the universal scaling behavior of atomic scales in the scalar theory of gravity may suggest to use the conformally rescaled metric as fundamental field of gravity, thereby eliminating the Minkowski metric from the description altogether, similarly to the procedure in the "flat approach to general relativity," where one starts with a mass $= 0$, spin $= 2$ field in Minkowski space. This "curved" interpretation of the scalar theory is possible. So what? It remains true that this theory has a formally consistent interpretation in Minkowski space. The curved interpretation is certainly not necessary in order to avoid formal inconsistencies, though some may find it more "natural."

---

[23] For a more in-depth analysis, see (Giulini 2008), in which the discussion with John Norton is also acknowledged.

John Norton Concerning Einstein urging us to regard clocks and rods as "solutions to differential equations," which you read as "obeying consistent dynamical laws": He says that later on, but doesn't do it. You've read Einstein's statements of his thought experiments pertaining to Nordstroem's theory, so you know as much about them as I do. My reading is that he assumes that a rod moved about in a gravitational field has its length fixed by the Minkowski metric, not by what mechanical equilibrium of dynamical systems yields.

Two answers to your "so what": First, if the length of real rods (and times of real clocks) responds to the conformal metric and not the Minkowski metric, then the Minkowski metric has become a kind of unknowable ether. The standard move in the "flat approach to general relativity" is to abandon the flat background, which is exactly what I take Einstein to be doing here.

Second, what if Einstein is establishing that real dynamical rods must vary in length according to the gravitational field (my original suggestion)? For purposes of a *reductio* argument he assumes they don't and ends up with a contradiction. That there is a consistent Minkowski metric theory is compatible with this contradiction, for the contradiction is just telling us that the consistent Minkowski metric theory must harbor rods that change in length with the field—which is Einstein's conclusion.

Domenico Giulini O.k., that I understand and I think there is little disagreement left. Would you then agree that (except for its experimental impossibility) the *only* flaw of scalar gravity is not an internal inconsistency, but a redundancy of its primitive elements which is precisely given by the conformal representative of Minkowski metric, which is not the one rods and clocks respond to? (I am not certain how general one may take the meaning of "clocks" and "rods" beyond that of "electromagnetically bound systems.")

John Norton Yes, I do think we are agreeing on this. The upshot of Einstein's thought experiment was that consistency required rods to respond, in effect, to the conformal metric and not the Minkowski metric, but nothing in the thought experiment spoke against the consistency of the resulting theory.

Perhaps the only difference left is one of emphasis. That the Minkowski metric has become inaccessible is generally taken as strong grounds for discarding it. It is the final step from flat spacetime to curved spacetime in the spin-2 field pathway to general relativity. The analogous step is taken when the Newtonian space and time background of Lorentz's consistent electrodynamics is discarded in favor of the Minkowski metric to which rods and clocks respond.

Acknowledgements

For their careful reading of earlier versions of this text, extensive discussions, and helpful commentaries, we would like to thank Michel Janssen, Domenico Giulini, Christopher Smeenk, and John Stachel.

References

Abraham, Max. 1912a. "Zur Theorie der Gravitation." *Physikalische Zeitschrift* 13: 1–4. (English translation in Renn 2007a, vol. 3).

———. 1912b. "Das Elementargesetz der Gravitation." *Physikalische Zeitschrift* 13: 4–5.

———. 1912c. "Der freie Fall." *Physikalische Zeitschrift* 13: 310–311. (English translation in Renn 2007a, vol. 3).

———. 1912d. "Das Gravitationsfeld." *Physikalische Zeitschrift* 13: 793–797.

———. 1912e. "Die Erhaltung der Energie und der Materie im Schwerkraftfelde." *Physikalische Zeitschrift* 13: 311–314.

———. 1913. "Eine neue Gravitationstheorie." *Archiv der Mathematik und Physik* 20: 193–209. (English translation in Renn 2007a, vol. 3).

———. 1915. "Neuere Gravitationstheorien." *Jahrbuch der Radioaktivität und Elektronik* (11) 4: 470–520. (English translation in Renn 2007a, vol. 3).

Barbour, Julian and Bruno Bertotti. 1977. "Gravity and Inertia in a Machian Framework." *Nuovo Cimento* 38B: 1–27.

———. 1982. "Mach's Principle and the Structure of Dynamical Theories." *Proceedings of the Royal Society London* 382: 295–306.

Barbour, Julian and Herbert Pfister (eds.). 1995. *Mach's Principle: From Newton's Bucket to Quantum Gravity. Einstein Studies*, vol. 6.

Behacker, Max. 1913. "Der freie Fall und die Planetenbewegung in Nordströms Gravitationstheorie." *Physikalische Zeitschrift* (14) 20: 989–992.

Cartan, Elie. 1986. *On Manifolds with an Affine Connection and the Theory of General Relativity.* Naples: Bibliopolis. (Excerpts reprinted in Renn 2007a, vol. 4).

Edward, Matthew R. (ed.). 2002. *Pushing Gravity: New Perspectives on Le Sage's Theory of Gravitation.* Montreal: Apeiron.

Einstein, Albert. 1907. "Über das Relativitätsprinzip und die aus demselben gezogenen Folgerungen." *Jahrbuch der Radioaktivität und Elektronik* 4: 411–462.

Einstein, Albert and Adriaan D. Fokker. 1914. "Die Nordströmsche Gravitationstheorie vom Standpunkt des absoluten Differentialkalküls." *Annalen der Physik* 44: 321–328.

Einstein Studies series, edited by Don Howard and John Stachel (Boston: Birkhäuser, 1989–).

Föppl, August. 1904. "Über absolute und relative Bewegung." *Königlich Bayerische Akademie der Wissenschaften, München, mathematisch-physikalische Klasse, Sitzungsberichte* 34: 383–395. (English translation in Renn 2007a, vol. 3).

———. 1905. "Über einen Kreiselversuch zur Messung der Umdrehungsgeschwindigkeit der Erde." *Königlich Bayerische Akademie der Wissenschaften, München, mathematisch-physikalische Klasse, Sitzungsberichte* 34: 5–28.

Friedlaender, Benedict and Immanuel. 1896. *Absolute oder relative Bewegung?* Berlin: Leonard Simion. (English translation in Renn 2007a, vol. 3).

Gans, Richard. 1905. "Gravitation und Elektromagnetismus." *Physikalische Zeitschrift* 6: 803–805.

——. 1912. "Ist die Gravitation elektromagnetischen Ursprungs?" In *Festschrift Heinrich Weber zu seinem siebzigsten Geburtstag am 5. März gewidmet von Freunden und Schülern.* Berlin: Teubner, 75–94.

Gentner, Dedre and Albert L. Stevens. 1983. *Mental Models.* Hillsdale: Erlbaum.

Giulini, Domenico. 2007. "Attempts to Define General Covariance and/or Background Independence." In I.O. Stamatescu and E. Seiler (eds.), *Approaches to Fundamental Physics.* Berlin: Springer.

——. 2008. "What is (not) wrong with scalar gravity." *Studies in the History and Philosphy of Modern Physics* 39: 154–180.

Grassmann, Hermann. 1844. *Die lineale Ausdehnungslehre, ein neuer Zweig der Mathematik.* Leipzig: Otto Wigand.

——. 1995. *A New Branch of Mathematics: The Ausdehnungslehre of 1844 and Other Works.* Translated by Lloyd C. Kannenberg. Chicago: Open Court. (Excerpts reprinted in Renn 2007a, vol. 4).

Hertz, Heinrich. 1894. *Die Prinzipien der Mechanik in neuem Zusammenhange dargestellt.* Leipzig: Barth.

Hilbert, David. 1916. "Die Grundlagen der Physik. (Erste Mitteilung.)" *Nachrichten von der Königlichen Gesellschaft der Wissenschaften zu Göttingen, Mathematisch-physikalische Klasse* 1915, 395–407. (English translation in Renn 2007a, vol. 4).

——. 1917. "Die Grundlagen der Physik. (Zweite Mitteilung.)" *Nachrichten von der Königlichen Gesellschaft der Wissenschaften zu Göttingen, Mathematisch-physikalische Klasse* 1916, 53–76. (English translation in Renn 2007a, vol. 4).

Hofmann, Wenzel. 1904. *Kritische Beleuchtung der beiden Grundbegriffe der Mechanik: Bewegung und Trägheit und daraus gezogene Folgerungen betreffs der Achsendrehung der Erde und des Foucault'schen Pendelversuchs.* Leipzig: M. Kuppitsch Witwe.

Isenkrahe, Caspar. 1879. *Das Räthsel von der Schwerkraft: Kritik der bisherigen Lösungen des Gravitationsproblems und Versuch einer neuen auf rein mechanischer Grundlage.* Braunschweig: Vieweg.

Jaumann, G. 1911. "Geschlossenes System physikalischer und chemischer Differentialgesetze." *Sitzungsberichte der math.-nw. Klasse der Kaiserl. Akademie der Wissenschaften,* Wien 120: 385–530.

——. 1912. "Theorie der Gravitation." *Sitzungsberichte der math.-nw. Klasse der Kaiserl. Akademie der Wissenschaften,* Wien 121: 95–182.

Lange, Ludwig. 1886. *Die geschichtliche Entwicklung des Bewegungsbegriffes und ihr voraussichtliches Endergebnis: ein Beitrag zur historischen Kritik der mechanischen Principien.* Leipzig: Engelmann.

Laue, Max von. 1917. "Die Nordströmsche Gravitationstheorie." *Jahrbuch der Radioaktivität und Elektronik* 14: 263–313.

Le Sage, Georges-Louis. 1784. "Lucrèce Newtonien." *Mémoires de l'Académie royale des Sciences et Belles Lettres de Berlin,* pour 1782.

Levi-Civita, Tullio. 1916. "Nozione di parallelismo in una variet qualunque e conseguente specificazione geometrica della curvatura riemanniana." *Circolo*

Matematico di Palermo. Rendiconti 42: 173–204. (English translation of excerpts in Renn 2007a, vol. 4).

Lorentz, Hendrik A. 1900. "Considerations on Gravitation." *Proceedings Royal Academy Amsterdam* 2: 559–574. (English translation in Renn 2007a, vol. 3).

——. 1910. "Alte und neue Fragen der Physik." *Physikalische Zeitschrift* 11: 1234–1257. (English translation in Renn 2007a, vol. 3).

Lunteren, Frans Herbert van. 1991. *Framing Hypotheses: Conceptions of Gravity in the 18th and 19th centuries.* Dissertation, Utrecht: Rijks University.

Mach, Ernst. 1883. *Die Mechanik in ihrer Entwickelung: historisch-kritisch dargestellt.* Leipzig: Brockhaus.

Mie, Gustav. 1912. "Grundlagen einer Theorie der Materie I, II und III." *Annalen der Physik* 37: 511–534; 39 (1912): 1–40; 40 (1913): 1–66. (English translation of excerpts in Renn 2007a, vol. 4).

——. 1914. "Bemerkungen zu der Einsteinschen Gravitationstheorie." *Physikalische Zeitschrift* 15: 115–122. (English translation in Renn 2007a, vol. 4).

——. 1915. "Das Prinzip von der Relativität des Gravitationspotentials." In *Arbeiten aus den Gebieten der Physik, Mathematik, Chemie: Festschrift Julius Elster und Hans Geitel zum sechzigsten Geburtstag.* Braunschweig: Vieweg, 251–268. (English translation in Renn 2007a, vol. 4).

Minkowski, Hermann. 1908. "Die Grundgleichungen für elektromagnetische Vorgänge in bewegten Körpern." *Nachrichten der Königlichen Gesellschaft der Wissenschaften zu Göttingen, Mathematisch-Physikalische Klasse,* 53–111. (English translation of excerpts in Renn 2007a, vol. 3).

Minsky, Marvin. 1988. *The Society of Mind.* New York: Simon and Schuster.

Neumann, Carl. 1870. *Ueber die Principien der Galilei-Newton'schen Theorie.* Leipzig: Teubner.

Nordström, Gunnar. 1912. "Relativitätsprinzip und Gravitation." *Physikalische Zeitschrift* 13: 1126–1129. (English translation in Renn 2007a, vol. 3).

——. 1913a. "Träge und schwere Masse in der Relativitätsmechanik." *Annalen der Physik* 40: 856–878. (English translation in Renn 2007a, vol. 3).

——. 1913b. "Zur Theorie der Gravitation vom Standpunkt des Relativitätsprinzips." *Annalen der Physik* 42: 533–554. (English translation in Renn 2007a, vol. 3).

Poincaré, Henri. 1906. "Sur la dynamique de l'électron." *Rendiconti del Circolo Matematico di Palermo* 21: 129–175. (English translation in Renn 2007a, vol. 3).

Reißner, Hans. 1914. "Über die Relativität der Beschleunigungen in der Mechanik." *Physikalische Zeitschrift* 15: 371–375.

——. 1915. "Über eine Möglichkeit die Gravitation als unmittelbare Folge der Relativität der Trägheit abzuleiten." *Physikalische Zeitschrift* 16: 179–185.

Renn, Jürgen (ed.). 2007a. *The Genesis of General Relativity* (4 vols.). *Boston Studies in the Philosophy of Science,* vol. 250. Dordrecht: Springer.

vol. 1: *Einstein's Zurich Notebook: Introduction and Source.* Michel Janssen, John Norton, Jürgen Renn, Tilman Sauer, and John Stachel.

vol. 2: *Einstein's Zurich Notebook: Commentary and Essays*. Michel Janssen, John Norton, Jürgen Renn, Tilman Sauer, and John Stachel.

vol. 3: *Gravitation in the Twilight of Classical Physics: The Promise of Mathematics*. Jürgen Renn and Matthias Schemmel, eds.

vol. 4: *Gravitation in the Twilight of Classical Physics: Between Mechanics, Field Theory, and Astronomy*. Jürgen Renn and Matthias Schemmel, eds.

Renn, Jürgen. 2007b. "Classical Physics in Disarray: The Emergence of the Riddle of Gravitation." In Renn 2007a, vol. 1.

———. 2007c. "The Third Way to General Relativity." In Renn 2007a, vol. 3.

Renn, Jürgen and Peter Damerow. 2007. "Mentale Modelle als kognitive Instrumente der Transformation von technischem Wissen." In H. Böhme, C. Rapp and W. Rösler (eds.) *Übersetzungen und Transformationen*. Berlin: De Gruyter.

Renn, Jürgen and Tilman Sauer. 2007. "Pathways out of Classical Physics: Einstein's Double Strategy in Searching for the Gravitational Field Equation." In Renn 2007a, vol. 1.

Roseveare, N. T. 1982. *Mercury's Perihelion from Le Verrier to Einstein*. Oxford: Clarendon Press.

Schemmel, Matthias. 2005. "An Astronomical Road to General Relativity: The Continuity between Classical and Relativistic Cosmology in the Work of Karl Schwarzschild." *Science in Context* 18(3): 451–478.

Schrödinger, Erwin. 1925. "Die Erfüllbarkeit der Relativitätsforderung in der klassischen Mechanik." *Annalen der Physik* 77: 325–336.

Schwarzschild, Karl. 1897. "Was in der Welt ruht." *Die Zeit*, Vienna Vol. 11, No. 142, 19 June 1897, 181–183. (English translation in Renn 2007a, vol. 3).

———. 1900. "Über das zulässige Krümmungsmaass des Raumes." *Vierteljahresschrift der Astronomischen Gesellschaft* 35: 337–347.

Seeliger, Hugo von. 1895. "Über das Newton'sche Gravitationsgesetz." *Astronomische Nachrichten* 137: 129–136.

———. 1909. "Über die Anwendung der Naturgesetze auf das Universum." *Königlich Bayerische Akademie der Wissenschaften, München, mathematisch-physikalische Klasse, Sitzungsberichte*, 3–25.

Stachel, John. 2007a. "The First Two Acts." In Renn 2007a, vol. 1.

———. 2007b. "The Story of Newstein or: Is Gravity just another Pretty Force?" In Renn 2007a, vol. 4.

Thomson, William (Lord Kelvin). 1873. "On the ultramundane corpuscles of Le Sage." *Phil. Mag.* 4th ser. 45: 321–332.

Whitrow, G. J. and G. E. Morduch. 1965. "Relativistic Theories of Gravitation: A comparative analysis with particular reference to astronomical tests." *Vistas in Astronomy* 1: 1–67.

Zenneck, Jonathan. 1903. "Gravitation." In Arnold Sommerfeld (ed.) *Encyklopädie der mathematischen Wissenschaften*, vol. 5 (Physics). Leipzig: Teubner, 25–67. (English translation in Renn 2007a, vol. 3).

Zöllner, Friedrich, Wilhelm Weber, and Ottaviano F. Mossotti. 1882. *Erklärung der universellen Gravitation aus den statischen Wirkungen der Elektricität und die allgemeine Bedeutung des Weber'schen Gesetzes*. Leipzig: Staackmann.

2

The Newtonian Theory of Light Propagation

Jean Eisenstaedt

SYRTE, Observatoire de Paris, CNRS, UPMC, France

2.1 From Newton to Einstein

At the end of the 18th century, a Newtonian theory of the propagation of light was developed as an application of Newton's *Principia* but quickly forgotten. A series of works completed the *Principia* with the formulation of a Galilean relativistic optics of moving bodies, a gravitational physics of light, and the discovery of the analog of the Doppler–Fizeau effect, as well as many other effects and ideas that are a fascinating preamble to Einstein's special and general relativity.

It is generally thought that light propagation cannot be treated in the framework of Newtonian dynamics. However, at the end of the 18th century and in the context of Newton's *Principia*, several papers, published and unpublished, offered a new and important corpus that represents a detailed application of Newton's dynamics to light. In it, light was treated in precisely the same way as material particles. This most interesting application—foreshadowed by Newton himself in the *Principia*—constitutes a relativistic optics of moving bodies, of course based on what we nowadays refer to as Galilean relativity, and offers a most instructive Newtonian analogy to Einsteinian special and general relativity (Eisenstaedt, 2005a; 2005b). These several papers, effects, experiments, and interpretations constitute the Newtonian theory of light propagation. I will argue in this paper, however, that this Newtonian theory of light propagation has deep parallels with some elements of 19th century physics (aberration, the Doppler effect) as well as with an important part of 20th century relativity (the optics of moving bodies, the Michelson experiment, the deflection of light in a gravitational field, black holes, the gravitational Doppler effect).

Surely, the Newtonian theory of light is not at all part of the context of discovery of relativity: it is not the road that Einstein or any of his predecessors took. Moreover, due to the incommensurable distance between Newtonian and Einsteinian concepts, the link between the two is not conceptual. As Thomas Kuhn clearly put it, "The transition from Newtonian to Einsteinian mechanics illustrates with particular clarity the scientific revolution as a displacement of the conceptual network through which scientists view the world" (Kuhn [1962] 1996, 102–103).

This paper will first of all show that the Newtonian scheme was able to—and did—treat light propagation in a consistent and powerful way. In fact, there are two different fields wherein Newtonian dynamics was used to understand light.

First, as I have shown elsewhere in detail (Eisenstaedt, 2005a), Newton's theory of refraction predicts a physical effect that has much to do with what is called a Doppler–Fizeau effect. More precisely, more than sixty years before Fizeau it was understood that a measure of refraction is a measure of the velocity of the incoming light—and thus of the relative velocity between the emitting star and the observer.

Second, the 18th century analysts obtained different effects actually later rediscovered by Einsteinian relativity, although there are many similarities and analogies between these effects. For example, Newtonian light deflection is qualitatively the same as Einstein's, but quantitatively it has half the Einsteinian value. As well the gravitational Doppler effect (also called the gravitational displacement of line rays) discovered by Michell in 1784 is quantitatively the same as Einstein's. Although the differences are more important in the case of "dark bodies" and "black holes," nevertheless, the physical effect is the same: gravitation acts on light and implies "dark bodies."

Such a comparison between two theoretical structures may have a historical and pedagogical interest in that it helps reveal the underlying physical meaning of the interaction between matter and light. It also makes visible the analogical relationships between the problems and solutions in both conceptual frameworks. It reveals what we can call "a physical context."

In the following, we will first recall Newton's treatment of light as a corpuscle in the *Principia* and then move on to the little-known but fundamental contributions of Michell (Michell, 1784) and Blair (Blair, 1786). In conclusion we will come back to the question of the relationship between these works which provide a complete Newtonian theory of light propagation and see its structural analogy to Einstein's relativistic treatment of light. Let us now look at the reasons why only a monochromatic theory was possible in this context.

2.1.1 Newton's Corpuscular Theory of Light

Newton's *Principia* established a dynamics that determines how a particle behaves when subject to a force, predicting its trajectory. We are here interested in two forces—actually two fields—that engaged our 18th century optical theorists: the long-range force of gravitation and the short-range refracting force[1] that was posited by Newton's corpuscular theory of light.

Let us first look at the corpuscular theory of light, a very brief discussion of which appears in the first book, section XIV, of the *Principia*, wherein Newton accounts for the Snell–Descartes law of refraction (Newton, [1687] 1999, pp. 622–629). Newton's

---

[1] The "refringent force"—as it will be called later on—because it is the force responsible for refraction.

calculations deal with "The motion of minimally small bodies [...] tending toward each of the individual parts of some great body" (Newton, [1687] 1999, p. 622). The dynamical model he has in mind applies to a (material) "body" but it is also valid for rays [corpuscles] of light, as Newton claims in the first line of his demonstration (and justifies later on):

> Therefore because of the analogy that exists between the propagation of rays of light and the motion of bodies, [...] meanwhile not arguing at all about the nature of the rays (that is, whether they are bodies or not), but only determining the trajectories of bodies, which are very similar to the trajectories of rays. (Newton, [1687] 1999, p. 626)

The model simply consists of a corpuscle of light incident on a transparent body. The constant force of refraction acts at the surface—in the "atmosphere" as Clairaut (Clairaut, 1741) would later name it—of the body. Between two planes, "in all its passage through the intermediate space let [the body] be attracted or impelled toward the medium of incidence, and by this action let it describe the curved line" (Newton, [1687] 1999, p. 622).

Newton not only obtained Descartes' sine law, he also gave a physical reason for refraction, which neither Descartes nor Snell had. Newton went on to deal with a light corpuscle that passes successively through several spaces—with different indices of refraction—bordered by parallel planes, a model that he will also use for astronomical refraction.[2] Newton's corpuscular theory of light is in fact little more than a ballistic theory of light.[3]

At first there was not much interest in Newton's corpuscular theory of light. In 1741 Clairaut was the first one to take it seriously, complaining that Newton did not carry it further. Clairaut himself developed algebraic calculations that directly produced the Descartes–Snell law of refraction. He also derived a relation that is implicit in Newton's demonstration, according to which the difference between the square of the refracted and incident velocities is a constant equal to the square of the refringent force.[4] The incident velocity determines the refraction: the greater the incident velocity, the smaller the angle of refraction, an extremely important point that John Michell uses later on. Thus, the corpuscular theory tells us how light behaves at short range while pouring through glass.

2.1.2 The Velocity of Light as the Parameter of Color

Newton himself tried to explain chromatic dispersion on the basis of his particle theory, according to which white light consisted of a stream of corpuscles of different kinds, with each kind corresponding to a specific spectral color. He hoped that

---

[2] Concerning astronomical refraction, see (Whiteside, 1980) and (Eisenstaedt, 1996, p. 127: note 39).

[3] Concerning the ballistic theory of light, see (Bechler, 1973) and (Eisenstaedt, 1996).

[4] See (Eisenstaedt, 2007, p. 743) in which all the equations for light propagation in Newton's theory are derived; see also (Eisenstaedt, 2005a, p. 357).

it would be possible to make a connection between color and velocity: each corpuscle of some specific color being endowed with a particular velocity. Velocity would then be the parameter associated with color. As red is the least refracted color of the spectrum, a red corpuscle would have the greatest velocity; a blue corpuscle, more refracted by a prism, should be slower. Consequently, the extreme red corpuscles were supposed to travel faster in air than the extreme violet ones. This is a consequence of the fact that Newton's is a ballistic theory of light. The analogy with a gun is enlightening: the faster a bullet at the exit of the gun, the smaller its angular deviation due to the gravitational field of the Earth, and conversely.

In the 1690s, Newton thought that a possible model for chromatic dispersion and a possible explanation of the spectrum of light was at hand.[5] But color being related to velocity—and vice versa—many effects are expected and could be tested. Thus, Newton's hypothesis of the different velocities of the components of light implied that a moon of Jupiter would change color as it moves behind the planet just before or after an eclipse; over a short period of time, the colors of the spectrum would disappear in turn behind the planet beginning with the fastest (red) rays.

In August 1691 and again in February 1692, Newton wrote to John Flamsteed, Royal Astronomer at Greenwich Observatory and his longtime correspondent, to ask if he had observed any change of color in Jupiter's satellites before they disappeared. Flamsteed replied that he "never saw any change to a bluish color or red but duskish" (Turnbull et al. 1959–1977, vol. 3: 202). As Alan Shapiro put it: "Newton justly respected Flamsteed's skill as an observer and took this as a definite judgment, for he never again adopted the assumption of different velocities for different colors" (Shapiro 1993, 218; Eisenstaedt 1996).

Newton did not publish anything regarding his failed velocity model for chromatic dispersion, nor did he discuss the reason for its failure. Only in the middle of the 18th century did the question of the color of Jupiter's satellites arise once again, and it remains possible that its reemergence derived from Newton via David Gregory or Jean-Dominique Cassini. Several natural philosophers then reconsidered the question of the color of Jupiter's satellites in the same context: Jean-Jacques d'Ortous de Mairan, Thomas Melvill, the Marquis de Courtivron, and Alexis-Claude Clairaut (Eisenstaedt, 1996, Ch. 5; Shapiro, 1993, p. 218). James Short, a well-known optician in London, performed the observations but he did "not perceive the least alteration in the color of the light reflected by the satellite, except in quantity" (Short, 1754, p. 268). Thus, Newton's approach to a theory of chromatic dispersion was refuted once again. His corpuscular theory of light was not, however, altogether destroyed: it was instead limited to issues that did not involve color and was reworked into a system that became known as "the emission theory," which dominated optics in the 18th century.[6]

---

[5] For an analysis of Newton's interest on this subject see (Bechler, 1973, pp. 14–23).

[6] Concerning optics as a branch of dynamics and its institutionalization see (Cantor, 1983, ch. 2).

2.1.3 Michell on the Velocity of Light

A new era—largely neglected by historians[7]—started with Michell's analysis of light propagation in the *Principia*. John Michell was a convinced Newtonian and a most inventive astronomer. In 1767, assuming that the stars are randomly distributed, he developed a probabilistic argument showing that nearby stars could be physically connected; and as a consequence he predicted the existence of "double stars" some twenty years before William Herschel actually observed them.

Then, in a paper published in 1784, Michell tried to measure the distances of stars. In order to determine the light trajectories, he used the dynamics of the *Principia*. Following Newton's own procedure, he made no distinction between a material particle and a light corpuscle. For refraction at short range, he used the corpuscular theory of light; at long range, he was simply to "suppose the particle of light to be attracted in the same manner as all other bodies with which we are acquainted; [...] gravitation being, as far as we know, or have any reason to believe, an universal law of nature" (Michell, 1784, p. 37).

As a consequence, the velocity of a light corpuscle emitted by a star would be diminished by the effect of gravitation: light was "retarded." In his impressive paper, Michell developed the concept—as well as the theory—of "dark bodies" (Michell, 1784, p. 42) as Laplace (Laplace, 1796, vol. 2, pp. 304–306) later named those strange (and of course then unobserved) stars whose light would return to them after traveling for some distance.[8]

When incident on a transparent body, a light corpuscle was subject to the short-range forces of the emission theory, with the consequence that the greater the incident velocity, the smaller the angle of refraction. By measuring the angle of refraction, one could in principle measure the change in the velocity of the corpuscle. Michell understood that a prism would be "very convenient" for this purpose. Indeed, it was clear to him that a prism was a good tool with which to measure the velocity of light. He discussed the point with Henry Cavendish, who showed great interest in the method, which he called "a very good one" in a letter to Michell.[9]

In 1848 (Fizeau, 1870a)[10] Fizeau understood that measure of a frequency shift would permit calculation of the relative velocity of the emitting object. A frequency shift can be interpreted as a refraction: what is seen as a frequency shift in the context of the wave theory can be interpreted as a refraction in the context of Newton's emission theory. Thus, the same effect that is explained by the frequency in the wave theory is explained by the angle of refraction in Newton's theory. So Doppler's effect—when limited to a monochromatic view—is precisely the same as Michell's (which as we will see was elaborated by Blair two years later); only the parametrization differs. In 1870, Fizeau described this double effect (frequency

---

[7] Except by Russell McCormmach who wrote quite an interesting paper on John Michell (McCormmach, 1968); see also (Jungnickel and McCormmach, 1999), (Eisenstaedt, 1991; 2005a; and 2005b, ch. 8 and 9), (Gibbons, 1979), and (Schaffer, 1979).

[8] Concerning dark bodies see (Eisenstaedt, 1991).

[9] (Cavendish to Michell, 27 May 1783) in (Jungnickel and McCormmach, 1999, p. 567).

[10] The article was read on December 23, 1848, but not published before 1870.

shift/refraction) as follows: "This modification of the wave length will imply a greater deviation produced by refraction through a prism" (Fizeau, 1870b, p. 1063). Most often this effect, the measure of a velocity through frequency shift, is—erroneously—named after Doppler (Doppler, 1842). Let us call it the "Doppler–Fizeau effect."

Though Michell and Blair had certainly calculated the effect, they did not publish the details. From the Newton–Clairaut results (the index of refraction of the prism being known), it is simple to derive the variation of the angle of refraction as a function of the variation of the velocity of the incident light. It correctly shows that, when the velocity of light increases, the refraction angle decreases.[11]

In his 1784 paper, Michell proposed a sophisticated experiment (Eisenstaedt, 1991; 2005b, ch. 8) to determine the distances of stars. He thought that it would be possible to observe a (still hypothetical) double star in the Pleiades, and hoped that his method would allow the measurement of the velocity difference between light rays coming from both components of the system. Light emitted by the central, heavier component of the star system would be much slower than that emitted by the lighter star component. Using a prism, it seemed possible to measure the difference of refraction angle between the two rays and thus the difference between the two velocities. This would provide additional data to determine the distance of the star system (Eisenstaedt, 1991, pp. 343–350). Michell's experiment was eventually performed by William Herschel and Nevil Maskelyne but—not surprisingly—they could not detect any differential refraction.[12]

2.1.4 Blair on the Velocity of Light

On November 27, 1783, Michell's article was read at the Royal Society of London; but for some months it had been discussed in a circle that included Henry Cavendish, William Herschel, Nevil Maskelyne, and Joseph Priestley, and it must have circulated among a larger group. Thus, "in June or July 1783," Robert Blair, the first professor of practical astronomy at the University of Edinburgh—not a very well-known astronomer indeed and mainly interested in constructing achromatic prisms—heard of it "accidentally" (Blair, 1786, p. 11). Actually, in the following letter, Cavendish

---

[11] Such a demonstration was never published in Michell's article, in Blair's manuscript, or in Arago's paper. But, as is often the case, it is implicit in their exposition. For the derivation of these equations, see (Eisenstaedt, 2007). Calculations are made explicit in a manuscript left by Arago (Arago 1806); a detailed analysis of this manuscript has been recently published (Eisenstaedt & Combes, 2011).

[12] Actually, the effect implied here does exist; it is nothing but what, in the context of general relativity (even quantitatively), is often called the "gravitational–Doppler effect": light subject to a gravitational field exhibits a "Doppler shift." See (Eisenstaedt, 2005a; 2005b, pp. 303–304). The question of the "distance" separating Newton's *Principia* and Einstein's theories in terms of the effects implied by each is most interesting; I only point out here the evident "analogy" between the effects.

objected to Michell's secrecy on scientific matters[13] and more precisely to his "method":

> But in the present case I can not conceive why you should wish to have it kept secret for when you was last in town you made no secret of the principle but mentioned it openly at our Monday meeting & if I mistake not at other places & I have frequently heard it talked of since then. [. . .] On the whole I think that instead of you desiring ⟨me⟩ to keep the princ[iple] of the paper secret you ought rather to wish me to show the paper to as many of your friends as are desirous of reading it.[14]

On April 6, 1786, a paper written by Blair was read at the Royal Society of London. It offered a "proposal for ascertaining by experiments whether the velocity of light be affected by the motion of the body from which it is emitted or reflected; and for applying instruments for deciding the question to several optical and astronomical enquiries" (Blair, 1786).

In his 1784 article Michell had only been interested in dynamics, in measuring the effect of gravitation on light. He had not seen that what he called "his method" would be convenient to measure the radial relative velocities of stars. Blair understood this point and realized how important Michell's method was. He reasoned that the velocity of a light corpuscle received by an observer on Earth was the sum of the emission velocity of the corpuscle,[15] of the velocity of the source, and of that of the Earth. He dealt with light in the very same way that one deals with a material particle: the Galilean law of addition of velocities was presumed to be just as valid for light as for matter. He then examined the question of the motion of light relative to the "refractive body":

> I observe, that this difference of refraction arising from a difference in the incident velocity of light, will be precisely the same, whether the difference of incident velocity, be real or relative; or in other words, whether light moves with greater velocity towards the refracting body supposed to be at rest; or, the velocity of light remaining the same, the refracting body move directly towards the body, where the light is emitted. The same relative velocity of the refracting body & light will produce the same refraction, even if we are to go so far as to suppose the refracting body in motion, to meet with the particles of light at rest. This is demonstrable from the laws of motion. (Blair, 1786, pp. 3–4)

Here Blair takes a thoroughly *relativistic* view—in the Galilean sense—of the motion of light: what matters—what is physically significant—is only the *relative*

---

[13] On these issues see (McCormmach, 1968, ch. V).

[14] (Cavendish to Michell, 27 May 1783) in (Jungnickel and McCormmach, 1999, p. 567).

[15] The emission velocity is the (initial) velocity of light relative to its source at the time of its emission; it is supposed to be constant—independent of its source—but this does not mean that the velocity of light is constant.

velocity of the corpuscle and the observer.[16] He even goes so far as to suppose that light could be at rest, the relative velocity being that of the refracting body only (Eisenstaedt, 2005a, pp. 353–357).

Blair used, as it were, a (classical) relativistic kinematics of light together with Michell's method in order to measure the velocity of light:

> The intention of this paper is to demonstrate that [. . .] a difference in the incident velocity of light might produce a difference in the angle of refraction, to show how much this difference ought to amount to, to describe an instrument, which will make this difference very sensible; and to point out of the advantages which the sciences of optics and astronomy may be expected to derive from such an instrument. (Blair, 1786, pp. 1–2)

And Blair clearly pointed out what astronomers would be "capable of resolving" with his instrument: from measurements of angles of refraction he expected to determine the relative velocity between the source and the observer in the direction of observation:

> The motion of the observer being given, to determine at once by inspection, the motion with which any Planet, Comet or fixed star however remote, is approaching to, or receding from the observer and conversely, the motion of the body from which light is emitted or reflected being given to determine the velocity with which the Observer is approaching to, or receding from that body. (Blair, 1786, p. 10)

This is precisely the program that would be completed by Hippolyte Fizeau in the second part of the 19th century (Fizeau, 1870a). In fact, the Blair–Michell effect is analogous to the Doppler–Fizeau effect.

Of course Blair also applied his effect to the case of the annual motion of the Earth, so as to measure the velocity of the Earth relative to the Sun. He expected a differential refraction on the order of 10 seconds of arc in his achromatic prisms.[17]

Blair realized that the effect was in any case too small to be easily observed. In order to magnify it he devised a complicated instrument with twelve achromatic prisms. With so many prisms light absorption was so great that no measurement at all was possible, and Blair was never able to complete any observation.[18] Moreover, as a tool against the undulatory theory of light Blair would as well express there a thought experiment, what we call now a "Michelson experiment" (Eisenstaedt, 2005a, pp. 372–374). His paper was never published.

---

[16] Which is *not* the case in undulatory theories in which the ether "absorbs" the velocity of the source. Such theories are not "relativistic."

[17] The use of achromatic prisms was then general but was not—far from it—the best tool for that purpose: a dispersive prism is needed in order to observe shifts in emission and absorption lines, though at the time these lines were not understood. The Fizeau effect was observed in the 1860s. See (Eisenstaedt, 2005b, ch. 13).

[18] Blair also discussed there a thought experiment, intended as a tool against the undulatory theory of light: it was what we now call a "Michelson experiment." See (Eisenstaedt, 2005a).

2.1.5 The Constancy of the Velocity of Light

Crucial for the Newtonian theory of the propagation of light is the variability of the velocity of light. Nevertheless even in the late 17th and 18th centuries there were good reasons to believe that the velocity of light was constant.

Newton had early accepted Römer's conclusion (Römer, 1676) that the velocity of light was finite. From Bradley's 1728 article on aberration, the *observed* velocity of light seemed to be constant, though the precision of the observation did leave room for discussion.[19] Nevertheless, in his 1747 introduction to Newton's philosophy, s'Gravesande, commenting on Bradley's aberration, stated "that Light comes from all the Stars with the same Velocity" (s'Gravesande, 1747, vol. 2, p. 106). And in his "History of Opticks" Joseph Priestley, commenting as well on Bradley, wrote "that the velocity of the light of all the fixed stars is equal, and that their light moves through equal spaces in equal times, at all distances from them" (Priestley, 1772, p. 401). At the beginning of the 19th century, Delambre compared Römer's data to Bradley's, i.e., the data from the observations of Jupiter's satellites to those linked to aberration, and his conclusion was that the measurements of the velocity of light agreed:

> The phenomenon of aberration so happily and ingeniously explained by Bradley endows starlight with the same velocity as the light of the sun, when the latter is reflected by the satellites of Jupiter. This was a strong presumption in favor of the opinion that supposed the light from the sun has the same velocity as that of the stars, and that there was no difference in this respect between direct and reflected light. (Delambre, 1809)

Laplace was of the same opinion: "Generally, however the luminous ray arrives from the void within a transparent medium its speed is the same" (Laplace, 1808, p. 298).

Thus, the velocity of light seemed to be constant on observational grounds, even if the precision of the measurements was limited. At the beginning of the 19th century François Arago was working along these lines in a series of experiments on the velocity of light.[20] Arago's article was the very last—and most complete— work that was based on Newtonian physics for light propagation. In his article he had developed theoretical reasons for doubting the constancy of the velocity of light, the first one being, of course, that the velocity should depend on the motion of the observer, and the second that it should also depend on the proper velocity of the emitting star. Arago was moreover convinced by Michell's opinion that the velocity would be affected by gravity.[21] As I have shown elsewhere, the calculations he made

---

[19] The logic is most clear: the coefficient of aberration is proportional to the mean velocity of the Earth and inversely proportional to the velocity of light; as the angle of aberration is (observationally) constant, as is the mean velocity of the Earth, the velocity of light coming from different stars at different distances always had to be the same.

[20] (Arago, 1853, p. 40). Arago's paper was written in 1810 but not published before 1853.

[21] In his paper Arago refers to Michell and to Blair; most probably he never read Blair's manuscript but heard of his work through Robison's article. See (Robison, 1797, pp. 284–285).

for his experiments and for estimating the variation of light velocity are thoroughly based on Newtonian light dynamics.[22]

Arago made numerous measurements of the refraction of light from various stars whose motions are different relative to the Earth. His instrument was more sophisticated but simpler than Blair's: it was essentially an achromatic prism placed in front of a telescope, refraction being measured on a "mural," an equatorial sector, essentially a large and precise sextant fastened on a wall (Eisenstaedt, 2005b, pp. 192–195). Like Blair he was looking for an effect similar to the Doppler–Fizeau effect, and with a far better precision he would have been able to measure that effect. But the use of an achromatic prism made it impossible to achieve convincing results.[23] Anyway, Arago was confident of the precision of his measurements; as he put it, "the rays of all stars are subject to the same deviations" (Arago, 1853, p. 53). In such a result he saw "a manifest contradiction with the Newtonian theory of refraction" (Arago, 1853, p. 46).

His way to reconcile his strange experimental results with the Newtonian theory was to suppose "that the luminous bodies are only visible when their velocities are included between fixed limits [...] as these velocities will determine the amount of refraction, the visible rays will be always equally refracted" (Arago, 1853, pp. 46–48). Shortly after, in his observations on Newton's rings, he had more reasons to question the principles of the emission theory (Buchwald, 1989, p. 73). Later he would push Fresnel—who relied on aberration and on Arago's experiment in order to contest the theory of emission (Fresnel, 1814a, 1814b, 1815)—to develop his wave theory of light. As one of the turning points from the emission theory to Fresnel's wave theory of light, Arago's experiments are important for the history of physics as Hirosige wrote (Hirosige, 1976, p. 12), and as the correspondence between Fresnel and Arago around 1815 reveals (Fresnel, 1818), but it seems that they have always been misinterpreted. As we have seen these experiments were driven by the Newtonian theory of light propagation and Blair's idea of a Doppler–Fizeau effect; unfortunately, such a fundamental point seems not to have been understood.[24]

The wave theories of light resolved this question, at least for a while, but—to put it very simply—with the Michelson experiments the question was to be reopened at the end of the century. It is in fact one of the most intricate questions in 19th century physics, only to be fully resolved by Einstein's kinematics (Einstein, 1905).

2.1.6 The Newtonian Theory of Light Propagation

At the end of the 18th century, the Newtonian theory of the propagation of light as an application of the *Principia* was apparently taken to be obvious by many Newtonian philosophers. John Michell, Henry Cavendish, Robert Blair, Joseph

---

[22] Arago never explicitly published his calculations. They appear in his 1806 manuscript, analyzed in detail in (Eisenstaedt & Combes 2011).

[23] Such a "Fizeau effect" on star velocities would only be measured in the 1860–1870s with the help of a dispersive prism and techniques of stellar photography.

[24] Rosmorduc is as far as I know the only historian of science who understood that Arago's experiment was a Doppler–Fizeau experiment (Rosmorduc, 1981, p. 843, note 7).

Priestley, William and John Herschel, Nevil Maskelyne, Johann von Soldner, Nathaniel Bowditch, John Robison, Pierre-Simon Laplace, Jean-Baptiste Biot, François Arago, Siméon-Denis Poisson, all used it, worked on it, believed in it, or at least alluded to it. Clearly such a view was the "normal" way to deal with light propagation; and even Euler, when working on aberration, hesitated between two conceptions, the wave and the corpuscular (Eisenstaedt, 2005a, pp. 353–357). But shortly after Arago's article—which was not published at the time of his experiments—there was very little interest in it. Quite soon the Newtonian theory of light propagation was forgotten. Nevertheless all these works had quite important consequences.

Finally, let us survey the views that the Newtonian theory of light propagation offers.

- A relativistic optics of moving bodies: a corpuscle of light is subject to Galilean kinematics, and thus to its principle of relativity as well as to the corresponding theorem of the addition of velocities. The velocity of a light corpuscle is the sum of the velocity of its source, of its emission velocity, and of the velocity of its observer; as a consequence it cannot be constant. Such an optics of moving bodies has quite the same structure as Einstein's special relativity—the Galileo transformations having of course to be replaced by Lorentz transformations. Blair imagined a thought experiment which a century later became a reality through Michelson's experiment (Eisenstaedt, 2005a, pp. 372–374).
- A Newtonian gravitational optics: in the very same way as a material particle, a corpuscle of light is subject to the long-range force of gravitation. And its velocity has therefore to vary from this cause alone: it will be diminished (or augmented) by the influence of the force of gravitation.
- Most physical ingredients of Einstein's relativity are there (light, gravitation, and—Galilean—relativity) as well as qualitatively important effects: the deviation of light by gravitation,[25] "dark bodies" as cousins of "black holes," the gravitational Doppler effect.
- An equivalent to Doppler's effect: incident on a transparent body, light is refracted as a function of its incident velocity. As a consequence, a prism was thought of and used as a tool to measure the velocity of light.

Notwithstanding that, much later, these results would become important in a different context, the theory itself was forgotten because very few scientists were closely interested in questions specifically involving the relation between the velocities of light and those of the source and the observer. There were many more long-standing, important and interesting problems to be dealt with concerning light. As we mentioned above, Newton himself was unable to deal with color within the framework of his corpuscular theory. A century later, Young and Fresnel dealt with these and other questions as well (diffraction, interference, polarization), questions that Fresnel would explain through his wave theory of light. Moreover, thanks to the

---

[25] The deviation of light by gravitation was derived on the basis of the same theory by (Soldner, 1800). Concerning this history see (Jaki, 1978); Einstein independently predicted the very same effect, see (Einstein, 1907, p. 461; 1911, p. 908).

ether, the wave theory had the virtue of solving—to first order and in a nonrelativistic way—the question of the independence of the velocity of light relative to its source, and the problem of aberration.

Afterwards there was no interest at all in Blair's idea of an optics of moving bodies, and very little in Michell's gravitational optics; only Doppler's effect was at stake—but without any reference to Michell and Blair. Actually, within the context of wave optics there was hardly any way to develop a dynamics for light, and so it was not really possible—and uninteresting[26]—to think of gravitation acting on a light-wave. As wave optics gained widespread acceptance very few physicists referred to the older Newtonian theory of light propagation: the only ones to do so, and then only as a matter of historical record, were Hippolyte Fizeau, James Clerk Maxwell, and Eleuthère Mascart (Eisenstaedt, 2005a, pp. 262–264); not to forget François Arago, whose work was at the frontier between the two theories.

It is important to emphasize that the body of work devoted to the Newtonian theory of light propagation *in no way* provides a context of discovery for Einstein's relativity. Clearly, Einstein's kinematics, his "special relativity," evolved out of his concern with problems in electrodynamics. It was little different with general relativity, Einstein's theory of gravitation. But as Einstein's 1911 paper on the deviation of light in a Newtonian context shows, the Newtonian issue was a way, not to find the (Lorentzian) relativistic theory of gravitation, but rather to have a more precise idea—at least quantitatively—of the action of gravitation on light. In 1913, Einstein wrote Erwin Freundlich "that the idea of a bending of light appeared rather natural at the time of the emission theory"[27] But as far as we know it was the only concern of Einstein in this respect.

Not so surprisingly, neither the possibility of a Newtonian optics of moving bodies nor that of a Newtonian gravitational theory of light has been easily "seen," neither by relativists nor by historians of physics; most probably the "taken-for-granted fact" of the constancy of the velocity of light did not allow thinking in Newtonian terms.

Acknowledgements

I would like to thank Jed Buchwald, Yves Gingras, Dominique Hirondel, Anne Kox, and unknown referees for comments, discussions, and linguistic assistance.

References

Arago, D. F. J. (1806). Sur la vitesse de la lumière. *Bibliothèque de l'Institut* (Paris). Manuscrit 2033, folios 117–121.

---

[26] Because the effect was so small as Michell and Soldner have already shown.

[27] Einstein to Freundlich, August 1913; quoted in (Eisenstaedt, 1991, p. 378).

——. (1853). Mémoire sur la vitesse de la lumière, lu à la première classe de l'Institut le 10 décembre 1810. *Académie des Sciences* (Paris). *Comptes Rendus*, 36, 38–49.

Bechler, Z. (1973). Newton's Search for a Mechanistic Model of Color Dispersion: a Suggested Interpretation. *Archive for History of Exact Sciences*, 11, 1–37.

Blair, R. (1786). A proposal for ascertaining by experiments whether the velocity of light be affected by the motion of the body from which it is emitted or reflected; and for applying instruments for deciding the question to several optical and astronomical enquiries. *Royal Society* Manuscript L & P, VIII, 182.

Buchwald, J. Z. (1989). *The rise of the wave theory of light: optical theory and experiment in the early nineteenth century*. Chicago: University of Chicago Press.

Cantor, G. N. (1983). *Optics after Newton. Theories of light in Britain and Ireland, 1704–1840*. Manchester University Press.

Clairaut, A.-C. (1741). Sur les explications cartésiennes et newtoniennes de la réfraction de la lumière. *Académie Royale des Sciences* (Paris). *Mémoires pour 1739*, 259–275.

Delambre, J.-B. (1809). Procès-verbal de la Séance du Lundi 4 Septembre 1809 et Rapport de Delambre à la Classe des Sciences. Sur le mémoire de M. Arago concernant la *vitesse de la lumière*. In *Procès-verbaux des séances de l'Académie tenues depuis la fondation de l'Institut jusqu'au mois d'août 1835*. Hendaye : Imprimerie de l'observatoire d'Abbadia, (1913), vol. 4, 245.

Doppler, C. (1842). On the coloured light of the double stars and certain other stars of the heavens. *Proceedings of the Royal Bohemian Society of Sciences*, 2, 103–133.

Einstein, A. (1905). Zur Elektrodynamik bewegter Körper. *Annalen der Physik*, 17, 891–921.

——. (1907). Über das Relativitätsprinzip und die aus demselben gezogenen Folgerungen. *Jahrbuch der Radioaktivität und Elektronik*, 4, 411–462; 5, 98–99.

——. (1911). Über den Einfluss der Schwerkraft auf die Ausbreitung des Lichtes. *Annalen der Physik*, 35, 898–908. Translated as, On the influence of gravitation on the propagation of light. In Lorentz, H. A., *et al. The Principle of Relativity*. A. Sommerfeld, ed. W. Perrett and G. B. Jeffery, trans. London: Methuen, 1923; reprint New York: Dover, 1952, 97–108.

Eisenstaedt, J. (1991). De l'influence de la gravitation sur la propagation de la lumière en théorie newtonienne. L'archéologie des trous noirs. *Archive for History of Exact Sciences*, 42, 315–386.

——. (1996). L'optique balistique newtonienne à l'épreuve des satellites de Jupiter. *Archive for History of Exact Sciences*, 50, 117–156.

——. (2005a). Light and relativity, a previously Unknown Eighteenth-Century Manuscript by Robert Blair (1748–1828). *Annals of Science*, 62, 347–376.

——. (2005b). *Avant Einstein Relativité, lumière, gravitation*. Paris: Seuil.

——. (2007). From Newton to Einstein: a forgotten relativistic optics of moving bodies. *American Journal of Physics*, 75, 741–746.

Eisenstaedt, J. and Combes, M. (2011). Arago et la vitesse de la lumière (1806–1810), un manuscrit inédit, une nouvelle analyse. *Revue d'histoire des Sciences*, 64, 59–120.

Fizeau, A.-H. (1870a). Des effets du mouvement sur le ton des vibrations sonores et sur la longueur d'onde des rayons de lumière. *Annales de Chimie et de Physique*, 19, 211–221. [Read at the Société Philomathique on December 23, 1848].

——. (1870b). Remarques concernant le déplacement des raies spectrales par le mouvement du corps lumineux ou de l'observateur. *Académie des Sciences* (Paris), *Comptes Rendus*, 70, 1062–1066.

Fresnel, A. (1814a). Augustin Fresnel à son frère Léonor. In H. de Sénarmont, É. Verdet and L. Fresnel (Eds.), *Œuvres Complètes d'Augustin Fresnel*, (1866–1870), vol. 2, 820–828. Paris: Imprimerie impériale.

——. (1814b). Lettre d'Augustin Fresnel à son frère Léonor. In H. de Sénarmont, É. Verdet and L. Fresnel (Eds.), *Œuvres Complètes d'Augustin Fresnel*, (1866–1870), vol. 2, 848–851. Paris: Imprimerie impériale.

——. (1815). *Premier Mémoire sur la Diffraction de la Lumière*. In H. de Sénarmont, É. Verdet and L. Fresnel (Eds.), *Œuvres Complètes d'Augustin Fresnel*, (1866–1870), vol. 1, 9–37. Paris: Imprimerie impériale.

——. (1818). Lettre d'Augustin Fresnel à François Arago. In H. de Sénarmont, É. Verdet and L. Fresnel (Eds.), *Œuvres Complètes d'Augustin Fresnel*, (1866–1870), vol. 2, 627–636.

Gibbons, G. (1979). The man who invented black holes. *New Scientist,* June 28 1979, 82 (1161), 1101.

Hirosige, T. (1976). The ether problem, the mechanistic worldview, and the origins of the theory of relativity. *Historical Studies in the Physical Sciences*, 7, 3–82.

Jaki, S. L. (1978). J.G. von Soldner and the gravitational bending of light with an English translation of his essay on it published in 1801. *Foundations of Physics*, 8(11–12), 927–950.

Jungnickel, C., and McCormmach, R. (1999). *Cavendish: The Experimental Life*. Cranbury, NJ: Bucknell.

Kuhn, T.S. (1962). *The structure of scientific revolutions*. Chicago: University of Chicago Press, 3rd ed., 1996, 102–103.

Laplace, P.-S. (1796). *Exposition du système du monde*. Ed. originale. 2 vols. Paris: Imprimerie du Cercle-Social.

——. (1808). *Exposition du système du monde*. 3° ed. Paris: Courcier.

McCormmach, R. (1968). John Michell and Henry Cavendish: weighing the stars. *The British Journal for the History of Science*, 4, 126–155.

Michell, J. (1784). On the means of discovering the distance, magnitude, &c. of the fixed stars, in consequence of the diminution of the velocity of their light, in case such a diminution should be found to take place in any of them, and such other Data should be procured from observations, as would be farther necessary for that purpose. By the Rev. John Michell, B. D. F. R. S. In a letter to Henry Cavendish, Esq. F. R. S. and A. S. *Royal Society of London. Philosophical Transactions*, 74, 35–57.

Newton, I. (1687). *The Principia: mathematical principles of natural philosophy. A New Translation.* Sir I. B. Cohen and A. Whitman. University of California Press, 1999.

Priestley, J. (1772). *The history and present state of discoveries relating to vision, light and color.* London. Kraus Reprint Co. Millwood, NY.

Robison, J. (1797). Optics. *Encyclopædia Britannica,* 3rd ed., vol. 13, 231–364.

Römer, O. (1676). Demonstration touchant le mouvement de la lumière. . . *Journal des Savants,* 7.XII 1676, 233–236.

Rosmorduc, J. (1981). L'expérience de Fizeau. *Bulletin de l'Union des Physiciens,* 632, 841–857.

Schaffer, S. (1979). John Michell and Black Holes. *Journal for History of Astronomy,* 10, 42–43.

s'Gravesande, W. J. (1747). *Elémens de Physique ou Introduction à la philosophie de Newton.* Trad. C. F. R. de Virloy's. Paris: C. A. Jombert. 2 vol.

Shapiro, A. E. (1993). *Fits, Passions, and Paroxysms: Physics, Method and Chemistry and Newton's Theories of Colored Bodies and Fits of Easy Reflection.* Cambridge: Cambridge University Press.

Short, J. (1754). Report to the Society. *Royal Society of London. Philosophical Transactions for the year 1753,* 48, 268–270.

Soldner, J. G. von. (1800). Etwas über die relative Bewegung der Fixsterne; nebst einem Anhange über die Aberration derselben. *Astronomisches Jahrbuch für das Jahr 1803,* 185–194.

Turnbull, H. W., Scott, J. F., Hall, A. R. and Tilling, L., eds. and trans. (1959–1977). *The Correspondence of Isaac Newton.* 7 vols. Cambridge: Cambridge University Press.

Whiteside, D. T. (1980). Kepler, Newton and Flamsteed on Refraction Through a 'Regular Air': the Mathematical and the Practical. *Centaurus,* 24, 288–315.

3

Mach and Einstein, or, Clearing Troubled Waters in the History of Science*

Gereon Wolters

University of Constance, Germany

3.1 Introduction: A Chilling Health Bulletin

In this paper I would like to analyze the relationship between Mach and Einstein. Ernst Mach was born in 1838 and died in 1916 aged seventy-eight. The last eighteen years of his life were heavily overshadowed by the consequences of a stroke which occurred in 1898, i.e., seven years before Einstein's *annus mirabilis*. Mach suffered paralysis on the right side of his body, heavy speech impairment, and had a painful bladder condition which required him to be catheterized at least twice-a-day. In addition, he suffered from a severe sleeping disorder, neuralgia, and the sometimes severe consequences of frequent falls. The reason for mentioning Mach's poor state of health is that it has much to do with his relationship to Einstein, or, for that matter, to relativity theory.

But Mach was not just an old and very sick man when relativity theory emerged. He was not even a professional theoretical physicist, i.e., he was not particularly well equipped to understand relativity. Rather, he was a sense physiologist and an experimental physicist. There is only one field of his teaching and research that related Mach to relativity: his so-called "historical-critical" studies in the history of physics, in particular, the history of mechanics, which culminated in the 1883 publication of *Die Mechanik in ihrer Entwickelung—historisch-kritisch dargestellt*, which Mach revised for the last time in its seventh edition in 1912. And it was this *opus*

*Work for this paper was done during two short research stays in 2002 and 2003 at Max Planck Institute for the History of Science in Berlin. This support is gratefully acknowledged. Earlier versions of the paper were presented at the Einstein celebration of the Società Filosofica Italiana in Messina in March 2005, at the Univeristy of Bari, and at an Einstein Seminar in the context of the *dottorato di ricerca* in philosophy at the University of Rome II (Tor Vergata) in October 2005. I am grateful to an anonymous referee for very useful suggestions, and particularly for greatly improving the existing English translations of Mach texts I have used.

magnum which greatly influenced Einstein. It constituted an important part of the reading program of the "Academia Olympia" of Einstein and friends in pre-relativity Bern.

From what has been said so far one may conclude that Mach's position with respect to relativity, if such a position existed at all, would certainly be of some historical interest, but its systematic significance would be rather low. It would reflect the opinion of an old and very sick man about the development of a theory in a field he himself had hardly ever worked in, although he had commented on some parts of it from a historical and epistemological point of view.

In the following I would like to outline Mach's influence on Einstein and to investigate whether and to what extent Einstein was aware of this (Section 3.2). I will then look at Mach's reaction to relativity (Section 3.3). This is followed by Section 3.4, which deals with the arrival on the scene of people who began to make trouble, first someone who forged texts in order to turn poor Mach into an antirelativist, and second, philosophers and historians of science who, beginning in the 1970s, took these texts seriously and tried to malign Mach both as a person and as an epistemologist. There were at least ten good reasons not to do this, even if the documentary evidence I uncovered in the 1980s was not at disposal. This documentary evidence led me to claim that the antirelativist texts, which carry Mach's name, are simply forged. A very brief account of this thesis constitutes the final part of the paper.

3.2 Mach's Influence on Einstein and How Einstein Himself Saw It

As is well known and has never been doubted so far, Einstein himself did not have the slightest doubt as to the influence Mach exercised on the shaping of both special and general relativity.[1] One may distinguish the following Machian positions, which influenced Einstein: (1) Mach's critique of the mechanical world picture, (2) Mach's principle,[2] (3) what I would like to call Mach's research program "General Relativity Theory" (GTR), (4) Mach's critique of basic concepts of physics, (5) Mach's methodology, (6) Mach's personality. In the frame of this paper, I can say here only a few words about each of these fields of influence.

(1) Mach's critique of the mechanical world picture is at the very heart of his philosophy. Mach rejects, from an *ontological* perspective, the mechanistic idea that the world, finally, consists of nothing but matter in motion. He rejects also the *methodological* consequence of this world picture, namely, reductionism as the requirement

---

[1] It is, of course, a different question whether—in an abstract systematic sense—Einstein was "justified" in tracing back certain of his ideas to Mach. Norton (1995), e.g., convincingly shows that Mach himself had no clear conception of what Einstein later coined "Mach's principle."

[2] In the sense of the preceding footnote Earman (1995, 158, fn. 16) remarks: "In retrospect it seems that the extent to which GTR incorporates Mach type principles is much less than Einstein originally thought."

to give mechanical models in all areas of natural science. Mechanical reductionism is particularly at odds with Mach's phenomenalist epistemology which takes sensations as irreducible elements of experience.[3] As is well known, Mach was wrong in rejecting mechanical models in the case of thermodynamics, but his critique of the mechanical world picture as a general position was soon proved to be right since all attempts to establish mechanical models for electrodynamics had failed. And, as early as 1910, Mach could have taken the following quote as support

> I need not emphasize that the mechanical view of nature is incompatible with this conception [i.e., the principle of relativity, G.W.]. Whoever regards the mechanical view of nature as a postulate of thinking in physics will never be able to make friends with relativity theory. (Planck 1969, 61, my translation)

This quote, in which an antimechanistic position (as vigorously defended by Mach) is described as a necessary condition for embracing the most modern theory of relativity, Mach was able to read in Max Planck's talk ("The New Physics and the Mechanical View of Nature") given at the 82nd "Assembly of German Scientists and Physicians," which was attended by almost everyone in German science. For Mach, the dissociation of the mechanistic world picture and relativity theory was even more important since the very same Planck had attacked him shortly before, again in a widely distributed talk ("The Unity of the Physical World Picture"), as a "false prophet," who should be recognized by the fruits of his labor.[4] But one has to keep in mind that this attack criticized Mach's attitude towards the reality of atoms, which he, in fact, had denied.

In order to adequately assess the historical situation, we have to also keep in mind that in the first years after 1905, relativity theory was not regarded as the revolution in physics it is today, but rather as the final part of the efforts to arrive at what was then called the electromagnetic world picture. As McCormmach (1970, 488) has shown, in the years immediately after 1905 most people did not distinguish between what they called 'Lorentz theory,' 'Lorentz–Einstein theory,' or 'relativity theory.'

The electromagnetic world picture, based on whichever of these theories, is a sort of alternative to the mechanical world picture. Physical reality in the simplest versions of the electromagnetic world picture consists basically in the electromagnetic ether and in electric particles that were understood as singularities in the ether or the field, respectively. In a methodological perspective this implies the reduction of mechanics to electrodynamics. This, in turn, was an idea on which Mach had spoken out positively on several occasions,[5] one of them as late as 1910 when he republished his famous 1871 talk on the conservation of energy, adding a few additional notes.

---

[3] For Mach mechanistic physics presents "a diagram of the world, in which we do not know reality again. It happens, in fact, to men who give themselves up to this view for many years, that the world of sense from which they start as a province of the greatest familiarity, suddenly becomes, in their eyes, the supreme 'world-riddle'" (Mach 1943, 159).

[4] The talk is in (Planck 1969, 28–51). An English translation of the last section of the talk is in (Blackmore 1992, 127–132).

[5] See, e.g., (Mach 1911, 95; Mach 1943, 131; Blackmore 1992, 125).

Later I will return to this republication as prime evidence for Mach's position on relativity.

(2) As to Mach's Principle one has to keep in mind that in classical mechanics inertial motion is regarded basically as a *kinematical* phenomenon. It is motion relative to absolute space and as such it is seen as motion without the imposition of forces. Mach, in his critique of Newton's interpretation of the bucket experiment, tentatively puts forward a *dynamical* conception of inertia: "The comportment of terrestrial bodies with respect to the earth is reducible to the comportment of the earth with respect to the remote heavenly bodies" (Mach 1989, 285). What "reducible" means here becomes clearer when we realize that causal interaction for Mach is nothing but the functional dependency between observed phenomena. And what one observes with respect to inertial motion is that there exists, in the last analysis, a functional relation between inertially moved bodies and the fixed stars. It is this idea, which Einstein first called "relativity of inertia," that influenced greatly the formation of general relativity. In a letter of June 25, 1913, Einstein let Mach know that his idea of the relativity of inertia had turned out to be a consequence of general relativity:

> For it follows of necessity [from the equivalence in general relativity of the acceleration of the reference system, on the one hand, and the gravitational field on the other] that *inertia* has its origin in some kind of *interaction* of bodies, exactly in accordance with your [i.e., Mach's] argument about Newton's bucket experiment. (Einstein 1995, 340)

(3) Let us now have a quick look at what I would like to call "Mach's research program GRT." Keep in mind that Mach was *not* a professional theoretical physicist, but took a historical and methodological interest in theoretical physics.

Mach was convinced that what might be called "strict empiricism" was a necessary condition for scientific progress. With "strict empiricism" I mean that every physical concept must refer to something observationally demonstrable. For Mach concepts without such reference are just "metaphysics," mostly of the sort that had impaired scientific progress in the past.[6]

In taking Mach as having designed a research program "GRT" I would like to point to four themes, which were important for the historical development of that theory. I do not claim, though, that this applies also to the theory as most present-day physicists understand it. The four themes are: (a) Mach's Principle, (b) the general relativity principle, (c) the equivalence principle, (d) the need for a theory that replaces Newtonian action at a distance by action by contact.

(3a) As is well known, Einstein dismissed what he had termed "Mach's principle" soon after he had explicitly stated it in 1918. But this principle nonetheless played a decisive role in the genesis of GRT. With Mach's principle Einstein had

---

[6] But one has to see also that according to Mach's instrumentalist conception of physical theories even theories without direct and complete observational reference can be useful in explaining and predicting (as, e.g., the theory of atoms), as long as one refrains from ascribing physical reality to theoretical concepts that, as a matter of principle, lack such reference.

tried to give Mach's rather loose and informal idea of inertia as resulting from a kind of interaction with the fixed stars a sufficient theoretical and formal expression. As mentioned above, Einstein first dealt with this idea under the heading "relativity of inertia" and then, from 1918 onwards, as "Mach's principle." Mach's principle consists basically in conceiving the gravitational field as generated by matter, or, in a formulation that refers to the field equations of general relativity: curved space-time generated by matter-energy. Matter and energy in the perspective of Mach's principle have ontological priority over space. In this sense Einstein wrote Mach at the turn of 1913/14:

> It seems to me absurd to ascribe physical properties to 'space.' The totality of the masses generates a $G\mu\nu$-field (gravitational field), that, in turn, regulates all processes, the propagation of light rays and the behavior of measuring sticks and clocks included. (Einstein 1995, 371)

(3b) The general relativity principle is the postulate that all laws of physics have the same form in *all* reference frames, and not only in inertial frames—as stated in special relativity and classical mechanics. We need not go into detail here, suffice to say that Einstein in his *Grundlagen* paper of 1916 ("The Foundation of General Relativity"), in which he gave the first comprehensive exposition of general relativity as well as in the corresponding popular booklet of 1917 (*On the Special and General Theory of Relativity (A Popular Account)*) explicitly credits Mach with the idea that the restriction of special relativity and classical mechanics to inertial reference frames is unfounded (Einstein 1997, 147 and 324f.).

(3c) Classical mechanics as well as special relativity distinguish two different concepts of mass, namely, inertial mass and gravitational mass. The inertial mass of a body should be conceived—according to a proposal of Mach of the 1860s—as its resistance to accelerating forces. The ratio of the masses of two bodies is thus expressed by the equation: $m_1/m_2 = -a_2/a_1$.

On the other hand, the ratio of the forces that are imposed on two bodies in the same gravitational field defines gravitational mass. Gravitational mass is, as is well known, what one measures with scales. Inertial and gravitational mass of a body had turned out at the turn of the twentieth century to be numerically the same in high-precision measurements. General relativity gives a *theoretical account* of their *actual* numerical equality. Einstein (1955, 57) speaks of "equality of the real nature" (*Gleichheit des Wesens*).

Mach had pointed in the same direction in several thought experiments and observations. He claimed, for example, that the sensations we have when we jump down from an elevated position, i.e., when we sense acceleration, should be the same as the ones we would have "on a celestial body smaller than the earth, if we were transferred there instantaneously," i.e., when we sense differential gravitation. According to Mach, corresponding sense experiences hold in the case of elevation and instantaneous transfer to a bigger celestial body (Mach 1989, 252). In the same section of his *Mechanics* he points out that the weight of a mass on a table would be zero if the table were accelerated towards the center of the earth with an acceleration equal to free fall. What these and other remarks of Mach have in common is that they claim sensations and observations to be the same whether or not they relate to accelerations or gravitational effects. This in turn ought to be a strong epistemological argument for an empiricist like Mach to take the two phenomena to be, in fact, the same. The equivalence of inertial and gravitational mass in general relativity can thus be regarded as a theoretical account of Mach's epistemological observations.

(3d) In a reflection in *Mechanics* on how an "ideal" theory of mechanics might look, Mach states that this ideal should be a theory "from which accelerated and inertial motions result in the *same* way" (Mach 1989, 296). Mach's own conception remained Newtonian in the sense that his making inertia relative by relating it to the fixed stars still took place in a Newtonian action-at-a-distance framework. As Einstein remarked, to actually theoretically realize Mach's conception "within the limits of the modern theory of action through a medium, the properties of the space-time continuum which determine inertia must be regarded as field properties of space, analogous to the electromagnetic field" (Einstein 1955, 56). Mach basically was of the same opinion, but lacked the instruments to theoretically realize it. For, in continuation of his vision of the ideal mechanics just quoted, he says: "The progress from Kepler's discovery to Newton's law of gravitation, and the impetus given by this to the finding of a physical understanding of the attraction in the manner in which electrical actions at a distance have been treated, may here serve as a model" (Mach 1989, 296).

(4) How influential Mach's empiricist critique of basic concepts for Einstein was has been shown to a large extent already in the preceding sections and is so well documented by Einstein himself that I need not deal with it in greater detail here. I would like to mention only that Einstein, in addition to what has been said already, links his reading of Mach also to his redefinition of simultaneity. Also Einstein's critique of the concept of ether is in part truly Machian (cf. Wolters 1987, 96ff.).

(5) As is generally agreed young Einstein has to be regarded as an avid follower of Mach in methodological matters. The majority of researchers believe, furthermore, that Einstein later fundamentally disagreed with Mach. This view seems rather exaggerated, but it has the advantage of being well founded by the one-sided interpretation Einstein gave of Mach's conception of scientific theory. The catchword here is Mach's notorious idea that science, at its best, was a "catalogue" that consisted in the description of constant functional relations between sensations. Einstein, however, in his critique of Mach, claims that theories are what he calls "free creations of the human mind."

If one presupposes something like the distinction between the context of discovery and the context of justification, Mach's and Einstein's positions, however, do not occur as completely contradictory. Note that for Mach's catalogue conception such a catalogue is (1) only the final state a scientific theory may reach in its historical development, and that (2) this catalogue is not just a description of what there is, but it ought to be a description as unified as possible by applying Mach's principle of economy. Consequently, the ideal, final state of a scientific theory as a unified descriptive catalogue has to be distinguished both from its earlier historical states and from the way one discovers scientific theories. Scientific theories and ideas are not just the result of careful observational bookkeeping. Rather, as Mach emphasizes over and over again, new scientific ideas arise from the workings of intuition and imagination. For example, Newton's discovery of the law of gravitation is for Mach in a decisive way an "achievement of imagination" (*Phantasieleistung*) (Mach 1989, 228). Mach's own term for the discovery of important scientific ideas is *erschauen*, which means an intuition that goes well beyond what is just observable.[7] In short, the difference between Mach's and Einstein's conception of scientific theories is not as big as Einstein would have us believe. Things are different, however, with respect to the scientific realism that Einstein began to embrace sometime in the early twenties, leaving behind Mach's biologically oriented instrumentalism.

(6) Finally, I would like to point out that Mach's ever-skeptical, antinationalistic, anti-anti-Semitic, leftist and liberal personality left a strong impression on Einstein, as one can gather, e.g., from his obituary of Mach. Mach was one of the very few professors in the German-speaking countries with whom Einstein could easily feel congeniality. And he might have deeply consented when he read in one of Mach's popular lectures ("The Relative Educational Value of the Subjects of Science Instruction"):

> I [i.e., Mach] would restrict drastically the number of school lessons. [. . .] I don't know of anything more terrible than those poor people who have learned *too much*. Instead of a healthy and forceful judgment that might have arisen, if they hadn't learned *anything,* their thoughts creep timidly and mesmerized on ever the same roads behind a few words, phrases, and formulas. [. . .] To put youngsters into a uniform is certainly suited for the military, but doesn't do any good to the brain.[8]

I think it a fair summary of what I have been saying so far that Einstein held Mach in high esteem, not only as a philosopher-scientist, but also for his personality.

Three quotations may suffice to document Mach's well-known influence on Einstein and the development of relativity theory. The first is from Einstein's long and moving obituary on Mach in 1916, published in *Physikalische Zeitschrift.* Among other things Einstein writes:

> [. . .] it is not an idle play, when we are trained to analyze the entrenched concepts, and point out the circumstances that promoted their justification and usefulness and how they evolved from the experience at hand. [. . .] Nobody can deny that epistemologists paved the road for progress; and for myself, I know at least that Hume and Mach have helped me a lot,

---

[7] See, e.g., (Mach 1989, 171) with respect to Galileo's intuition that "it is *accelerations* which are the immediate effects of the circumstances that determine motions, that is, of the forces." The usual English rendering of *erschauen* as *perceive* does not seem adequate to me.

[8] Mach (1923, 343, 344, 346), my translation. The text is not contained in Mach (1943). Mach, on the other hand, would have read with great pleasure Einstein's newspaper article "The Nightmare" (*Der Albtraum*) which concludes with the exclamation "away with the final secondary school exam!" (Einstein 1997, 449).

both directly and indirectly. [...] These quotations [from Mach's *Mechanics*, G.W.] show that Mach clearly recognized the weak points of classical mechanics, and thus came close to demand a general theory of relativity – and this almost half a century ago! It is not improbable that Mach would have hit on relativity theory when in his time – when he was in fresh and youthful spirit – physicists would have been stirred by the question of the meaning of the constancy of the speed of light. [...] The contemplations on Newton's experiment with the pail demonstrate how close his mind was to the demands of relativity in the wider sense (relativity of acceleration). (Einstein 1997, 142–144)

The second quotation is from Einstein's "Autobiographical Note" that was written in 1948 for the Schilpp volume on Einstein:

It was Ernst Mach who, in his History of Mechanics, shook this dogmatic faith [in classical mechanics as firm and definite foundation for all physics]; this book exercised a profound influence upon me in this regard while I was a student. (Schilpp ed. 1970, 21)

The third and last quotation is from an interview Einstein gave in 1955 to I. B. Cohen, only two weeks before his death.

Although Einstein did not agree with the radical position adopted by Mach (with respect to the existence of atoms), he told me he admired Mach's writings, which had a great influence on him. (Cohen 1955, 72)

3.3 How Mach Reacted towards Relativity Theory

Now I would like to ask how Mach himself received relativity. From what I have said so far, one thing seems clear: However Mach received relativity, if he did receive it at all, it had to be the reaction of an old and sick man about a development in a field of physics he had not worked in, and for which he possessed only some historical and epistemological expertise.

Given this background it cannot come as a surprise that after 1905 it took quite a while, until Mach came across relativity theory. All existing evidence suggests that Mach first learned about it through Hermann Minkowski's widely publicized and famous talk "Space and Time" at the 80th Assembly of German Scientists and Physicians in September 1908.[9] It was this talk that made relativity theory transcend the small circles of specialists.

Mach was obviously unable to understand Minkowski's group-theoretical, four-dimensional presentation of the theory which soon became standard. So he asked the theoretical physicist Philipp Frank (1884–1966) at the turn of the year 1909/10

---

[9] There is no mention of relativity or of its creator in the existing Mach material prior to 1909.

to explain relativity to him.[10] Frank gives a report in a letter to the East German historian of science Friedrich Herneck:

> He [i.e., Mach] particularly wanted to know more about the use of four-dimensional geometry. As you [i.e., Herneck] perhaps know from his book *Knowledge and Error* it was one of Mach's favorite ideas that a four-dimensional geometry is more practical for explanations in physical science than three-dimensional geometry. I, on that occasion, had the impression that he was in complete agreement with Einstein's 'special' theory, and also particularly with its philosophical basis. Mach asked me to give him my representation in written or printed form. I did this, and therefore the representation of Einstein's theory that Mach agreed with exists also in a printed text.[11]

Special relativity and his own positive role in its development became for Mach at the same time particularly important. As already mentioned, Planck had attacked Mach severely and personally in his talk "The Unity of the Physical World Picture" (1908, published 1909), apostrophizing him in the last sentence a "false prophet," who should be known by the fruits of his labor.[12] Clearly, Planck was here referring to Mach's opposition to atomism. Still, as an old man in his "Personal Recollections from the Old Days" (1946) he resented Mach because he was of the opinion that the sound arguments of his own early work on thermodynamics, particularly on irreversibility

> fell on deaf ears. It was simply impossible to be heard against the authority of men like W. Ostwald, Ch. Helm and E. Mach. [. . .] the annoying thing was that I was not to have at all the satisfaction of seeing myself vindicated. (Planck 1969, 12, my translation)

Mach felt very much offended by Planck's outburst. As far as I can see, he never felt so bitter about a colleague during his whole life. So for Mach it came as a gift from heaven, as it were, that there was a new and fundamental theory whose coming into being had not only not been impaired by Mach's philosophical ideas, but could be regarded as a sort of physical implementation of them. So Mach could build a defense line against Planck's criticism by making clear that relativity theory in any case was a *valuable* fruit from the tree of his earlier methodological ideas.

But Einstein had not mentioned Mach in his 1905 paper. Thus, in the context of his line of defense against Planck, it became essential for Mach to point out to Einstein that he himself had developed similar 'relativistic' ideas. So it was Mach

---

[10] Unfortunately Frank did not report his visit in his Einstein biography. Even more unfortunate is that after Frank's death nobody at Harvard, where Frank had been a "Permanent Faculty Lecturer," seems to have thought it worthwhile to secure his *Nachlass*. Only thanks to the quoted letter from Frank Friedrich Herneck, do we know about the visit.

[11] Herneck (1966, 49), my translation. Frank's "printed text" is Frank (1910).

[12] There is an English translation of the decisive (and last) section 4 of the paper in (Blackmore 1992, 127–132).

who contacted Einstein in 1909 and sent him the just published second edition of his 1871 (first published 1872) talk in Prague on the "History and Root of the Law of the Conservation of Energy," which contained a new preface and additional footnotes. The republication of the booklet was—as Mach writes in a letter to Paul Carus— "caused by Planck's attack."[13] This means, of course, that the republication had in Mach's eyes functioned to demonstrate that he was *not* the false prophet that Planck had depicted. Mach could reach this goal only by proving that his ideas had been scientifically fruitful. And there was only one field in which this could be done: the emergence of relativity and—as Mach and many others thought—connected with this, the replacement of the mechanical world picture by an electrodynamical one.

Einstein reacted exactly as Mach might have hoped he would. In a letter of August 9, 1909, he tells Mach that he has "already carefully read" the Conservation of Energy Paper. Then he confesses to know Mach's major works, of which he "especially admire(s) the one on mechanics." He continues:

> You [i.e., Mach] have had such an influence on the epistemological views of the younger generation of physicists that even your current opponents, such as, e.g., Mr. Planck, would undoubtedly have been declared to be 'Machists' by the kind of physicists that prevailed a few decades ago. (Einstein 1995, 130)

In answering Mach's thank-you letter,[14] in which Mach had obviously also mentioned his illness, Einstein says that he is "very happy that the theory of relativity gives you [i.e., Mach] pleasure." He closes his letter with the phrase "your admiring student" (Einstein 1995, 130).

There are three footnotes in his published writings of this time, in which Mach in an almost hesitant way tries to take advantage of the theoretical continuity between his ideas and the winning horse "relativity theory." That it is footnotes, in which Mach wants to point to this continuity, does not seem to be accidental. When in 1909/10 the seventy-two-year old, ailing Mach called in Philipp Frank to have him explain relativity theory, it was perhaps the first time in his life that Mach had asked for help in order to fully understand a physical theory. Sincerity and modesty prevented him, in my view, from trumpeting the continuity between his own ideas and those of Einstein, whose mathematical details he was unable to master by himself.

The first and most hesitant footnote one finds already in the republication of the "Conservation of Energy," whose explicit objective was to show that Mach's methodology bore healthy fruits:

> Space and time are not here conceived as independent entities, but as forms of the dependencies on one another. I subscribe then to the principle of relativity which is also firmly upheld in my *Mechanics* and *Wärmelehre*. Cf. "Raum und Zeit physikalisch betrachtet" in "Erkenntnis und Irrtum" 1905, H. Minkowski, "Raum und Zeit 1909" (Mach 1911, 95).

---

[13] Written on January 7, 1910 (OCA, my translation).

[14] Unfortunately, Mach's letters to Einstein all seem to be lost.

In the second footnote, Mach seems somehow encouraged by Einstein's gentle reaction to Mach sending him the "Conservation of Energy." It is in Mach's direct answer to Planck's attack. In this context it is, of course, particularly important to refer to the continuity between him and relativity theory:

> Even if the kinetic physical world picture, which in any case I consider hypothetical without intending thereby to degrade it, could 'explain' *all* physical appearances, I would still hold that the diversity of the world had not been exhausted, because for me *matter, space,* and *time* are also *problems*, which moreover, the physicists (*Lorentz, Einstein, Minkowski*) are also moving closer toward. (Blackmore 1992, 139)

The third and last footnote can be found in the paper "Sensory Elements and Scientific Concepts" (1910):

> Similarly, one will have to distinguish between metrical and physical space, with time included in the latter. I have already carried this out in my book *Erhaltung der Arbeit* (1872), p. 35, suggested on p. 56, and in *Erkenntnis und Irrtum* (1906), p. 434ff.; it is also a direction in which essential progress has been made by the work of A. Einstein and H. Minkowski. (Blackmore 1992, 125)

I think there can be no reasonable doubt that relativity came to the old Mach as a gift from heaven. Apart from these footnotes Mach did not publish a word about relativity. The reasons for such restraint are obvious, and I have mentioned them already. First of all, he was not a theoretical physicist; second, his time of commenting on the course of physics from an epistemological point of view, based on proper self-understanding, was over, as is evident from the episode with Philipp Frank who had to explain relativity to him; third, he was an old and sick man who did not feel fit to immerse himself in the quarrels of the scientific community of his day.

3.4 Mach's Alleged Refusal of Relativity Theory

Everyone would have viewed the situation as I have described it in the last section if Mach's *History of Optics* had not been published in 1921—five years after the death of its author. There to everyone's surprise, one can find in the preface, dated July 1913, a curt rejection of relativity and of all attempts of having counted himself among the "forerunners" of that theory:

> In this book I have set myself a task similar to that of the books on Mechanics and Heat, but the method of presentation corresponds mostly to that of the latter. I hope that I have laid bare, not without success, the origin of the general concepts of optics and the historical threats in their development, extricated from metaphysical ballast. Results of historical research have not, however, been accumulated here, and certain chapters, treated exhaustively in other quarters, are only cursorily surveyed. [...]

On account of my old age and illness I have decided, yielding to pressure from my publisher, but contrary to my usual practice, to hand over this part of the book to be printed,[15] while radiation, the process of light emission, Maxwell's theory, together with relativity, will be briefly dealt with in a subsequent part. The questions and the doubts arising from the study of these chapters formed the subject of tedious researches undertaken conjointly with my son, who has been my colleague for many years. It would have been desirable for the collaborated second part to have been published almost immediately, but I am compelled, in what may be my last opportunity to touch on my views on relativity theory.

I gather from the publications which have reached me and especially from my correspondence, that I am gradually becoming regarded as a forerunner of relativity. I am able even now to picture approximately what new expositions and interpretations many of the ideas expressed in my book on Mechanics will receive in the future from this corner.

It was to be expected that philosophers and physicists should carry on a crusade against me, for, as I have repeatedly observed, I was merely an unprejudiced rambler, endowed with original ideas, in varied fields of knowledge. I must, however, as assuredly disclaim to be a forerunner of the relativists as I withhold from the atomistic belief of the present school or church.[16]

The reason why, and the extent to which I refuse for myself the present-day relativity theory, which I find to be growing more and more dogmatical, together with the particular reasons which have led me to such a view – considerations based on the physiology of the senses, epistemological reservations, and above all the insight gained from my experiments – must remain to be treated in the sequel.

The ever-increasing amount of thought devoted to the study of relativity will not, indeed, be lost; it has already been both fruitful and of permanent value in mathematics. Will it, however, be able to maintain its position in the physical conception of the universe in some distant future as a theory which has to find a place in a universe enlarged by a multitude of new ideas? Will it prove to be more than an ingenious side note in the history of this science?
[...]

München-Vaterstetten, July 1913
Ernst Mach

A similar rejection one finds in quotations, allegedly from Ernst Mach, in the preface to the 9th edition of Mach's *Mechanics*, which was published in 1933, and

---

[15] Footnote in the original text: The printing was commenced in the summer of 1916, but at the wish of the author further experiments were to be tried and completed. The delay in the publication of the present book is due to the long absence of the person to whom this task was entrusted, as a result of his mobilization during the same summer, and to a series of adverse circumstances resulting from the conditions of the times.

[16] Footnote in the original text: "Scientia," Vol. 7, 4th year (1910), No. 14.

which, in a sense wisely enough, was not included in the English translation (Mach 1989)[17]:

Already in my younger years my attitude was relativistic, as one calls it today, because of my continual self-analysis and self-critique. Perhaps I could have pursued these things further, but, less occupied by my own thoughts, I was interested first in the prospect, in leaving behind the barriers of the past and the spell that great thinkers generally have around them. Therefore, I have always been actively preoccupied with sense physiological and psychological questions.

And then, one lives only once, and I wanted to take from the world, as far as it was accessible to me, as much as possible and was not able to fill my life completely with just one thought; I was unable to be economical. [Then Ludwig Mach continues:]

The founder of "Critical Physics," as he [i.e., Ernst Mach] is called by H. Dingler, was substantially involved in carving out the foundations on which the theory of relativity was erected later; but given the immensely quick development of that theory there was so much that did violence to the facts that he could not join in unconditionally. [Then Ludwig allegedly quotes again his father:]

"I did not view the Newtonian Principles as perfect and final, but in my old age I can as little accept relativity as a dogma as the existence of atoms and other things. Nothing could have been further from my thoughts than to produce acolytes, quite to the contrary this was prevented by my irresistible desire to keep off the beaten paths, which also caused a far reaching sympathy for dissenters, although not following them, for nothing is more beneficial than doubts raised by dissent." [Ludwig continues:]

As far as my recollection of these very early times goes, and as is also shown by his [i.e., Ernst Mach's] correspondence with Ludwig Lange, he was exclusively devoted to his work and driven by an uncompromising elemental desire for personal clarity. Pointing out the weaknesses of classical mechanics required in those days as much courage and selflessness as rejecting more modern trends today.

The economist of thought, who regarded death only as "destruction of a unity of economy of thought" could grant mathematics and theory only the role of instruments of enlightenment in the physical world picture.

Circumstances that need not be discussed here force me to be short and refrain from further elaboration. But on the basis of experimental insights that we have secured collaboratively after years of cooperation I am sure that E. Mach would have approved P. Lenard's original and ground breaking study (Über Äther und Uräther, Leipzig 1923)[18] as well G. von Gleich's

---

[17] I would like to thank Nicholas Rescher (Pittsburgh) for correcting my translation.

[18] This is hardly imaginable for at least two reasons: Lenard's central concepts are "Äther" and "Uräther," both prime examples for metaphysical 'conceptual monsters' (*Begriffsungetüme*), which Mach had fought all along. Apart from that, Lenard had become the

book that is imbued with the most serious struggle for reality and fully
appreciates mathematical investigations.

München-Vaterstetten, January 1933

Here is a list of ten good reasons that should have made people right away
suspicious about the authenticity of the two antirelativistic texts quoted above—even
without the benefit of hindsight that my forgery thesis provides.

(1) Einstein himself and the other physicists of his time were well aware of the
continuity between certain ideas of Mach and both special and general relativity theory.[19] Thus the harsh rejection of relativity in the *Optics* preface came to
Einstein as well as others as a great surprise. Nobody saw himself in a position
to systematically reconstruct in the context of Mach's work the rejection of relativity. Consequently, the standard reaction to the *Optics* preface, if there was any
reaction at all, until the 1970s was to ascribe it to Mach's illness and old age.[20]
People who saw things like that could not know, of course, that all extant documentation proves that Mach does not show any decline in his basic intellectual
or moral capacities—despite severe physical handicaps.

(2) When one takes into account that the rejection of relativity in the two texts
quoted came as a surprise to everyone until the 1970s, it should have stirred
up serious suspicion that both were published posthumously, and that in both
cases the same person—Ludwig Mach, i.e., Ernst Mach's eldest son—took care
of the publication.

(3) Now to the texts themselves. It is hardly imaginable that a scholar of Mach's
stature would use half of the preface to a book on the history of optics in order
to issue a harsh rejection of relativity, only to announce that the respective arguments would follow in a volume that was still to be published. Apart from that,
two weeks before his death in February 1916 Mach wrote to the Leipzig physicist
Otto Wiener that he had no intention of publishing his history of optics, which
he had begun to write more than twenty years previously, because its results
were outdated by new developments in the field. When Mach wrote this, he was
obviously unaware of the fact that at this point, initiated by Ludwig Mach, the
printing of the book had already been under way for several months.

(4) In the first section of the *Optics* preface we learn that this book, although it is on
the history of optics, has not accumulated "results of historical research." Well,
what else could it have "accumulated"? The answer is: nothing. The book is,
in fact, an accumulation of historical research, if one wants to express it in this

---

spearhead of an anti-Semitic campaign against Einstein, and the book was part of that campaign. Mach, however, throughout his life unshakably fought anti-Semitism.

[19] I have devoted most of the first chapter of my 1987 book to demonstrate and analyze these
continuities. I will touch upon some details in the next section. Generally, I would like to
state here that adequate evidence for all *factual* statements and claims I make in this paper
without giving evidence for them here is to be found in my book.

[20] The latest example of this reaction I know of is (Cohen 1998/99), who, as many others,
obviously has not yet taken account of my forgery thesis.

clumsy way. The nonsense actually written came about because Ludwig Mach made a little mistake when he practically copied from the preface to Mach's *Theory of Heat*. But *there* Mach wrote that "the reader should not expect to find here the results of a *search through the archives* [emphasis G.W.]" (Mach 1986, 1). Ludwig seems to have been rather hasty, or he had no clear idea about the difference between historical and archival work.

(5) Note the queer logic in the last sentence of the third section ("It would have been desirable...."): after the first part of the sentence, which talks about the desirability to contemporaneously also publish the collaborated (with Ludwig) second volume of the book, one would expect that in the second part of the sentence or later Mach would give some explanation for why this is not the case. But instead one finds the curious turn to his rejection of relativity which sounds perhaps even more curious in the German original.

(6) Further, the first footnote states only that "the printing commenced in the summer of 1916," i.e., several months after Mach's death in February, and that "at the wish of the author" there were still experiments to be "checked and supplemented." Thus, we are confronted with the strange situation that somebody sends out a manuscript to the publisher and tells him that it is not yet finished. Nonetheless, however, the publisher starts printing. I have never heard of such a procedure before or after. And it is unimaginable that Ernst Mach or any serious and decent publisher—and *Hofrat* (court counselor) Arthur Meiner, owner of the publishing house J. A. Barth, was serious and decent—could engage in a procedure of this sort.[21]

(7) In section four Mach seems to complain that in publications and correspondence he was "gradually becoming regarded as a forerunner of relativity." With respect to the alleged date of the preface (July 1913), this is nonsense. As far as I can see, up to July 1913 Mach had been mentioned only twice in the context of relativity, both times by Einstein: first in an article published in 1912 in a rather remote place, and second in the already mentioned Einstein/Grossmann paper of 1913, to which Mach—as we have seen—had reacted favorably in the undated letter to Einstein.[22]

(8) Look at the queer logic of the first section of the preface to the 9th edition of *Mechanik* quoted above. The "argument" that one has to reconstruct from this associative flight of ideas goes like this: Because of his "relativism" which was an expression of his "continual self-analysis and self critique" (!) Mach *could* have gone towards relativity theory. He did not do this, however, because he was "less occupied" by his own thoughts. One would now expect that Mach, instead,

---

[21] Apart from this Ludwig's claim in this footnote that the "printing commenced in the summer of 1916" (i.e., *after* Mach's death in February 1916) is simply a lie. As can be demonstrated from the correspondence, Ludwig had the printing begin—without the author's knowledge—already in 1915, when Ernst Mach was still alive and mentally alert.

[22] The 1912 paper ("Is There a Gravitational Effect Which is Analogous to Electrodynamic Induction?") (Einstein 1996, 126–129) was published in a quarterly of forensic medicine, the *Vierteljahrsschrift für gerichtliche Medizin und öffentliches Sanitätswesen,* which makes it unlikely that Mach had seen it.

stayed with the thinking of the physical tradition. But the text continues to the contrary: Mach wanted to rather transcend that tradition, and "therefore" he took an interest in sense physiology and psychology.

(9) It is hardly imaginable that Ernst Mach could have appreciated Lenard's partly anti-Semitic pamphlet, as is suggested by Ludwig.

(10) Finally, a general remark: When one reads these texts, visibly confused in style[23] and content as they are, one can hardly imagine that they were written by one of the greatest philosopher-scientists of the 19th century, whose other writings show stylistic security and even elegance.

As noted above in (1) the standard reaction to Mach's apparent rejection of relativity was more or less to not take it into account; and if one took account of it one ascribed it to Mach's poor physical and, supposedly, also mental health. To me this reaction is in a sense humane, because it takes into account the concrete living situation of the researcher.

Things began to change in the 1970s in the wake of the antiempiricist turn in history and philosophy of science that began in the 1960s with Thomas S. Kuhn's *Structure of Scientific Revolutions*. Now it is historians of science who start to trouble the waters again. Now is proudly presented what is missing in the forged texts, i.e., reasons, or better alleged reasons for Mach's apparent rejection of relativity as a consequence of his sensualistic epistemological and/or his empiricist methodological orientation. In each case these explanations have been at least induced by Mach's apparent rejection of relativity.[24]

The argumentation goes basically like this: Mach's epistemology/methodology bears rotten fruits. The main proof for this is: it led Mach to reject relativity and to stubbornly refuse to acknowledge the existence of atoms.[25] The authors, then, usually set out to provide arguments for the rejection of relativity from a supposedly Machian perspective.[26] In the end we are confronted with the historical hypothesis that it was exactly those arguments which induced Mach (or, at least, should have induced him) to reject relativity. The arguments, however, which are presented by various authors, are not in every case of the professional quality that is standard in the history of science. This is no surprise, because, among other things, they build upon and at the same time attempt to explain something that, in fact, did not occur.

---

[23] See, e.g., the embarrassing self-congratulation ("not without success") in the first section of the *Optics* preface.

[24] I conclude this from the fact that nobody who adduces such arguments fails to mention Mach's apparent rejection of relativity.

[25] This resumes Planck's aggressive strategy against Mach, mentioned above, in the notorious talk "Die Einheit des physikalischen Weltbildes" ("The Unity of the Physical World Picture") in 1909 that with respect to Mach quotes in its last sentence the biblical saying "ye shall know them by their fruits."

[26] Such arguments were, as we have seen, precisely missing in the two texts quoted above, in which Mach allegedly rejects relativity.

Apart from that they implicitly draw a picture of Mach's personality which does not have any historical foundation.[27]

3.5 The Forgery Hypothesis

Notwithstanding the ten warning signals I have identified in the last section nobody had ever doubted the authenticity of the antirelativity texts quoted above until I published in 1987 a book in German (*Mach I, Mach II, Einstein und die Relativitätstheorie: eine Fälschung und ihre Folgen*) in which I gave ample and what I still take to be convincing documentary and literary evidence for claiming that without reasonable doubt the two texts are forged. The forger is Mach's son Ludwig Mach (1868–1951).[28]

Documents that I have literally found in the attic, together with the already available material, show firstly how things really went and what circumstances led Ludwig Mach to forge those texts. Secondly, the evidence can explain every detail of the forged texts down to dating and to peculiarities of formulations. Thirdly, the evidence shows how poor a great deal of (English language) Mach scholarship has become.

To tell this story would take a little book. The story is at the same time sad and fascinating. At its center is Ludwig Mach's disturbed personality. And then the story is about money, or better: the lack thereof and of desperate attempts at procuring it, of ambition, drugs, lies, Nazism, anti-Semitism, and so on. Of the ingredients of a good story, alas, only intriguing women are missing.

Sad as this story is, it has a very satisfying result, which certainly would have pleased Einstein. It clears the troubled waters: *Ernst Mach ab omni naevo vindicatus*.

References

1. Archives

(OCA): Open Court Archive, Morris Library, Southern Illinois University at *Carbondale.*

---

[27] The main proponents among many others of this line of research are John Blackmore and Gerald Holton. Whereas my results on the whole have been received positively by researchers (cf., e.g., the essay review Di Salle 1990), Blackmore and Holton (cf., e.g., their respective papers in Blackmore 1992) have criticized them extensively. I will deal with this criticism on another occasion.

[28] A short account of the forgery thesis in English is (Wolters 1989). For those who prefer reading Chinese there is a Chinese translation of this text in the 1988 volume of *Zi ran bian zheng fa tong xung* (*Journal of Dialectics of Nature*), 16–26.

2. Publications

Blackmore, John (ed.) (1992). *Ernst Mach—A Deeper Look: Documents and New Perspectives*. Dordrecht: Kluwer.

Cohen, I. Bernard. (1955). "An Interview with Einstein." *Scientific American* 193 (July), 69–73.

Cohen, Robert S. (1998/99). "Mach and Einstein. A Posthumous Dialogue." *Philosophia Scientiae* (Nantes) 3(2), 167–182.

Di Salle, Robert. (1990). "Critical Notice." *Philosophy of Science* 57 (1990), 712–723.

Earman, John. (1995). *Bangs, Crunches, Whimpers, and Shrieks: Singularities and Acausalities in Relativistic Spacetimes*. New York: Oxford University Press.

Einstein, Albert. (1955). *The Meaning of Relativity*, 5th ed. Princeton: Princeton University Press.

——. (1995). *The Collected Papers*, vol. 5 (The Swiss Years: Correspondence, 1902–1914 – English Translation). Martin J. Klein, A. J. Kox and Robert Schulmann, eds. Anna Beck, trans. Princeton: Princeton University Press.

——. (1996). The Collected Papers, vol. 4 (The Swiss Years: Writings, 1912–1914 – English Translation). Martin J. Klein, ed. Anna Beck, trans. Princeton: Princeton University Press.

——. (1997). *The Collected Papers*, vol. 6 (The Berlin Years: Writings, 1914–1917 – English Translation of Selected Texts). A. J. Kox, Martin J. Klein and Robert Schulmann, eds. Alfred Engel, trans. Princeton: Princeton University Press.

Frank, Philipp. (1910). "Das Relativitätsprinzip und die Darstellung der physikalischen Erscheinungen im vierdimensionalen Raum." *Zeitschrift für physikalische Chemie, Stöchiometrie und Verwandtschaftslehre* 74, 466–495.

Herneck, Friedrich. (1966). "Ernst Mach and Albert Einstein." In *Symposium aus Anlaß des 50. Todestages von Ernst Mach. Veranstaltet am 11./12. März 1966 vom Ernst-Mach-Institut Freiburg i. Br.* Freiburg: Ernst Mach Institut, 45–59, 60–61 (discussion).

Mach, Ernst. (1911). *History and Root of the Principle of the Conservation of Energy*. Philip E. B. Jourdain, trans. and annot. Chicago: Open Court and London: Keagan Paul, Trench, Trübner & Co.

——. (1923). *Populär-wisssenschaftliche Vorlesungen*, 5th ed. Leipzig: J. A. Barth.

——. (1943). *Popular Scientific Lectures*, 5th ed. Thomas J. McCormack, trans. LaSalle, IL: Open Court.

——. (1953). *The Principles of Physical Optics: An Historical and Philosophical Treatment*. John S. Anderson and A. F. A. Young, trans. New York: Dover (reprint of the first Engl. ed. London: Methuen, 1926).

——. (1986). *The Principles of the Theory of Heat Historically and Critically Elucidated*. Brian McGuiness, ed. Dordrecht: Reidel.

——. (1988). *Die Mechanik in ihrer Entwicklung. Historisch-Kritisch dargestellt.* Gereon Wolters, ed. and introd. Darmstadt: Wissenschaftliche Buchgesellschaft (reprint of the 9th edition Leipzig: Brockhaus, 1933).

——. (1989). *The Science of Mechanics: A Critical and Historical Account of Its Development*. Thomas J. McCormack, transl. Karl Meger, introd. LaSalle, IL: Open Court.

McCormmach, Russell. (1970). "H. A. Lorentz and the Electromagnetic View of Nature." *Isis* 61, 459–497.

Norton, John. (1995). "Mach's Principle before Einstein." In *Mach's Principle: From Newton's Bucket to Quantum Gravity*. Julian B. Barbour, ed. Boston: Birkhäuser, 9–57 (*Einstein Studies*, vol. 6).

Planck, Max. (1969). *Vorträge und Erinnerungen*. Darmstadt: Wissenschaftliche Buchgesellschaft.

Schilpp, Paul Arthur, ed. (1970). *Albert Einstein. Philosopher-Scientist*. LaSalle, IL: Open Court, 3rd edition.

Wolters, Gereon. (1987). *Mach I, Mach II, Einstein und die Relativitätstheorie: eine Fälschung und ihre Folgen*, Berlin: de Gruyter.

——. (1989). "Phenomenalism, Relativity and Atoms: Rehabilitating Ernst Mach's Philosophy of Science." In *Logic, Methodology and Philosophy of Science VIII: Proceedings of the Eighth International Congress of Logic, Methodology and Philosophy of Science, Moscow 1987*. J. E. Fenstadt *et al.*, eds. Amsterdam: North-Holland, 641–660.

Part II

Contexts of the Relativity Revolution

4

Tilling the Seedbed of Einstein's Politics: A Pre-1905 Harbinger?

Robert Schulmann

Bethesda, MD, USA

Einstein's political interests only crystallized in 1919 after his most significant discoveries in physics lay behind him. In the critical years leading up to his greatest scientific achievements his political and social interests lay fallow, their moral roots unarticulated. One challenge this poses for the Einstein biographer is determining if and when latent political sensibilities emerged and what forms they assumed. Perhaps the earliest indication that he felt socially out of joint is provided in his comment on schoolmates in Munich:

> Among the children anti-Semitism was alive particularly at elementary school. It was based on the children's remarkable awareness of racial characteristics and on impressions left from religious instruction. Active attacks and verbal abuse on the way to and from school were frequent but usually not all that serious. They sufficed, however, to establish an acute feeling of alienation already in childhood.[1]

A comment by sister Maja adds weight to this autobiographical account of marginality. In the 1924 biography of her brother, she claims that it was this very sense of alienation early on and the resulting identification with the lot of other young Jews that first kindled his commitment to a Hebrew University in Palestine.

But what about the Swiss period which played such a decisive role in the development of Einstein's science? Are there trace elements here as well of a sensibility or at least of influences that inform Einstein's politics in Berlin?

In what follows, I very consciously have set aside the question of influences that his readings in the philosophies of Schopenhauer and Spinoza may have had on Einstein. I examine what I think is a promising though indirect set of connections which Einstein enjoyed in the period between 1905 until his true political awakening in the 1920s and beyond. These are connections that are admittedly oblique, that were not laid down by intellectual appropriation but by personal contact. I'll be speaking of personal linkages and congruities of thought, not of cause and effect. The exercise runs on two tracks:

---

[1] Einstein to Paul Nathan?, 3 April 1920, (CPAE 2004, Vol. 9., Doc. 366).

1) establishing biographical, if not hard-and-fast linkages between the young Einstein and members of a pacifist organization in Zurich;
2) more importantly, comparing Einstein's later political convictions with the aims of that society.

I begin on a personal note. Back in the palmy days of the Einstein Papers Project, the editorial writ was conceived broadly. John Stachel, the founding editor, envisioned the Project as a comprehensive editorial enterprise encompassing and integrating all aspects of Einstein's contributions and personal development. This was a very welcome approach for a nonscientist such as myself. On my first research trip to Switzerland, I found a letter in the director's copybook in the manuscript department of the ETH Library. In it Albin Herzog, Director of the Swiss Polytechnic (later the ETH) retained a copy of a letter of 1895 to a certain Gustav Maier. Responding to Maier's letter of the previous day, Herzog never mentioned Einstein by name, but urged Maier to encourage the "Wunderkind" in his charge to obtain a leaving-certificate before entering the Polytechnic. All the more so as the boy was well under the regular age of admission. Parenthetically, Einstein, at 16, was two years underage. Should Maier and the boy in question reject this advice, however, he will make an exception and allow him to sit for the entrance examination. The story of Einstein's failure—due probably to deficiencies in his knowledge of Swiss literary and political history—and the decision to spend the next school year in Aarau to finish his secondary-school education is well known and needs no further elaboration.

Stumbling over the Herzog/Maier letter, however, had a number of ramifications. One line of pursuit led us to discover that Maier had an extensive correspondence with Jost Winteler, Einstein's father figure in Aarau. Unfortunately, this exchange of letters, though rich in other regards, scarcely touched on their relationship to Einstein. One letter did, however, explain that it was at Maier's suggestion that Einstein arrived at the Winteler doorstep in neighboring Aarau in the autumn of 1895. Reading Winteler's memoir "Erinnerungen aus meinem Leben" of 1917 it also became clear that the friendship between Maier and himself had grown out of a common interest in political matters. Winteler, it turned out, was a frequent and avid participant in the meetings of a pacifist organization in Zurich—of which Gustav Maier was president.

I will mention another clue to Maier's continuing intellectual interaction with Einstein in a moment. More immediately I'd like to review briefly the evidence for Winteler's tutelage. The historian Jost Winteler, a native of the Toggenburg region of eastern Switzerland, had been trained at the University of Jena in Thuringia, where he developed a deep-seated aversion to the great-power triumphalism that accompanied the founding of the German Reich. He was also a freethinker who had fled the stifling atmosphere of ultramontane Fribourg in Switzerland, where he taught before taking an appointment at the cantonal school in Aarau. Here he taught in the humanistic Gymnasium, not in the technical school where Einstein was registered. Evenings were long and the increasingly close ties that bound the sixteen-year-old to the Winteler family provided a congenial atmosphere for wide-ranging discussions. Eldest daughter Anna recounted that:

Einstein was a pleasant, very solid member of the household and never a spoilsport. He enjoyed having scientific discussions often exhibiting a good deal of humor, occasionally bursting out in laughter. He seldom went out in the evenings, often he worked, but more often he sat with the family around the table, where one read to the others or discussions were held. He joined the family on Sunday walks, on which occasion he would philosophize with my father or articulate his thoughts on physics (Lüscher 1944).

To what degree Winteler influenced, even if only tangentially, latent political musings of the young Einstein remains unclear of course. Certain it is though that politics figured in conversations around the table and during their walks. After Hitler's seizure of power forty years later, Einstein wrote sister Maja: "I often think of Papa Winteler and the prophetic accuracy of his political views. I always felt it but never this clearly and forcefully." Broader social commentary seems also to have had its place. A hint of this is preserved in a contemporary letter of Einstein to Winteler, in which he delivers one of his most quoted epigrams: "What you have said about German professors is not at all exaggerated... Slavish obedience to authority is the greatest enemy of the truth" (July 1901).

After graduating in autumn 1896 from the cantonal school in Aarau, Einstein entered the Poly and renewed his relationship with Maier. What was the bond between the wealthy banker who had withdrawn from an active business role and the freshly minted Poly student? In a letter to his Swiss biographer Carl Seelig many years later, Einstein provided the answer: "On my arrival in Switzerland from Italy, I was advised by an old friend of my father, a Mr. Gustav Maier, who was the chairman of a society for Ethical Culture there." Writing to Gustav and his wife in 1922 on the occasion of their golden wedding anniversary, Einstein remembered his frequent visits to the Maier household during the fours years he was a student in Zurich: "Your hospitable home was always open to me during my days at the Poly, even when I came down from the Uetliberg with mud on my boots." As with Winteler in Aarau, details of what was discussed with Maier are unknowable. But what is evident from Maier's memoirs "Siebzig Jahre politischer Erinnerungen und Gedanken" of 1918 is that the author spent all his waking hours consumed by the affairs of the Ethical Culture Society in Zurich over which he presided. It is not, I think, too much of a stretch to assume that some conversations with Einstein in that hospitable home centered on the *Ideenwelt* of the Ethical Culture Society so dear to Maier's heart.

Given the tantalizing web of relationships, the Einstein editors tried actively in the mid-1980s—as Volume 1 of the *CPAE* was going to press—to demonstrate unequivocally an immediate connection between the "Wunderkind," Maier, Winteler, and the Ethical Culture Society. In vain. But with the luxury afforded me twenty years later of following such elusive leads, I return to the question of connections.

The case for latent interest in the Ethical Culture Society on the part of the young Einstein rests first of all on the potential of close personal associations and an affinity of attitude. More forcefully, it is based on a congruence of values that underlie the objectives of the Society and the goals of the politicized Einstein of later years.

In order to allay the kind of misunderstanding that frequently dogs the question of Einstein's political views, it is useful initially to point out that some critics have denied a political character to his focus on the moral individual. These critics have argued that Einstein's was an ethical stance and not a political one. While one can certainly debate whether such a position is adequate to effecting his ideal of a just society, I contend that Einstein's concentration is unquestionably a political one, perhaps most closely resembling that of a nineteenth-century liberal. His goal, after all, was a social system that rejected gross income inequality and the exploitation of the economically vulnerable, while placing equal if not more importance on the traditional liberal goal of self-realization of the individual.

As a way of setting the stage for considering the influence on Einstein of his other Swiss mentor, Gustav Maier, let us first discuss the latter's seminal role in the German and Swiss Ethical Culture Societies. Second, we need to lay out the fundamental tenets of the Society and compare them with the aims of the political Einstein decades later.

Gustav Maier was born in 1844 in Einstein's hometown of Ulm. Culturally assimilated and religiously indifferent like many Jews of the generation of Einstein's parents, he was eager to avail himself of the economic opportunities only recently and still grudgingly conceded to his *Stammesgenossen*. Far more successfully than Hermann Einstein, he mastered the economic upsurge generated by the industrial and financial revolution in the second half of the nineteenth century. By the age of thirty-two he was manager of the Deutsche Reichsbank in Ulm, where he made the acquaintance of the Einstein family. A year after the Einsteins left Ulm for Munich, Maier became Director of the Deutsche Reichsbank in Frankfurt. Unconventional he certainly must have been. Where many financiers and bankers of the Second Reich unselfconsciously rode the crest of wealth accumulation, Maier had his eyes on other goals. In 1886, to counter the saber-rattling mood that accompanied the founding of the Reich, he helped found the Frankfurt Peace Union and shortly after became a member of the German Society for Ethical Culture. At the tender age of fifty, he withdrew from his active business interests and moved to Switzerland to devote his considerable energies to social, economic, and pedagogic issues. In spite of writing, speaking, and travel commitments, he couldn't resist the offer of a bank directorship as a kind of peripheral challenge.

Maier's departure from Germany sprang from the same sense of resentment at arrogant authority as that which fueled Einstein's resolve only a few years later to leave the school disciplinarians of Munich behind. After relocating to a small town north of Zurich, the retired banker threw himself into a seemingly endless round of lectures and travel. One such foray took him to Berlin. While attending a party in high society, he reduced his audience to nervous giggles and suppressed hilarity by parodying a song that had recently been composed by no less a personage than the Emperor himself. Aside from many other quirks, William II fancied himself both a musical heir to his illustrious ancestor, Frederick II, and the father of a German fleet to rival England's. In a thinly veiled allusion to himself, the song took the form of a melodic homage to Aegir, the Norse god of the sea. Maier set his parody to the Emperor's melody and mimicked its lyrics in a piece entitled "To the Lord of Rain in the Tiergarten." To compound the insult, he proposed a toast, in which he made a prediction about the political future of Germany. Forty years hence, Germany would be a republic, a socialist and a liberal would share the chair of German chancellor, and William was president of the United States of Europe and threatening to declare himself dictator. As Maier happily recounted in his memoir, the audience was suitably shocked. But as a Swiss citizen, he was able to escape a charge of *lèse-majesté*.

In the early 1890s Maier wrote and published incessantly, for the most part in the weekly house organ of the Ethical Culture Society. He tackled every social subject under the sun, including freemasonry, anti-Semitism, and anarchy. He was particularly critical of the *Allgemeine Elektrizitäts-Gesellschaft* which he called a noxious hybrid of industrial and financial capital, whose monopolistic practices extended as far as Genoa. Could Einstein's father, the victim of such practices both in Munich and in northern Italy, have said it any better?

After helping to promote a Swiss branch of the Ethical Culture Society in Zurich, Maier became a member of its board in January 1896. The range of political opinion among the directors was broad, not surprising in that one of the fundamental principles of the Ethical Culture Society was the equality of all parties. A number of members were socialists. Maier stood more in the tradition of the German Liberal Party, which condemned nationalist prejudices, advocated free trade and religious tolerance, and was ill-disposed toward state socialism. At the third meeting of the Society in April, midway through Einstein's year in Aarau, Maier gave a speech urging Swiss authorities to offer political asylum not so much to individuals as to the authentic German *Geist* under siege from reactionary forces in his home country. Judging from the description of the lecture in the organization's weekly, the reception was very favorable: "The audience greeted the address with enthusiasm, particularly Prof. Winteler of Aarau."

In 1898, he completed his most important work, *Social Movements and Theories up to the Modern Labor Movement*, which went through nine editions in the prestigious Teubner Verlag. Believing that his antimilitarist position needed a more explicit political platform, Maier joined the Zurich Peace Society in 1900, remaining active in its ranks until 1914. In doing so, he also cemented a bond with an individual who had been present at the creation of both the Ethical Culture Society and the Peace Society in Germany. Astronomy professor Wilhelm Julius Foerster bridged the worlds of establishment academia and outspoken political dissidence. He had been rector of the University of Berlin in the early 1890s and concurrently active in the organized peace movement. He enters our saga not only because of his collegial friendship with Maier, but also because of a special relationship with Einstein. It was he after all who was to join Einstein at the outbreak of the First World War in signing a condemnation of the German Army's incursion into neutral Belgium, the so-called Counter-Manifesto. To get a sense of the appeal of the Ethical Culture movement to Maier and Foerster—and indirectly Einstein—let us turn now to a quick sketch of that society's development and principles.

The Ethical Culture Society, begun in 1876, is of American origin. Its founder Felix Adler was Professor of Hebrew and Oriental Literature at Cornell University. Though his father was head of Temple Emanu-El, one of the most influential

Jewish Reform congregations in New York City, Adler early on advocated ridding religion of its superstitious traditions in order to better focus on the ethics which he felt was central to any human community. This concept found great resonance in post-*Kulturkampf* Germany with Foerster as its chief proponent. In a talk at the founding meeting of the Society Foerster stated its aim as overcoming the struggle for life between human beings and its detrimental effects on the fate of nations and of their populations. War was quite simply "mankind's most serious infantile disorder." A staunch advocate of the Enlightenment's vision of the perfectibility of man, Foerster observed that "wars will certainly become impossible once we attain a higher level of civilization." His faith in the progress of mankind blended easily with the observation that social organizations must be governed by scientific principles.

In a speech given in Eisenach in autumn 1893, Maier elaborated on the movement's goals. The relations between nations must be based on peaceful understanding. Without it, hostility arising out of wounded national pride will inevitably lead to perdition. Two pillars support the goal of *Verständigung*. The first is a universal education which makes short shrift of rote learning and the discipline of the barracks in the classroom. The aim after all is to foster critical, freethinking individuals not to encourage herd mentality. This universal education must be available to all regardless of social standing. Here Maier stressed the importance of adult education, evening classes for the working man, and full educational equality for women.

The second pillar was a religious sensibility which stressed respect for all persons and valued ethical behavior over theological rigor. Confessional differences were irrelevant. The individual, family, and community ought to be encouraged to pursue what Maier called "a cult of the natural, of truth, and beauty." In the explicitly political sphere, partisan conflict should be avoided as much as possible. As with tolerance of confessional differences, the movement should stand above party wrangling.

A number of aspects would certainly have appealed to a headstrong young man only beginning to find his way in Switzerland. While followers of the Ethical movement generally shared common beliefs about what constituted ethical behavior and the good, individuals were encouraged to develop their own personal understanding of these ideas. Ethical principles were viewed as being related to deep, not arbitrary truths about the way the world worked. At the same time members recognized that complexities rendered understanding of ethical nuances subject to continued dialogue. Paramount was the observation that too often disputes over religious or philosophical doctrines distracted people from actually living ethically and doing good. "Deed before creed" lay at the heart of the movement's teachings.

By many, the organization was accused of a starry-eyed optimism. Its refusal to take partisan sides in the overheated political climate at the end of the nineteenth century also came at a high cost. Social democrats considered the Ethical Culture Society a bourgeois, elitist enterprise which by emphasizing personal morality diverted energies from the workers' movement. Conservative circles regarded it as crypto-socialist. The churches dismissed the idea of faith without theological and liturgical underpinnings. For obvious reasons nationalists despised their ideal of pacifism. In the end the movement was doomed to ineffectiveness. In an age of mass mobilization it organized its core appeal around the concept of personal moral engagement with the world. The very absence of a practical political program almost certainly sealed its fate. The outbreak of the First World War crippled the Ethical Culture Society as it did much of the German peace movement.

Yet I would argue that what most distinguished Einstein's early nonscientific writings are those very features which characterized the movement of Maier, Winteler, and Foerster. It was a profoundly ethical, and unprogrammatic view of society and the relations between nations. Some of the charges leveled against the Society can similarly be brought against Einstein. In what follows let us look briefly at excerpts of some of the writings that I think bear this out. Because of constraints of space, I will only delve into the first decade after 1914.

Einstein arrived in Berlin in spring 1914 at the age of thirty-five to take up a position in the Prussian Academy of Sciences. The outbreak of the war less than four months later was a rude awakening for him and led to his first tentative grappling with political issues. What he retained from an earlier period was the sensibility and vocabulary of a citizen of the republic of letters, not of someone prepared to engage in public debate or political action at close quarters. He expressed his sense of moral outrage to his friend Paul Ehrenfest, writing that he was experiencing "a mixture of pity and disgust," an instinctive recoiling from "Europe in its madness." Two months later insult was added to injury when ninety-three German intellectuals published a "Manifesto to the Civilized World" declaring that "the German Army and the German people are one." Einstein was shocked by this appeal to a narrow nationalism as advocated by the very cultural elite to which he had recently been recruited. He responded, as noted above, by co-signing the counter-manifesto with Foerster that reached out to all Europeans:

> The struggle raging today will likely produce no victor; it will probably leave only the vanquished behind. Therefore, it seems not only *ethically fitting*, but rather bitterly *necessary, that intellectuals of all nations* marshal their influence such that the *terms of peace shall not become the cause of future wars*.

As heirs of Germany's greatest poet, Einstein and other German cultural paladins were called upon by the Goethe Society in 1915 to contribute to a patriotic commemorative volume. Einstein seized the opportunity in a piece called "My Opinion on the War" to display his visceral feelings on the phenomenon of war in general. When the editors of the volume deemed the essay too unpatriotic and requested that Einstein delete the harshest passages, he agreed. Yet he could not resist pointing out that he shared Tolstoy's view of patriotism as a mental disorder. The essay was published in 1916 without the offending paragraphs:

> The psychological roots of war are—in my opinion—biologically rooted in the aggressive characteristics of the male creature. We "lords of creation" are not the only ones who sport this crown: some animals—the bull and the rooster—surpass us in this regard. This aggressive tendency comes to the fore whenever individual males are placed side by side... I will never forget what honest hatred my classmates felt for years for the first-graders

of a school in a neighboring street. Innumerable brawls occurred, resulting in many a gash in the heads of the boys... The thirst for power and greed should, as in the past, be treated as despicable vices; the same applies to hatred and contentiousness. I do not suffer from an overvaluation of the past, but in my opinion we have not made progress on this important point; on the contrary, we have declined. Every well-meaning individual should work hard at improving himself and his personal surroundings in this regard. Then the grave afflictions which plague us in such terrible fashion today will vanish... Subtle intellects of all times have agreed that war is one of the worst enemies of human development, and that everything must be done to prevent it... How can a powerless individual creature contribute to reaching this goal? Should everyone perhaps devote a considerable portion of his abilities to politics? I really believe that the intellectually more experienced people in Europe have sinned in their neglect of general political questions; yet I do not regard the pursuit of politics as the path to an individual's greatest effectiveness in this area. I rather believe that everyone should act privately in such a way that those traits, which I have discussed in detail earlier, can no longer represent a curse to society.

The root of the problem in Germany as Einstein defined it in a letter to Europe's foremost pacifist, Romain Rolland, was the worship of force:

Through military victory in 1870 and successes in the fields of commerce and industry this country has arrived at a kind of religion of power... This religion holds almost all intellectuals in its sway; it has driven out almost completely the ideals of Goethe and Schiller's time. I know people in Germany whose private lives are guided by a virtually unbounded altruism, but who were waiting with the greatest impatience for the declaration of unlimited submarine war.

Five years after his arrival in Berlin, the German Empire lay in ruins and a hopeful new Weimar Republic struggled to find its footing. Meanwhile Einstein had vaulted to international fame. His cautious testing of the political waters during the war gave way to an increasingly urgent engagement with social and political issues. Two factors determined his greater access to and interest in the political realm. One was thrust on him in late 1919 after a British solar expedition confirmed a prediction of his general theory of relativity. Now a world figure, pronouncements on public affairs came to be expected of him. The other was a conscious redefinition for himself of the role of the intellectual with access to the media in the mass society of the twenties. Though old elites had not been swept away in Germany, Einstein now recognized that the ineffectiveness to which the intellectual had been relegated under the Empire could be overcome by appeals to an international press that created his popular fame and continued to further it. His views on moral and political issues, confined to his correspondence and the occasional publication during the war, would now be broadcast throughout the world and carry great weight. It is under these circumstances that Einstein joined the political fray while always maintaining his detachment from it.

After he became ever more engaged in political concerns, Einstein remained without partisan political affiliations. Though he called for the founding of the liberal German Democratic Party, he took pains to deny publicly that he was a member. Eager to educate the German public about the events of the war and to counter feelings of revenge against the Allies, he joined a nonpartisan private commission to evaluate German war guilt. He proved equally evenhanded in assessing blame for the turmoil of the early Weimar years, expressing his distrust of extremes on the right and left, though he saw the greater danger from the former, particularly after the military putsch of March 1920. Throughout the war years and after, the one constant in Einstein's attitude toward public life was his unwavering commitment to the international character of all intellectual activity.

About to embark on a trip to the United States to raise money for a Hebrew University in Palestine in spring 1921, he told an interviewer that

> the internationalism that existed before the war—the internationalism of culture, the cosmopolitanism of commerce and industry, the broad tolerance of ideas—was essentially right. There will be no peace on earth and the wounds inflicted by the war will not heal until this internationalism is restored.

It took three and a half years after the war for the bitterness to subside sufficiently between the archenemies, Germany and France, for Einstein to pay a visit to the western neighbor.

On his return he addressed a peace congress of fifteen pacifist organizations, including a French delegation, on the floor of the Reichstag:

> I believe that the condition in which the world finds itself today makes it not only a matter of idealism but one of dire necessity that a greater unity of spiritual and material cooperation among nations be attained. Those of us who are aware of these needs must no longer think in terms of 'What can be done for my country?' Rather one should ask: 'What should my country do to lay the groundwork for a larger world community?' For without that greater community no individual nation will be able to endure for long. I believe that only a person who is constantly aware of this and who strives to meet every situation in life with this thought in mind will be able to break through the frozen barriers that divide cultures. I consider it extremely important that, whenever the possibility arises, men of different languages and holding different political and cultural ideas should communicate with each other across their frontiers—not with the feeling that something can be squeezed out for their own benefit or for that of their country, but in an attempt to bridge the gulf between different, relatively independent intellectual groups. Only thus can we hope to achieve a political unity—at least in Europe—that will enable us to survive economically and safeguard our intellectual heritage. Only then will life be worth living.

Whatever his initial debt to the Ethical Culture movement, Einstein retained a remarkable consistency in his political views throughout his life. I end with an excerpt from an essay he published in 1930, entitled "What I Believe," republished with his permission in 1954 as "The World as I See It."

> The really valuable thing in the pageant of human life seems to me not the political state with its panoply of parties, but the creative, sentient individual, the personality; it alone creates the noble and the sublime, while the herd as such remains dull in thought and dull in feeling... This topic brings me to that worst outcropping of herd life, the military system, which I abhor. That a man can take pleasure in marching in fours to the strains of a band is enough to make me despise him. He has only been given his big brain by mistake; unprotected spinal marrow was all he needed. This plague-spot of civilization ought to be abolished with all possible speed. Heroism on command, senseless violence, and all the loathsome nonsense that goes by the name of patriotism—how passionately I hate them! How vile and despicable seems war to me! I would rather be hacked in pieces than take part in such an abominable business. My opinion of the human race is high enough that I believe this bogey would have disappeared long ago, had the sound sense of the peoples not been systematically corrupted by commercial and political interests acting through the schools, parties, and press.

Written several decades after his brushes with Winteler, Maier, and Foerster, the tenor and content of this statement hauntingly evokes their legacy. One cannot of course draw a direct causal link nor claim that Einstein's thinking on political issues grew out of an institutional affiliation with the Ethical Culture Society, whether in Europe or America. As a member of the board of leaders of the New York Society wrote to Einstein's Swiss biographer, Carl Seelig, in July 1953: "I cannot say that we have any special relationship to Albert Einstein. Some of us have corresponded with him and visited him. We are in great sympathy with his viewpoint and his work."[2] A congratulatory message that Einstein prepared for the Society's 75th anniversary two years earlier suggests the same. While praising the Society's ideals ("There is no salvation for humanity without 'ethical culture'"), he in no way attributes to it a formative influence on him.[3]

Lacking any exposure to the working-class movement in the waning decades of the German Empire, Einstein maintained a lifetime wariness of the mass organizations of the left which sprang up or developed a new lease on life in the Weimar republic, though he expressed his solidarity with some of their goals. In later life he demonstrated philosophical sympathy for socialism with a human face, including many of its economic principles, but he never identified with the intellectual tradition of the European labor movement or the Marxist legacy. Instead, he continued to place his faith in appeals to reason by a liberal intelligentsia, which, in availing itself of the decorous and principled use of manifestos, might best guide the fortunes of republican Germany. The search for a harmonious society was to reinforce

---

[2] ETH, *Wissenschaftshistorische Sammlungen*, 304: 200.
[3] 5 January 1951, in Einstein Archives, 28–904

his unease with "a political culture of problem-solving by negotiation, dispute, and majority vote" as well as with the Marxist concept of class struggle (Goenner 2003).

It is certainly true that various strands of late nineteenth-century free-thinking may have had an impact on Einstein's budding political sensibilties during his Swiss years. Other movements than that of the Ethical Culture Society placed a premium on the moral, self-determining, creative individual, rejected the herd mentality and militarism, called for internationalism, and abjured the narrow hatreds of nationalism. It is perhaps even true that in browsing through the *Parerga and Paralipomena* of Schopenhauer, which Einstein frequently refers to in letters to Mileva Marić during their years at the ETH, he may have come upon that philosopher's defense of limited government and his plea that the state "leave each man free to work out his own salvation." Weighing more heavily than these conjectures, however, it is Einstein's close friendship with his early mentors and their passion for the Ethical Culture movement that present us with the only concrete case of personal influence in the realm of politics during his Swiss period.

References

CPAE. (2004). The Collected Papers of Albert Einstein. Bd. 9: The Berlin Years: Correspondence, January 1919 – April 1920. D. Kormos Buchwald, R. Schulmann, J. Illy, eds. Princeton, N.J.: Princeton University Press.

Goenner, Hubert. (2003). Albert Einstein and Friedrich Dessauer: Political Views and Political Practice. Physics in Perspective 5: 21–66.

Lüscher, Edgar. (1944). Albert Einstein in Aarau. Schweizerische Lehrerzeitung 89: 622–623.

5

The Early Reception of Einstein's Relativity among British Philosophers

José M. Sánchez-Ron

Universidad Autónoma de Madrid, Spain

I've become very good friends with the Ehrenfest's children and play with them very much. I also have to study Cassirer's manuscript, which is less amusing. These philosophers are peculiar birds.

Albert Einstein[1]

The theory of relativity is primarily a physical theory. But anyone who wishes on that account (as has sometimes been done) to deny the philosophical character and scope of the theory, has failed to realize that the physical and philosophical viewpoints can by no means always be rigorously separated from each other; that each, on the contrary, passes into the other as soon as they address themselves to dealing with the highest and most general concepts of physics.

Moritz Schlick (1922).[2]

5.1 Introduction

The early reception of Einstein relativity theories among British philosophers is studied, paying special attention to the debate idealism versus realism, and to the ideas and contributions of individuals such as Viscount Haldane, Charles D. Broad, H. Wildon Carr, T. Percy Nunn, Dorothy Wrinch, Bertrand Russell, and Alfred North Whitehead, as well as Arthur Eddington, the scientist most influential in this field, and the somewhat philosophically minded scientist Alfred A. Robb. Foremost among the conclusions which can be derived from this case study is that no matter how different some of their basic philosophical standpoints could be, most British

---

[1] Albert Einstein to Elsa Einstein, 19 May 1920, reproduced in Kormos Buchwald, Sauer, Rosenkranz, Illy, and Holmes, eds. (2006: 264–265). For the English translations I have used the versions included in Hentschel, transl. (2006: 164).

[2] English translation in Schlick (1979b: 343).

philosophers argued that Einstein's new theories favored their philosophical positions. Einstein's relativity, both the special and the general theories, has maintained an intimate relationship with philosophy, and thousands of pages have been written dealing with its philosophical implications. More problematic is the question of the geographical distribution (i.e., the nationalities) and ideas of those who wrote those pages.[3] Were they mainly German-speaking individuals, people like Schlick, Carnap, Reichenbach, Cassirer, and Popper? Were they scientists turned, briefly or more permanently, philosophers (as, for example, Poincaré, Mach, or Bridgman), or philosophers who recognized the fecundity and relevance to their discipline of Einstein's contributions (as in the cases of Bergson and Russell)?

I will not attempt to answer here these questions. Instead, I will deal with the early reception of Einstein's relativity among British philosophers. My study suggests that it is not true what John Passmore (1966: 334) wrote in his splendid, although partial ("it is written," confesses its author in the preface, "from an English point of view"), book *A Hundred Years of Philosophy*:

> Professional philosophers, as distinct from philosophical journalists, have been singularly little affected by the revolution in physics. They have been inclined to suspect that, like a great many other revolutions, the revolution in physics raised no new philosophical problems and settled no old ones, for all the dust and fury. As well, it must be confessed, professional philosophers have been intimidated by the mathematics into which philosophical physicists so gratefully sink at crucial points of their reasoning; nor has the philosophical crudity of what they could understand led philosophers to expect any considerable degree of illumination from what passes their comprehension.

It is true, however, that immediately after having said so, Passmore weakened his previous affirmation by admitting that: "There are, of course, exceptions. That very remarkable philosopher-statesman, R. B. Haldane, in his widely read *The Reign of Relativity* (1921), attempted to incorporate Einstein's theory within the Hegelianism to which he so faithfully adhered; Alexander welcomed what he took to be a partial confirmation of his theory of space-time; Russell wrote popular expositions of the new physics and shows traces of its influence; and attempted to make philosophical sense out of contemporary developments in physics. On the whole, however, one must turn to the philosopher-scientists, of whom there have been more than enough, for philosophically-toned accounts of recent science."

How many and how rapidly or slowly philosophers reacted to relativity is important, but there are other questions that deserved to be considered. One of them is the

---

[3] Contributions to these questions are Ryckman (2005) and Hentschel (1990); for the Italian case Maiocchi (1985) as well as Reeves (1987: 206–208), for the Soviet reaction Graham (1972) and Vucinich (2001); though not dedicated exclusively to the philosophers' reaction, some interesting details in the French case are contained in Biezunski (1981) and Paty (1987), as well as in Biezunski, ed. (1989), where the correspondence that Einstein maintained with the philosopher Émile Meyerson and the scientist interested in philosophy André Metz is reproduced.

nature of their reaction; were they favorable or not to the new physical theory? Again, of course, the answer to such question might perhaps be different in different countries or philosophical communities (thus, it seems that the German case was rather specific, as a glimpse of Einstein's already published correspondence suggests).

5.1.1 General Aspects of the Reception of Einstein's Relativity in Great Britain

Leaving aside questions like the preceding ones and concentrating on the early reception of relativity in Britain, I must make clear beforehand that my discussion is based mainly in the study of a number of books and of two journals, *Mind* and the *Proceedings of the Aristotelian Society*, both recognized as the leading philosophical journals in Britain. It is one of my contentions that this material covers a substantial part of the British philosophical landscape, especially that related to Einstein's relativity contributions, at least as it concerns the period 1915–1930.

On the basis of such information, several facts emerge. First, the bulk of the philosophers' reaction was concentrated in a very narrow period, 1920–1922, with not very many contributions before, mainly Whitehead's *An Enquiry Concerning the Principles of Natural Knowledge* (1919), as well as a couple of articles by Wildom Carr, one read at the Aristotelian Society in the Session of 1913–14 (*Proceedings of the Aristotelian Society*, Vol. XIV), and another ("The Metaphysical Implications of the Principle of Relativity") published in the *Philosophical Review* (1915).[4] Of course, what made the difference was the announcement of the results of the 1919 British eclipse expedition, made public at the now legendary November 6 joint meeting of the Royal Society and the Royal Astronomical Society, at Burlington House, London. That event made Einstein a world celebrity. Prior to that, very few philosophers paid attention to the special theory of relativity and almost none to the general theory, nor to Einstein's previous attempts to cope with the gravitational field in a relativistic framework.[5] Take, for instance, an article published in 1915 by Charles Dunbar Broad (1887–1971), the Cambridge Trinity College man who became in 1920 (after having been associated with St. Andrews University) Professor of Philosophy at Bristol University and who, as we shall see later on, participated actively in the philosophical treatment of Einstein's relativity in Britain. The paper, which had the suggestive title of "What do we mean by the question: is our space Euclidean?" (Broad 1915), does not include any mention of Einstein's works on gravitation; actually, Einstein's name is not mentioned at all, even though there is a passing reference to the "modern theory of Relativity in Electrodynamics." The scientists named were Lorentz, FitzGerald and, especially, Minkowski and Robb. It might be of interest to quote the following comment made by Broad: "I do not say that the facts of electrodynamics do force us to conclude that either Euclidean space or Newtonian bodies

---

[4] We must take also into account the two books published by Robb (1911, 1914), but he was a scientist, although, as I will argue later, with a "philosophical sensitivity."

[5] It must be taken into account in this sense that early in 1915 Adriaan D. Fokker (1915) published an article in *Philosophical Magazine* discussing Einstein and Grossmann's 1913 theory.

are unreal in the present sense; but I take this as an illustration of the sense of reality under discussion, and remark in passing that these facts have actually led certain mathematicians and philosophers, e.g., Minkowski and Mr. Robb of Cambridge,[6] to elaborate a new system of geometry and a new system of physics which shall consistently fit all the facts." That is, Broad, as many others at the time, was particularly attracted by Minkowski's four-dimensional space-time and Robb's presentations of Einstein's special theory of relativity.

As a matter of fact, Broad's 1915 article contained assertions which will prove to be incompatible with general relativity, as well as with Einstein's post-1913 metric theories. He said, for instance, that "space and time cannot be conceived as capable of causal action on matter."

Broad's paper was commented on the following year, 1916, by J. E. Turner; but neither in his note nor in Broad's reply—both entitled "The nature and geometry of space" (Turner 1916, Broad 1916)—were Einstein's name or the general theory of relativity mentioned.

A second point that should be stressed is that, as far as journals are concerned, it was *Mind* where most articles dealing with the philosophical aspects of relativity—mainly general relativity—appeared. Based at Oxford and edited by the famous G. E. Moore, this philosophical journal was the organ of the Mind Association, which, at the time we are considering, had 137 members, among them men like Samuel Alexander (1859–1938), the author of the influential *Space, Time, and Deity* (1920),[7] F. H. Bradley, H. Wildon Carr, J. S. Mackenzie, J. M. E. McTaggart, Bertrand Russell, as well as A. J. Balfour, the tory politician who became Prime Minister between 1902 and 1905. Richard Burdon Haldane was also a member of the association and, as we shall see, an important figure in the reception of relativity in Britain—one of the great spokesmen of science in the country.

Mind was important for the philosophical discussion of relativity, not only because of the articles it published but also because it gave fast, detailed, and informed reviews of many books dealing with Einstein's relativity theories, Charles Broad being, by far, the most prolific reviewer.[8] Thus, among its pages we find reviews of: Einstein's *Relativity, the Special and the General Theory: A Popular Exposition*, Eddington's *Space, Time and Gravitation*, Erwin Freundlich's *The Foundations of Einstein's Theory of Gravitation*, Wildon Carr's *The General Principle of Relativity in Its Philosophical and Historical Aspect*, Cassirer's *Zur Einstein'schen Relativitätstheorie*, Schlick's *Space and Time in Contemporary Physics*, Haldane's *The Reign of Relativity*, Ebenezer Cunningham's *Relativity, the Electron Theory, and Gravitation*, Alfred Robb's *The Absolute Relations of Time and Space*, Hermann Weyl's *Space, Time and Matter*, Broad's *Scientific Thought*, and Whitehead's *The Principles of Natural Philosophy*, *The Concept of Nature*, and

---

[6] The work Broad mentioned here is Robb (1914).

[7] I will say more about this work later on.

[8] See the bibliography of Broad's writings included in Schilpp, ed. (1959). Besides his reviews mentioned below, Broad reviewed another important philosophical book dedicated to relativity, that nevertheless was published a few years later: Émile Meyerson's *La déduction relativiste* (Meyerson 1925, Broad 1925).

The Principle of Relativity.[9] It is worthwhile to point out that, with one exception, all these reviews were published in 1921.

A further point to be remarked is that it was only a small number of philosophers who contributed, through the pages of *Mind* and the *Proceedings of the Aristotelian Society*, to the philosophical discussion of relativity; the most active of them being Broad, Wildon Carr, Turner, Dorothy Wrinch, and Whitehead, with others like Thomas Greenwood, Ross, and Taylor making only occasional appearances.

To the previous list, and besides Eddington and Robb, we must add the name of one physicist: the young F. A. Lindemann, who together with his father A. F. Lindemann, as John Stachel (1986) pointed out, had taken notice of Einstein's new theory of gravitation as early as 1916.[10] Together with Eddington, W. D. Ross, and Broad, Lindemann participated in a Symposium organized by the Mind Association during the International Congress of Philosophy that was held at Oxford in September 1920; however, his role in the philosophical discussion of relativity was negligible.[11]

Of course, it seems more than probable that some of the philosophers "joined the race," to say so, without really being prepared, as in the case of W. D. Ross, mainly remembered today for his magnificent editions of major works by Aristotle, and also, though less so, for his works in ethics. In the Oxford Symposium I have just mentioned, Ross limited his comments to the special theory, while the discussion was dedicated mainly to the general theory. He excused his more than probable ignorance of Einstein's new theory of gravitation by saying that "until one can be satisfied about the truth" of the restricted theory "it would be useless to discuss the general theory which is an extension and in some degree a correction of it."[12] The only argument that really concerned general relativity put forward by him was one that had been used several years before—even before the final version of the theory (1915)—by Max Abraham (1913). "Incidentally," Ross stated, "one's faith in the argument should surely be somewhat shaken by the fact that the constant relative velocity of light, which is asserted in the special theory, is denied in the general." It was Charles Broad who answered Ross's queries.[13]

Another remarkable fact concerning philosophy and general relativity in Britain refers to the temporal distribution of the articles and books published. As noted before, the discussion was initiated in full in 1919–1920 if we take into account Whitehead's *An Enquiry Concerning the Principles of Natural Knowledge*—but it reached its climax almost immediately, decaying very rapidly.

While, as pointed out before, the reviews of relativity books were especially numerous in 1921, in the case of articles the prime year was 1922: Between five and ten papers (depending on the criteria adopted) were published in *Mind*. That year

---

[9] These reviews appeared in Taylor (1921), Broad (1921a), Ross (1921), Broad (1921b), Carr (1921), Broad (1921c), Jeffreys (1923), Wrinch (1924), and Broad (1920, 1921d, 1923b).

[10] Lindemann and Lindemann (1916–1917). Their paper was submitted on December 4, 1916.

[11] Eddington, Ross, Broad, and Lindemann (1920). More will be said about this Symposium in the section dedicated to Eddington.

[12] Ross in Eddington, Ross, Broad, and Lindemann (1920: 423).

[13] Broad in Eddington, Ross, Broad, and Lindemann (1920: 430).

the Aristotelian Society organized an interesting discussion in London, published in the *Proceedings* of the Society the same year (Carr, Nunn, Whitehead, and Wrinch 1922), in a volume that also includes two papers, by Whitehead and Greenwood, dealing with relativity. I will analyze that meeting later.

If we focus on the temporal distribution of the publication of monographs, we find it equally concentrated: Whitehead's *An Enquiry Concerning the Principles of Natural Knowledge* (1919) and *The Concept of Nature* (1920), Eddington's *Space, Time and Gravitation* (1920), Wildon Carr's *The General Principle of Relativity in Its Philosophical and Historical Aspect* (1920), Haldane's *The Reign of Relativity* (1921), Robb's *The Absolute Relations of Time and Space* (1921), Whitehead's *The Principle of Relativity* (1922), Nunn's *Relativity & Gravitation* (1923), and Broad's *Scientific Thought* (1923).[14]

Several comments can be made in light of what has been said so far. First, in some aspects British philosophers reacted to the new theory in the same way as other collectives, although perhaps with more intensity. I am thinking of the fact that the years 1920–1921 saw the publication of a large number of books—and of general articles, of course—trying to explain the physics of general relativity. That boom, therefore, also affected British philosophers, or, better, some British philosophers.

Why, however, was that momentum not sustained?

To answer such a question, it would be necessary to know more about what happened in other countries. However, it can be argued that the loss of momentum suffered by the discussions dealing with the philosophical implications of relativity among British philosophers might have been due to the nature of the topics in which they were interested, the research programs they cultivated. Theirs was a rather academic philosophy, in most cases not particularly dependent on science. Nothing of the sort of interest aroused in Central Europe—remember the case of the Vienna Circle—took place in Great Britain. Perhaps because no Helmholtz, Mach, or Boltzmann—that is, scientists with deeply rooted philosophical interests—had existed in nineteenth-century Britain, there did not appear men like Schlick, Carnap, or Reichenbach. In other words, the sort of philosophy promoted by the Vienna Circle, and, we should add, thinking of Reichenbach and Popper, by people subjected to its influence, was particularly well-suited to continue being developed, as happened in fact through the methodology of science (the case of Popper), or through the different efforts dedicated to the axiomatization of scientific theories.

Related to that situation is the fact that in the 1920s not many of the British philosophers had the scientific training of their German and Austrian counterparts. In general, the British were by education just philosophers; citizens of an academic world dominated by the study of the ancient classics, morals and ethics, philosophy of religion and of politics; it is true that they were also interested in logic as well as in the foundations of mathematics (Russell and Whitehead), a topic which required a deep mathematical knowledge, but it was a science quite different from

---

[14] Something similar happened in other countries. In France, for instance, Henri Bergson's *Durée et simultanéité* was published in 1922 (Bergson 1922), though Meyerson's *La deduction relativiste* appeared in 1925 (Meyerson 1925).

physics, in its methodology as well as in its purposes. Not one of those who appear in the present study were physicists transformed in philosophers, as was the case with Moritz Schlick and, though to a lesser extent, Hans Reichenbach.

Significant in this sense is what Haldane (1921: 39–40) wrote in *The Reign of Relativity*. After pointing out that "Einstein's language is that of the mathematician, and mathematics is his chief instrument," he stated that "into the purely mathematical aspects of such doctrine as that of Einstein, few philosophers are rash enough to attempt to enter. Mathematicians' talk is an admirably lucid language which is exclusively their own." Also illuminating is what Charles Broad candidly wrote in the preface of his *Scientific Thought* (Broad 1923: 4): "In some. . . chapters the reader will find a number of mathematical formulae. He must not be frightened of them, for I can assure him that they involve no algebraical processes more advanced than the simple equations which he learnt to solve at his mother's knee. I myself can make no claims to be a mathematician: the most I can say is that I can generally follow a mathematical argument if I take enough time over it."

5.1.2 Einstein's Relativity and British Philosophy

The years when relativity was developed were rather special as far as British philosophy is concerned. During the thirty years prior to the advent of Einstein's relativity, the British philosophical landscape had tended increasingly to group itself around two central and directly opposite and contradictory positions. One took "the thing-hood of the thing," to use Wildon Carr's no doubt obscure expression, as the typical reality, and emphasized the objectivity of existence and the subjectivity of the knowing relations.[15] The other took the mind and its activity as the immediate intuition of reality, conceiving the fundamental universal reality as the original activity of which the individual mind is the type. The chief influences in consolidating the first, or realistic position (also "new realist" or "neo-realist"), came from American philosophers, although there were notable exponents in England— the already mentioned Samuel Alexander, for example, with his book *Space, Time, and Deity* (Alexander 1920)— while the most striking formulation of modern idealist theory, or "neo-idealist," came from Henri Bergson (1859–1941) in France and Benedetto Croce (1866–1952) in Italy.[16]

---

[15] Carr (1921: 464).

[16] In spite of what its title might suggest, Alexander's *Space, Time, and Deity* belongs to a philosophical tradition and interests very different from and with no connection, as far as I know, to Einstein's relativity. The product of a series of lectures (the "Clifford Lectures") Alexander delivered in Glasgow in 1917 and 1918, that book was a bold adventure in speculative philosophy. What this Australian-born philosopher installed in Manchester defended is that space-time (for him Space-Time) constitutes the primordial reality from which everything evolves and where everything is found. "Space-Time," he wrote (Alexander 1920: 342, vol. I), "the universe in its primordial form, is the stuff out of which all existents are made. . . But it has no 'quality' save that of being spatio-temporal or motion." As to Croce, it might be interesting to recall what Passmore (1966: 301) wrote: "reality, according to Croce, is 'spirit'; to be real, that is, is to play a part in one of mind's diverse activities.

At first, those new realists tried to turn away from the methods of their predecessors, and particularly from those of the idealists; they sought, for instance, to bring philosophy into close relation with science by endeavoring to adopt its modes of investigation. Thus, they gave to the non-mental world the status of being self-subsistent and completely independent of the mind of the observer. "Actual objects did not exist for them in the mind," pointed out Haldane (1921: 261), "but in a medium that is independent of the mind."

Given that situation, it is not surprising that when in 1919 relativity became a world celebrity, a debate arose among British philosophers—who had felt very vividly the polemic idealism versus realism—as to what were the philosophical implications, or meanings, of the new physical theories, especially of the gravitational one (general relativity). As we shall see, what characterized that debate is that almost everybody—realists and idealists being the main contenders, although other positions were also defended—tried to appropriate relativity to favor his/her own philosophical standpoint. A layman, even a scholar not versed on philosophical matters, would have been lost when reading the different—often philosophically incompatible—statements, counter–statements, and counter-counter-counter-statements made by British philosophers during that brief and intense debate. As the always perceptive Bertrand Russell (1926: 331) noted: "There has been a tendency, not uncommon in the case of a new scientific theory, for every philosopher to interpret the work of Einstein in accordance with his own metaphysical system, and to suggest that the outcome is a great accession of strength to the views which the philosopher in question previously held. This cannot be true in all cases; and it may be hoped it is true in none. It would be disappointing if so fundamental a change as Einstein has introduced involved no philosophical novelty."[17]

5.1.3 A Stranger in a Strange Country: Arthur Eddington

Before entering fully in the professional philosophical territory, I will briefly mention some of Arthur Stanley Eddington's (1882–1944) ideas. He was, as is well known, a prominent figure in the introduction and defense of Einstein's relativity in Great

---

Croce opposes any sort of 'transcendence', any suggestion that there is an entity which lies wholly outside the human spirit, whether it be Kant's thing-in-itself, or the Christian 'God', or the naturalist's 'Nature'. Whatever mind cannot find within itself Croce rejects as mythical."

[17] Not exactly in the same vein, Schlick (1922, 1979b: 344) also commented on the reception of relativity among philosophers: "No small number of philosophers have attempted to deny or to minimize this transforming influence; they have adopted a very radical stance towards the theory of relativity. Some, that is, have supposed themselves forced into a flat rejection of Einstein's ideas, as being opposed to their 'commonsense', while others have thought they could derive these ideas from old and long-familiar philosophical themes with as little trouble as if they were something quite obvious, and hence constituted no essential advance whatever in philosophical knowledge." The differences in Russell's and Schlick's comments might perhaps be due to the different situation in their respective philosophical communities.

Britain, as well as an important contributor to its development. His role in the 1919 British eclipse expedition and the *Report on the Relativity Theory of Gravitation* (Eddington 1919a) he wrote were particularly important in the diffusion of general relativity in Britain. However, and although less known, he played also a significant role in the philosophical debate concerning relativity in his country.

To *Mind* he made two contributions. The first, a paper entitled "The meaning of matter and the laws of nature according to the theory of relativity" (Eddington 1920b), was of not much philosophical interest.[18] With the second, a seminal contribution to the already mentioned September 1920 Oxford Symposium organized by the Mind Association during the International Congress of Philosophy, in which, as we already know, also participated Ross, Broad, and Lindemann, it was different.[19] Actually, both the Symposium in general, and Eddington's article in particular, were a sort of starting point for the debate on the philosophical implications of general relativity in the pages of *Mind*, where the contributions were published.

In his presentation, Eddington defended the physical dimension of the theory (Eddington, Ross, Broad, and Lindemann 1920: 415): "It is natural for a scientific man to approach Einstein's theory of Relativity with some suspicion, looking on it as an incongruous mixture of speculative philosophy with legitimate physics. There is no doubt that it was largely suggested by philosophical considerations, and it leads to results hitherto regarded as lying in the domain of philosophy and metaphysics. But the theory is not, as its nature or in its standards, essentially different from other physical theories; it deals with experimental results and theoretical deductions which naturally arise from them." However, he recognized also that there was a domain in it of philosophical interest (Eddington, Ross, Broad, and Lindemann 1920: 416): "I would emphasize then that the theory of relativity of time and space is essentially a physical theory, like the atomic theory of matter or the electromagnetic theory of light; and it does not overstep the natural domain of physics. But, speaking to an audience of philosophers, I shall not hesitate to trespass beyond the borderline on

---

[18] In the second paper, which I will consider immediately, Eddington summarized this article saying that in it he had tried to show "the exact method by which, starting from a relation undefinable in its absolute character, we arrive from a single source at the physical quantities which describe space and time on the one hand and the quantities which describe things on the other hand. If we describe the character (or geometry) of space and time throughout the world, we at the same time necessarily describe all the things in the world." Eddington, Ross, Broad, and Lindemann (1920: 420).

[19] Eddington, Ross, Broad, and Lindemann (1920). Though his participation was not included in the published version, Thomas Greenwood also commented on relativity at the Oxford International Congress, as he pointed out in a paper he published afterwards in *Mind*. There he stated that he had "maintained, as against the crude subjectivism of Prof. Eddington and the extreme absolutism of Dr. Ross, that the Theory of Relativity cannot be taken as a crucial system to decide between idealism and realism. This view is generally shared by scientists, who cannot profess a great sympathy for the doctrines of those philosophers who endeavor to drive science on to a certain ground which is not its own. I now go further, and hold that the Theory of Relativity is rather a thing prejudicial to idealism, at least as Prof. Carr proposes it" (Greenwood 1920: 205).

my own account. I shall be a stranger in a strange country; and the lurking pits might well intimidate me, if I did not rely on your friendly hands to pick me out."

Actually, Eddington's ideas as put forward in the Symposium and in *Mind* were soon expanded in his classic book—very much referred to among British philosophers (and not only among them)–*Space, Time and Gravitation* (Eddington 1920a).[20]

It would take me far too long to discuss the contents of this wonderful book, in which Eddington showed his outstanding literary abilities, as, for example, in the famous sentences with which he ended the work (Eddington 1920a: 201): "We have found a strange foot-print on the shores of the unknown. We have devised profound theories, one after another, to account for its origin. At last, we have succeeded in reconstructing the creature that made the foot-print. And, Lo! it is our own."[21] Nevertheless, it is necessary to remark that *Space, Time and Gravitation* contained comments which could not pass unnoticed by a philosophical mind, as, to cite an example, the following one (Eddington 1920a: 197): "Our whole theory has really been a discussion of the most general way in which permanent substance can be built up out of relations; and it is the mind which, by insisting on regarding only the things that are permanent, has actually imposed these laws on an indifferent world. Nature has actually very little to do with matter; she had to provide a basis–point event; but practically anything would do for that purpose if the relations were of a reasonable degree of complexity. The relativity theory of physics reduces everything to relations; that is to say, it is structure, not material, which counts. The structure cannot be built up without material; but the nature of the material is of no importance."[22] And at that point, he linked his discussion with Bertrand Russell's *Introduction to Mathematical Philosophy*, from which he quoted a long paragraph.

Among philosophers, Haldane was one of the first who reacted to Eddington's ideas as explained in *Mind* and in *Space, Time and Gravitation*, dedicating several pages of his *The Reign of Relativity* to it (Haldane 1921: 104–121). Again,

---

[20] John Stachel (1986; 2002: 462–464) has made interesting comments on Eddington's *Space, Time and Gravitation*.

[21] I cannot resist quoting another, less known, of Eddington's statements. He made it in one of his first general papers dealing with general relativity, one he published in *The Contemporary Review*. Here are its last sentences (Eddington 1919b: 643): "It is not necessary to picture scientists as prostrated by the new revelations, feeling that they have got to go back to the beginning and start again. The general course of experimental physics will not be deflected, and only here and there will theory be touched. If our common view of nature is a dream, still our business is with the fabric of our vision. Search for truth may reveal that there is underlying the dream of a real world, and the true laws of the phantasms are to be expressed in terms of—let us say—indigestion. The discovery is epoch-making, and much that appeared inexplicable in the dream-world is now traceable to its source. But the dreamer goes on dreaming."

[22] Note how philosophically different what Eddington said here—"The structure cannot be built up without material; but the nature of the material is of no importance"—is from Alexander's idea mentioned in note 16, that "the universe in its primordial form, is the stuff out of which all existent are made."

it would take us too long to enter into this question here, though it must be pointed out that Haldane made efforts to relate some of what Eddington said to Kant's *Critique of Pure Reason* as well as to Whitehead's position. It is also of interest to point out that, after his "philosophical initiation," Eddington would make in the future further philosophical comments, no matter that they could be very brief, even in some of his more technical works. Thus, in the Introduction to *The Mathematical Theory of Relativity*, he wrote, in a philosophical vein (Eddington 1923: 3):

> The study of physical quantities, although they are the results of our own operations (actual or potential), gives us some kind of knowledge of the world-conditions, since the same operations will give different results in different world-conditions. It seems that this indirect knowledge is all that we can ever attain, and that it is only through its influences on such operations that we can represent to ourselves a 'condition of the world.' Any attempt to describe a condition of the world otherwise is either mathematical symbolism or meaningless jargon. To grasp a condition of the world as completely as it is in our power to grasp it, we must have in our minds a symbol which comprehends at the same time its influence on the results of all possible kinds of operations. Or, what comes to the same thing, we must contemplate its measures according to all possible measure-codes—of course, without confusing the different codes. It might well seem impossible to realize so comprehensive an outlook; but we shall find that the mathematical calculus of tensors does represent and deal with world-conditions precisely in this way. A tensor expresses simultaneously the whole group of measure-numbers associated with any world-condition; and machinery is provided for keeping the various codes distinct. For this reason the somewhat difficult tensor calculus is not to be regarded as an evil necessity in this subject, which ought if possible to be replaced by simpler analytical devices; our knowledge of conditions of the external world, as it comes to us through observation and experiment, is precisely of the kind which can be expressed by a tensor and not otherwise. And, just as in arithmetic we can deal freely with a billion objects without trying to visualize the enormous collection, so the tensor calculus enables us to deal with the world-condition in the totality of its aspects without attempting to picture it.

And he added: "Having regard to this distinction between physical quantities and world-conditions, we shall not define a physical quantity as though it were a feature in the world-picture which had to be sought out. *A physical quantity is defined by a series of operations and calculations of which it is the result*," a phrase that, of course, has an operationalistic dimension of the sort Percy Bridgman (1927) expressed in his famous *The Logic of Modern Physics*, but made long before the American physicist wrote it.

5.1.4 In Defense of Idealism: Wildon Carr and Viscount Haldane

It was pointed out before that it is reasonable to suppose that the philosophers'
reactions towards Einstein's relativity theory might depend on the country considered. Something similar can be said in connection with the philosophical traditions or schools, which also might be associated with countries. Such contentions
are supported when one considers the case of idealism, a theme presented intensely
in the philosophical discussions dealing with relativity in Britain. In this sense,
while studying the Italian case, Barbara Reeves (1987: 206–208) explained that the
philosophical debate there "must be situated in the context of the neo-idealist near-
hegemony in philosophy and the longstanding neo-idealist devaluation of scientific
and mathematical knowledge and research. Epitomized in the *Logic* of Benedetto
Croce, published as a book in 1908 but based on a paper first published in 1905,
this devaluation denied that science or mathematics were creative activities of the
human spirit, that they had philosophical value as knowledge, and they could lead
to truth. Scientific and mathematical research resulted not in genuine knowledge but
only in classification schemes or techniques useful for practice." And she added the
following comparative comment, which is particularly interesting for us:

> Therefore, most Italian neo-idealists just dismissed the question of the
> philosophical consequences of relativity, unlike the situation in England,
> for example. There such idealists as Wildon Carr, Collingwood, Eddington,
> and Whitehead made relativity considerations central to their writings in this
> period. Ugo Spirito, young neoidealist follower of Giovanni Gentile, found
> the efforts of Wildon Carr and his countrymen Alessandro Bonucci and
> Antonio Aliotta to advocate or at least discuss an idealist interpretation of
> relativity 'arbitrary' and the idea of an idealist science 'a contradiction in
> terms,' since an idealist science would no longer be science but philosophy.
> Furthermore, he denied any connection between the physical principle of
> relativity and philosophical relativity, finding the physical theory of relati-
> vity to be 'only pure idealism.'[23]

Having made these general comments, let us go to the British case, and in
particular to two philosophical books dedicated to relativity that were published in
Britain, namely, Wildon Carr's *The General Principle of Relativity in its Philosophical and Historical Aspect* (Carr 1920), and Richard (Viscount) Haldane's *The Reign
of Relativity* (Haldane 1921), which shared, in spite of their different contents and
tactics, the same philosophical position: a defense of idealism.[24] First, however, a
few words about who they were.

---

[23] See Spirito (1921).

[24] It might be helpful to recall what Bertrand Russell (1917: 588) wrote about idealism and
materialism: "Idealist say all matter is really mind; Materialist say all mind is really matter;
American Realists say mind and matter are the same thing, but neither mental nor material."
About the New Realism, he said: "In all these respects the New Realism has aimed at
inculcating a greater restraint...The New Realism has tried to invent a logical method by
which the legitimate conclusions and no more can be extracted from any body of data.
This modest and scientific spirit on the constructive side has perhaps been concealed from

Herbert Wildon Carr (1857–1931) was for many years Secretary of the Aristotelian Society. He had joined it in the second year of its existence (1881), when he was employed as a broker on the London Stock Exchange. By 1907 he had made enough money to retire and devote himself full-time to philosophy, serving as professor of Philosophy in the University of London from 1918 (he was then 61 years old) until he moved to the University of Southern California in the late 1920s.

More can be said about Richard Burdon Haldane (1856–1928), a great British statesman who made significant contributions to philosophy.[25] Between 1905 and 1912 he was War Secretary, and subsequently, in 1912 until 1915, Lord Chancellor (in 1911 he was created peer). In Ramsey MacDonald's first short Labour ministry (1924), he was again Lord Chancellor. He was also the first Chancellor of the University of Bristol and elected Lord Rector of the University of Edinburgh, and wrote a number of philosophical books, among them, and besides *The Reign of Relativity*, *Pathway to Reality* (1903) and *The Philosophy of Humanism* (1922). According to Frederick Maurice (1939: 98), Haldane's biographer, *The Reign of Relativity*—the work most important for the present discussion—was "the result of years of thought and of much reading and research. Einstein had acted as a spur to what was already moving in his mind." However, the book went far beyond the study of the physical theory of relativity, as he conceded in the Preface when he wrote: "The topics of this book are Knowledge itself and the relativity of reality to the character of Knowledge."

Haldane's *The Reign of Relativity* was a success: it went through three editions in six weeks, a large edition was required in America, and it was translated into German, French, and Russian. It might have had its origins not in Einstein's theories but in Haldane's philosophical thoughts, but its success surely owed much to the wide interest which Einsteinian relativity had aroused (also to the public personality of its author). Moreover, it was a good book, as Eddington pointed out in the Haldane Memorial Lecture ("The reign of relativity") he delivered at Birkbeck College, London, in 1937[26]:

> I have often been asked, How far did Lord Haldane understand Einstein's theory? I will try to answer that. To say that he understood it better than any other British pure philosopher at that time, would, I am afraid, be a poor compliment. For what it is worth it is undoubtedly true. Nor is it much of a testimonial to say that he understood better than many who are perfectly familiar with the mathematical calculus, and can manipulate tensors as easily as numbers. As I have already said, the heart of the theory is not in the mathematics, though the mathematics is very necessary in working out its physical consequences. It is all too common to see the formulae misapplied

---

readers by a certain arrogance on the critical side, for the New Realism holds that, though there are an infinite number of metaphysical theories which may be true and comparatively few which must be false, it so happens that all the ambitious systems with the exception of that of Leibniz, belong to those few that are demonstrably false."

[25] For Haldane's biography, see Maurice (1937, 1939).

[26] Quoted in Maurice (1939: 104–105).

because the conditions which determine their applicability are misunderstood. Perhaps we may come to the point this way: if you think that Lord Haldane, who (as he tells us) has been a philosophical relativist for forty years, jumped at the superficial resemblance of the new physical theory and distorted it to support his views, you are probably mistaken. In the first place, that was not Lord Haldane's way—I am sure those who are familiar with his other activities will bear me out. He had a wonderful power of concentration which enabled him to get to the essentials of an unfamiliar subject. In the second place, that was not his approach. He discussed Einstein's theory not for the purpose of claiming its support, but to see whether he could conscientiously support it himself. If you read the long fifth chapter of his book, you will see that it is the Einsteinians who are arraigned and cross-examined to see whether, beneath their unphilosophical language, beneath their glib use of the word 'relativity', they have any true perception of the relativistic outlook.

It happens, moreover, that Haldane had good relations with Einstein. When the famous physicist visited England in June 1921, he offered Einstein to stay at his London home—the offer was accepted—and organized a private dinner party in honor of his guest. It was on that occasion when Einstein met the Archbishop of Canterbury, Davidson, who, as J. J. Thomson (1936: 431) publicized, asked Einstein what was the relevance of the theory of relativity for religion, receiving the hasty reply: "None! Relativity is a purely scientific matter and has nothing to do with religion."[27]

Apparently, Haldane—who was fluent in German, having studied in Göttingen— and Einstein went carefully through the chapters of *The Reign of Relativity* that dealt with the relativity theory, with the result that Haldane made several important alterations in the third edition of the book.[28] Besides, Haldane was well aware that Einstein's visit would help the sales of his book, which was published, let us remember, that same year. In this sense, on May 12, he wrote to his editor, John Murray (Clark 1972: 337): "Einstein arrives here in the early days of June, and his advent will make a market for us which we must not lose." Philosophy, physics, and business meld in a very productive way.

Coming back to philosophy, and in particular to Haldane and Carr's defense of idealism, we have that Haldane (1921: 261) declared that "the New Realist's point according to which actual objects do not exist in the mind, but in a medium that is independent of mind, is apparently taken to be that of self-subsistent space and time,

---

[27] The story, as well as the details of Einstein's visit (he went also to Manchester and Oxford), is fully described in Clark (1972: 336–344).

[28] It seems that Einstein enjoyed his visit to England as well as his relationship with Haldane. Thus, on leaving England he wrote to Mrs. Haldane that "one of the most memorable weeks of my life lies behind me. Visiting this country for the first time I have learned to marvel at its splendid traditions and treasures of knowledge... The scientific talk with Lord Haldane has been for me a source of pure stimulation, and so has the personal intimacy with him and his remarkable knowledge." Quoted in Maurice (1939: 107).

or of their union in a foundational space-time continuum. For space and time may prove in the end to be only two inseparable forms of a general and self-subsistent externality. Some New Realists go so far as to call space-time the final substance of the phenomenon of experience." On the contrary, he left no doubt that "if the principle of relativity is well-founded the very basis of the New Realism seems to disappear into vapour" (Haldane 1921: 273).

Nevertheless, Haldane's defense of idealism was not so explicit and repetitive as was the case with Wildon Carr, the real champion of that philosophical doctrine in Britain, at least concerning the use of Einstein's relativity theories for its support. In *The General Principle of Relativity in its Philosophical and Historical Aspect*, Carr (1920: vi) explained how he came to be interested in Einstein's theories:

> My interest in the principle of relativity is purely philosophical, but it is not casual or accidental. I first became acquainted with it at the International Congress of Philosophy at Bologna in 1911, when M. Pierre Langevin, Professor of the Collège de France, revealed its philosophical importance in a remarkable paper entitled "L'évolution de l'espace et du temps."[29] I introduced the subject to the Aristotelian Society in a paper read in the Session of 1913–14 (*Proceedings of the Aristotelian Society*, Vol. XIV), and I contributed an article, "The Metaphysical Implications of the Principle of Relativity," to the *Philosophical Review* of January 1915. Since then the philosophical importance of the principle has received full recognition. It was not, however, until the preparation of my courses of lectures on the 'History of Modern Philosophy' delivered on 1918 and 1919 at King's College, London, led me to read anew the works of Descartes and Leibniz that the quite special historical interest of the main problem impressed me. It is this historical aspect of the principle to which I have tried to give expression in this study. The main idea was developed in a course of lectures on 'Historical Theories of Space, Time and Movement' delivered at King's College in the spring of this year.

It is also worth quoting what Carr (1920: 160–162) had to say in the final paragraphs of his book:

> It seems to me... that the principle of relativity is a philosophical principle which is not only called for by the need of mathematical and physical science for greater precision in the new field of electro-magnetic theory in which it is continually advancing, but is destined to give us a new worldview. It will be found, as it has always been found, that the poets with their mythical interpretations, and the philosophers with their speculative hypothesis, have led the way in this new advance...

---

[29] Paul Langevin's paper at the Fourth International Conference of Philosophy in Bologna (April 1911) was published afterwards in *Scientia* (Langevin 1911), and in *La physique depuis vingt ans* (Langevin 1923: 265–300). For Langevin and philosophy, see Bensaude-Vincent (1988).

I conclude, then, that in every reflection on our actual experience we are directly conscious of an objectivity which we distinguish from our subjective activity of knowing. Whether we approach the problem of that objectivity from the abstract standpoint of physical science or from the concrete standpoint of philosophy, the result is the same. Ultimately, in spite of its claim to independence, all that an object or event is, in substance or in form, it derives from the activity of the life or mind for which alone it possess the meaning which makes it an object or event. This is not a mystical doctrine, nor is it esoteric. If we adopt in mathematics and physics the principle of relativity (and have we any choice?) the obstinate, resistant form of the objectivity of the physical world dissolves to thin air and disappears. Space and time, its rigid framework, sink to shadows. Concrete four-dimensional space-time becomes a system of world-lines, infinitely deformable. And these world-lines, do not they at last bring us in sight of an irreducible minimum of self-subsistent objectivity? No. The world-lines are not things-in-themselves, they are only the expression for what is or may become common to different observers in the relations between their stand-points. Carried to its logical conclusion the principle of relativity leaves us without the image or the concept of a pure objectivity. The ultimate reality of the universe, as philosophy apprehends it, is the activity which is manifested in life and mind, and the objectivity of the universe is not a dead core serving as the substratum of this activity, but the perception-actions of infinite individual creative centres in mutual relation.

Passing now to Haldane's *The Reign of Relativity*, from which I have already quoted several times, the least that can be said of it is that it was a well-informed work. In the Preface, Haldane (1921: x) explained its purpose: "The remarkable ideas developed by Einstein, as the result of his investigation of the meaning of physical measurement, have provided fresh material of which philosophy has to take account... The advantage which the methods of science possess is that by them results can be reached and formulated with a precision that is unrivalled, so far as they can go. A price for this advantage has, however, to be paid, and science is apt to find itself in strange regions if it does not limit its scope with genuine self-denial. The inquiry entered on by Einstein has, perhaps because of the presence to his mind of something like this reason, stopped short in his hands of the general problem of the Relativity of all Knowledge. The question that remains is whether the investigation of that problem can be carried further, and if so, whether the philosophical method which appears to be required is a reliable one."

That he was a philosopher of a different sort than, say, a Reichenbach, is made evident in most of the pages of the book. Take, for example, what he had to say about the principle of relativity (Haldane 1921: 425–426):

The real lesson which the principle of relativity of knowledge teaches us is always to remember that there are different orders in which both our knowledge and the reality it seeks have differing forms. These orders we must be careful to distinguish and not to confuse. We must keep ourselves aware that the truth in terms of one order may not necessarily be a sufficient guide in the search for truth in another one. As an aid to our practice, the principle points us in a direction where we may possess our souls with tranquility and courage... The real is there, but it is akin in its nature to our own minds, and it is not terrifying.

Perhaps more clearly, we have what he wrote (Haldane 1929: 345–346) in his autobiography referring to his works, beginning with *The Pathway to Reality* (the Glifford Lectures that he delivered at St. Andrews University in 1903–4), and later on in *The Reign of Relativity* and subsequent essays:

> These works are concerned with what human experience seems ultimately and essentially to imply. They describe an outlook to which, after more than half a century of meditation, I found myself finally compelled. That outlook is not now likely to change substantially.
>
> As against it the physicist may think that he can succeed in resolving the Universe into an assemblage that can be most properly expressed in quantitative equations. For some at least of his critics the question will at once arise as to how the physicist, for whose reflection such an assemblage is present, is himself to be accounted for along with his reflection. For his mind seems to lie at the very foundation of the experience with which he is concerned, actual or possible.
>
> Meditation in this direction led me away from the facile postulates of scientific method in the Victorian period, here and on the Continent. That method of scientific approach does not appear to have been rendered more easy by the recent revolutions in outlook on it which are arising with much insistence in the physics and biology of the twentieth century.

5.1.5 Relativity Debated at the Aristotelian Society

One of the philosophical debates—perhaps the most interesting—in which Einstein's relativity theories were involved took place on February 20, 1922, in the Aristotelian Society, the home ground of the "London Philosophy" as it was called. The debate, presided over by Haldane himself, was dedicated once more to the possibility of an idealistic interpretation of Einstein's theories, and had as participants Wildon Carr, the educationalist and philosopher Percy Nunn, Alfred North Whitehead, and Dorothy Wrinch. The main purpose of it was to discuss a thesis put forward by Carr, which reads as follows[30]: "Einstein's theory is a scientific interpretation of experience based upon the principle of relativity. This principle is in complete accord with the neo-idealist doctrine in philosophy, and in complete disaccord with the fundamental standpoint of every form of neo-realism."[31] This thesis made no distinction between the special and general theories of relativity.

---

[30] Carr, Nunn, Whitehead, and Wrinch (1922). The quotations which follow belong to this reference.

[31] Commenting on these words in her obituary of Carr, Hilda Oakeley (1930–1931: 290) wrote: "According, however, to the papers of Sir Percy Nunn, and Professor Whitehead, it

For Carr, "neo-realism" was the philosophical standpoint according to which "knowledge requires us to presuppose existence, and that in some sense a universe exists in space-time, the entities within which are discoverable by minds, which themselves are accorded a place therein on equal terms with the entities they discover," while "neo-idealism" meant that "reality in its fundamental and universal meaning, is not an abstract thing opposed to nature, or an entity with its place among other entities in space and in time, it is concrete experience in which subject-object, mind-nature, spirit-matter, exist in an opposition which is also a necessary relation."

His arguments in favor of neo-idealism were rather simple: "The principle which Einstein follows in physics is based in the recognition that the phenomena which constitute its subject-matter are presented in the form and only in the form of sense-experience. Ultimately and fundamentally the qualities of physical objects are sensations. In this he avows himself the follower of Mach." Classical mechanics, on the contrary, was based in the neo-realist affirmation of an existence independent of sense experience to which the subject of experience referred his sensations. Thus, "the principle of relativity... rejects in physics the metaphysical principle of materialism which presupposes an objective transcendent cause of experience."

Moreover, Carr did not stop here, with physics; he also thought that the principle of relativity equally rejects in mathematics "the metaphysical principle of intellectualism which presupposes pure reason, enlightenment, discernment, as the transcendent subjective cause of experience. That is to say, it rejects the view that in mathematics the mind, endowed with reason, contemplates eternal truth." Finally, as his conclusion, he claimed that all he has been saying was "in essentials the Leibnizian conception. The principle of relativity proposed in science precisely the methodological reform which Leibniz proposed in philosophy when he said, 'The monads are the real atoms of nature.'"

Carr's arguments were contested on different grounds by the other participants in the discussions: Nunn, Whitehead, and Wrinch.

Sir (since 1930) Thomas Percy Nunn (1870–1944) was a professor of Education in the University of London from 1913 till 1936, and although best known as an educationalist, he was also an active member of the Aristotelian Society (he served as its President in 1923–1924). He wrote little on philosophy, but that little was rather influential. In particular, his contributions to a symposium on "Are secondary qualities independent of perception?" (*Proceedings of the Aristotelian Society* [1909]; see also his book *The Aims and Achievements of Scientific Method* [1907]) were widely studied both in England, where it struck Bertrand Russell's roving fancy, and in the United States. There, Nunn sustained two theses: (1) that both the primary and the secondary qualities of bodies are really in them, whether they are perceived

---

was by no means necessary for the realist to modify his position, on account of the principle of relativity. Their arguments, following on those of Professor Carr, suggest the inference that the philosopher's attitude to relativity will depend upon the position from which he starts." Again, as we see, the same philosophical malleability we found before using a text of Russell.

or not; and (2) that qualities exist as they are perceived. Nunn was not ignorant of Einstein's ideas; in 1923, he published a book on *Relativity and Gravitation*. What he wrote (Nunn 1923: 5) in the Preface gives an idea of its content: "Books upon the Theory of Relativity which are not philosophical in aim generally fall into one of two classes. They are either popular expositions intended for readers who have next to no mathematics, or else serious treatises presupposing in the student a considerable technical equipment. The present work seeks to fill a modest place between the two subgroups."

In his reply to Carr, Nunn was not concerned "to assert that the principle of relativity is incompatible with idealism; only to deny that it is incompatible with neo-realism." He argued that, like Einstein, neo-realists "have taught explicitly that the varying appearances of the 'same thing' to different observers are not diverse mental reactions to an identical material cause, but are correlated sense-data or 'events' belonging to a single historical series." "In other words," he added, the neo-realists "have professed the view which Professor Carr seeks to convey by his explanation, that what makes an object 'common' is point to point correspondence between the experiences of different observers." In fact, Nunn explained that "before the doctrine of relativity had risen above their horizon, neo-realists had already gone a long way to meet it," and that if some were restrained from going farther it was "because they shrank from breaking with the traditional beliefs about space and time which they shared with the idealists of their day," a limitation which, however, he considered Bertrand Russell had avoided in 1914, in his book *Mysticism and Logic*, in which he had "carried neo-realism right into the camp of the relativists" by declaring that "two places of different sorts are associated with every sense-datum, namely the place at which it is and the place *from* which it is perceived."[32] Thus, Carr's refutations were transformed, according to this line of argumentation, into confirmations; neo-realism was even a precursor of relativity![33]

As for Whitehead (of whom I will say more later on), he saw no problem for the realists to associate themselves with the concepts of the new physics: "Why should a realist," he asked, "be committed to an absolute theory of space or of time?" For him, relativity "actually removes a difficulty from the way of the realist. On the absolute theory, bare space and bare time are such very odd existences, half something and half nothing," that it reminded him of Milton's account of the Creation, "with the fore-paws of the lions already created and their hinder quarters still unfinished."

---

[32] According to Haldane (1921: 273), Bertrand Russell was "another brilliant exponent of the doctrine of the New Realism school in philosophy."

[33] Nunn also made some efforts to clarify what he thought were Carr's misunderstandings: "The physics of Einstein takes no more account of the 'subjective' in experience than did the physics of Newton," he pointed out, although it was true that "expositions, especially popular expositions, of the doctrine of relativity make numerous references to the 'observer,' an expository device to keep, before the reader's mind, the 'vital fact' that what distinguishes the older physicists from the modern ones is that the former believed that 'all observations, by whomsoever made, could be referred to a single space-time framework,'" while the latter "know they cannot."

In fact, he held that "so far as modern relativity has any influence on the problems of realism, it is all to the advantage of such philosophical systems."

We must note in this last sentence, Whitehead's *finesse*: he was not a realist, but a constructive metaphysician, with a solid and wide mathematical and physical background, Indeed, he took the opportunity granted to him by the Aristotelian Society debate to remark that the "realist's main difficulty is, however, not removed [by relativity]. Nature is the apparent world; but after all, appearance is essentially appearance for knowledge, and knowledge is a different order of being from mere nature."

The remaining participant in the discussion was Dorothy Wrinch, a fascinating though rather controversial—and certainly difficult—woman who is not as well known as she should be. Let me, for this reason, say something about her before considering her presentation at the Aristotelian Society meeting.[34]

Dorothy Maud Wrinch (1894–1976) entered Girton College, Cambridge, in 1913, choosing mathematics as her subject. She was the only Girton woman Wrangler (highest ranking among those graduating with honours by grades in the final examinations) in the Cambridge 1916 Mathematical Tripos. However, her interests were not only mathematics, but also philosophy (specially, symbolic logic). Thus, in her first year at Cambridge she attended Bertrand Russell's lectures at Trinity College on "Our knowledge of an external world," and took Part II of the Moral Sciences Tripos in 1917.

During some years, she was close to Russell and his circle, which provided her with both intellectual and social excitement (when in 1916 Russell lost his Trinity College position due to his refusal to serve his country in World War I, Wrinch was one of the few students—together with men such as Jean Nicod and Victor Lenzen—to whom Russell continued privately to teach mathematical logic). In 1924, she married the physicist and mathematician John William Nicholson (1881–1955), who had recently been appointed fellow and director of studies in mathematics and physics at Balliol College, Oxford, a marriage that did not bring her much happiness (it was dissolved in 1938). In Oxford, Wrinch was able to get a position (beginning in 1923–24) as lecturer at Lady Margaret Hall, an arrangement that prevailed throughout the 1920s and 1930s, working and publishing on subjects such as real and complex variable analysis, Cantorian set theory, transfinite arithmetic, and mathematical physics (especially applications of potential theory in electrostatics, electrodynamics, vibrations, elasticity, aerodynamics, and seismology).

However, in the early 1930s Wrinch began to change her interests from mathematics and physics to biology (in 1932, she was a founding member of the Biotheoretical Gathering, a group of philosophically and ideologically minded scientists keen on developing a theoretical biology). Indeed, by 1934 she was absorbed with the biological applications of potential theory, and between 1936 and 1939 made her most important contributions, namely, the first theory of protein structure, or the cyclol theory (Wrinch 1937). In 1941, she married the American biologist Otto Charles Glaser and settled in the United States, teaching at Smith College, Massachusetts.

---

[34] For more information about Wrinch, see Abir-Am (1987).

In philosophy, besides her interest on the philosophical status of relativity, which prompted her participation at the 1920 symposium of the Aristotelian Society, in the early 1920s Wrinch wrote papers on the scientific method, especially on inference, inductive logic, and philosophy of probability, often in collaboration with Harold Jeffreys (1891–1989), while in the late 1920s she addressed topics such as the philosophical principles of theories like electron theory and quantum theory, continuing her interest on relativity theory, which she considered a "kind of achievement which stands alone in Scientific Theory" (Wrinch 1927: 163), and venturing into "embryonic sciences" (Wrinch 1929–1930) such as physiology, genetics, psychology, and sociology. It deserves to be mentioned that it was Wrinch who made the necessary arrangements to have Ludwig Wittgenstein's *Tractatus Logico-Philosophicus* published, both in German (in Wilhelm Ostwald's *Annalen der Naturphilosophie*; 1921) and in English (Kegan Paul, London 1922).

As to her participation in the Aristotelian Society debate, in some sense Wrinch did not really enter into the discussion because for her to decide between "realistic" and "idealistic" one would have to recur to the mind, and in this regard she thought it "unjustifiable to conclude that the new concepts of relativity allow any deductions whatever to be made as to the nature of mind." She thought that relativity was simply a theory which had taken "as its data the particular facts of the external world and arranging and collating them in general propositions—by means of probability inference—interprets its results in terms of the deductions which can be made from certain facts which are known, to other particular facts which may or may not be known." And she added, in a Machian vein: "Correlation between different facts is the only aim of science."

The nature of Wrinch's remarks can be understood more easily if we take into account some of her other works. Thus, in the February 17, 1921, issue of *Nature*, in an article authored jointly with Jeffreys, she made clear her opposition to Eddington's presentation of Einstein's theory of gravitation, especially as it appeared in his already mentioned book *Space, Time and Gravitation.*[35] What Wrinch and Jeffreys could not accept was Eddington's view, according to which coordinate systems are arbitrary, having no real physical importance, and always eliminating themselves in any actual physical process. Eddington was charged with the idea that "the only thing that has physical importance is space-time," an idea that, Wrinch and Jeffreys thought, was not the point of view of physics, where the "co-ordinate systems actually chosen are adopted entirely because they give specially simple forms to relations between measured quantities, and thus are not chosen arbitrarily"; moreover, "the properties of space-time never appear in physical laws; thus it is space-time that eliminates itself when the problems are reduced to terms of measurement." In support of their point of view, Wrinch and Jeffreys mentioned Einstein's book *Relativity, the Special and the General Theory* (Einstein 1917, 1920). There, they wrote, "Einstein's attitude towards 'space' is closer to ours than to Eddington's... [because] he appears to regard space, not as a primary entity of Nature, but merely as a conventional construct, composed of the aggregate of all

---

[35] Wrinch and Jeffreys (1921).

possible values of the three position co-ordinates. In this form the notion may be useful in theoretical work, but we cannot attribute any ultimate physical importance to a thing we have constructed ourselves."

As to Wrinch's own notion of "space," let us see what she wrote in a paper published in *Mind* (Wrinch 1922: 204):

> to ask 'What is Space' is not significant. We want to investigate what consequences can be deduced from the 'space properties' of terms; we want to establish as many propositions as possible about the further properties which necessarily belong to any term possessing this set of properties. We want to trace the various alternative sets of space properties which are logically possible and to see how far results which are verifiable can be obtained. It will be very advantageous if propositions are discovered which make it possible to apply tests to decide between alternative theories. It will be of great importance, for example, if the further development of Weyl's theory yields some deduction as to the shift of the lines in the spectrum of the sun which would enable us to make a decision between it and other theories which yield other results as to the shift of the lines. But whichever part of the general investigation is being undertaken, not in the case of any of them is it significant to ask, 'What is Space.' It is the properties and not the intrinsic nature of the space which is the subject of investigation.

It is difficult to think of a more concise statement of Wrinch's analytic and logic-based view of science, than this final sentence: "It is the properties and not the intrinsic nature of the space which is the subject of investigation." On the basis of such a statement, it is not unfair to think that an important element in Wrinch's interest in the theories of relativity is that they were famous theories, important to physicists, mathematicians, philosophers, and the public at large, which she thought could be subjected to her philosophical analytic methods.

5.1.6 J. E. Turner versus Carr and Haldane

The debate regarding the possible idealist or non-idealist meaning of relativity did not end, by any means, with the Aristotelian Society discussion. An interesting example in this sense is the paper written by the Liverpool philosopher J. E. Turner (1922a) commenting on Carr's and Haldane's books.[36]

For Turner, it was "fundamentally important to recognize that the scientific theory [of relativity] in itself has at bottom very little bearing on any form of the philosophic principle of relativity... [for] the scientific theory is concerned, and is concerned only, with certain definitely *limited aspects* of space and time—a limitation which is, of course, perfectly legitimate from the scientific standpoint, but

---

[36] Turner, the most prolific Liverpool philosopher of the time, wrote books such as *An Examination of William James's Philosophy: A Critical Essay for the General Reader* (1919), *Personality and Reality* (1926), and *The Nature of Deity* (1927). He was, as we see, not particularly dedicated to science, something that, of course, also says something about the impact of Einsteinian relativity in philosophy.

which must none the less be carefully borne in mind in any discussion of the theory's philosophical implications. To pass from these 'physical conceptions' to the philosophical aspects of time and space is to 'cross the Rubicon;' and much of the prevailing confusion is due to the transition being undertaken as though it were of no significance, so that what is true of the scientific concept is also regarded as true of time and space within philosophy." Turner thought that Einstein could not be accused of confusing the terms "scientific" and "philosophic," and in this sense he quoted from his book *The Theory of Relativity* the statement that "there was nothing specially, certainly nothing intentionally, philosophical about" his theories, adding that neither Carr nor Haldane had adequately recognized "the crucial importance of the distinction" between philosophy and science.

One of the specific points of the idealist conception that Turner could not accept was the role it assigned in relativity to the mind, to the observer's mind, or, in other words, the implications idealism drew from the role played by observers in relativity. In *The Reign of Relativity*, Haldane (1921: 83) had written in this sense that "To understand Einstein's principle as it applies... it is necessary to get out of our heads the persistent assumption that when we look out on the universe of space and time we are looking at something which is self-subsistent. For him spatial and temporal relations in that universe depend on the situations and conditions of observers. The character of space and time is therefore purely relative, and so is their *reality*." But that kind of assertion Turner could not accept because it meant to pass "directly from Einstein's physical concepts to time and space in any philosophical sense," something that he, faithful to his distinction between science and philosophy, denied. The fallacy behind the identification of time and space, philosophically considered, with "their physical conceptions," was, according to him, a consequence of the fact that scientific relativity "is not concerned with categories as *categories*; it introduces no new ones and it dispenses with none of our earlier ones; it merely accepts the basic categories of space and time and then renders more precise their application to natural phenomena—corrects our measurements of time and space intervals, our estimates of mass and energy. If, at the outset of experiences, sensations present themselves as we distinguish them, in relations of space and time, then science, for its more special and limited purpose *must* presuppose them; but its increased accuracy in the employment of its fundamental categories is altogether independent of their nature and validity *as* category."

Summing up his arguments, Turner ends his paper with a clear and concise statement in which the intrinsic value of pure philosophy was estimated at its highest:

> All this implies, finally, that what philosophy has to recognize in scientific relativity is simply an increased degree of accuracy due to the greater exactitude of physical concepts; which means, again, that little, if indeed anything, truly metaphysical is in question at all. The established conclusions of the Theory will contribute to the future Philosophy of the universe; but this involves neither a complete revolution in fundamental concepts, nor any substantial advance in the Idealist view of experience and knowledge. 'Change in standpoint,' once more, gives no change in the actual.

Wildon Carr (1922) did not lose much time in replying, "amazed, at what seems to me," as he put it, "their [that is, his fellow philosophers; Turner, of course, in particular] short-sightedness in imagining that philosophy can be indifferent to this stupendous revolution in science." However, his arguments were not new; he only insisted on what he took to be "the special and important work of Einstein in so far as it affects philosophy," in particular his view that the principle of relativity rejected an absolute which is independent of experience.[37] Neither was Turner's (1922b) own reply to Carr especially illuminating. It is obvious, however, that he had not liked Carr's accusation of "short-sightedness," and thus he rapidly pointed out that by no means could he be charged with being indifferent to the new theory (let us stress once more that, as we are seeing, very few significant British philosophers, whatever their philosophical standpoints, were prepared to be accused of being indifferent to Einstein's new theories), fully admitting "both its outstanding scientific value and its philosophic importance." Still, Turner thought that Carr had misread the significance of those theories "as regards the philosophical principle of relativity."

Turner argued that Carr had made an assertion that relativity could not support when he stated that the absolute is not in the object of knowledge, taken in abstraction, not in the external world, but that it is in the observer or subject of knowledge. "The fallacy [of such arguments] becomes obvious," Turner pointed out, "[when one notes that] the relativist's system of reference (which is undoubtedly part of the external world) is transformed into the 'object of *knowledge*' and transferred from the external world to the observer... The Question becomes—Is the relativist's reference system a standpoint furnished by the observer? I venture to think that this epistemological problem is as foreign to many physicists as relativity mathematics is to the majority of philosophers. It is indeed a problem which can never be solved on any purely scientific basis such as underlies the theory. The only science which can be appealed to is the science of knowledge. The issue, that is, is epistemological; it cannot therefore be affected by the scientific Theory in any way." His final conclusion was unequivocal: "if any form of subjective idealism has been *already established*, or is *presupposed*, then the Theory amply confirms that philosophy. On the other hand, the Theory itself cannot substantiate it; it is indeed equally consonant with either objective idealism, realism or even materialism; it is, for philosophy, a benevolent neutral."

5.1.7 Bertrand Russell and Relativity

It is now time to consider the case of Bertrand Russell (1872–1970), one of the British philosophers traditionally associated with Einstein's theories, although it is

---

[37] He wrote, for instance, that the superiority of Einstein's scheme from the standpoint of philosophy was "that its construction and constitution are inherent and never transcend the conditions of actual individual experience... The principle of relativity is not the rejection of an absolute and the affirmation of universal relativity. That would be equivalent to the affirmation of universal skepticism. What the principle rejects is an absolute which is independent of experience, and therefore outside knowledge, an absolute which has to be postulated as the condition of knowledge."

true that more often than not he was more a general commentator on, if not divulgator of, them than a deep analyzer of their philosophical meanings or significance.

It is not clear when Russell first undertook a serious study of Einstein's theories of relativity. A remark he made in a paper he published in 1922 indicates that he was aware of the theory of relativity when he wrote, in the autumn of 1913, *Our Knowledge of the External World* (Russell 1914a), although he did not then appreciate its importance for the topics with which he was concerned in that book. "As I explained in my book on the *External World* (which, however, laid too little stress on relativity), we have to start with a private space-time for each percipient, and generally for each piece of matter," he wrote then (Russell 1922; 1988: 132).

Also, in a paper he published that same year in *Scientia*, "The relation of sense-data to physics" (Russell 1914b), that followed closely the position developed in *Our Knowledge of the External World*, he showed that relativity was not unknown to him. Thus, in section X, dedicated to time, he wrote (Russell 1914b, 1986: 18): "It seems that the one all-embracing time is a construction, like the all-embracing space. Physics itself has become conscious of this through the discussions connected with relativity." However, what it is really interesting—and for this reason I am dealing with this paper now—is what Russell said in a footnote added to the section titled "Time": "On this subject, compare *A Theory of Time and Space*, by Mr. A. A. Robb (Camb. Univ. Press), which first suggested to me the views advocated here, though I have, for the present purposes, omitted what is most interesting and novel of his theory. Mr. Robb has given a sketch of his theory in a pamphlet with the same title (Heffer and Sons, Cambridge 1913)." Therefore, we know that Robb, a name which has already appeared here (Broad also mentioned him in 1914), was influential if not in Russell's introduction to Einstein's relativity, certainly in some aspects of his comprehension of it. Thus, we must not forget Robb in the present paper; but before coming to him let us continue with Russell.

A fact that must be pointed out is that Russell lamented not having had general relativity at his disposal when he wrote *An Essay on the Foundations of Geometry* (Russell 1897), which arose from his 1895 Trinity College fellowship dissertation; and he could not, for the simple reason that relativity, both the special and general, still remained to be developed in the future. In *My Philosophical Development* (Russell 1959, 1975: 31) he wrote in this sense:

My first philosophical book, *An Essay on the Foundations of Geometry*, which was an elaboration of my Fellowship dissertation, seems to me now somewhat foolish. I took up Kant's question, 'how is geometry possible?' and decided that it was possible if, and only if, space was one of the three recognized varieties, one of them Euclidean, the other two non-Euclidean but having the property of preserving a constant 'measure of curvature.' Einstein's revolution swept away everything at all resembling this point of view. The geometry in Einstein's General Theory of Relativity is such as I had declared to be impossible. The theory of tensors, upon which Einstein based himself, would have been useful to me, but I never heard of it until he used it. Apart from details, I do not think that there is anything valid in this early book.

Further down in this book, Russell explained that two questions interested him specially in the philosophy of physics. The first was the question of absolute or relative motion. "Newton had an argument to show that rotation must be absolute and not relative. But although this argument worried people and they could not find an answer to it, the arguments for the contrary view, that all motion is relative, seemed at least equally convincing. This puzzle remained unsolved until Einstein produced his Theory of Relativity" (Russell 1959, 1975: 33). The other problem which concerned him was "whether matter consists of atoms separated by empty space, or of a plenum pervading all space." And at first he felt inclined to the former view, "of which the most logical exponent was Boscovich"; that is, he adhered to the action-at-a-distance point of view. Afterwards, at the urging of Whitehead, who had dedicated his fellowship dissertation to the study of James Clerk Maxwell's great book, *A Treatise on Electricity and Magnetism* (1873), he opted for Maxwell's approach, that is, for fields as vehicles for the transmission of interaction.[38] "When I adopted the more modern view, I gave it to a Hegelian dress, and represented it as a dialectical transition from Leibniz to Spinoza, thus permitting myself to allow what I considered the logical order to prevail over that of chronology."

Looking back in the 1950s on all those ideas and changes of position, Russell (1959, 1975: 33) sadly concluded: "On re-reading what I wrote about the philosophy of physics in the years 1896 to 1898, it seems to me complete nonsense, and I find it hard to imagine how I can ever have thought otherwise. Fortunately, before any of this work had reached a stage where I thought it fit for publication, I changed my whole philosophy and proceeded to forget all that I had done during those two years." Einstein's relativity was instrumental in that change.

Anyhow, relativity appeared and Russell learnt of its existence. We know that at least by the spring of 1919 (that is, before the November 1919 announcement of the eclipse expedition) he had studied both Einstein's special and the general theories. He was staying in the countryside when the measurements of the light bending due to a gravitational field were made during the total eclipse of the Sun on 29 May 1919. His friend, the Cambridge mathematician John E. Littlewood, had arranged with Arthur Eddington, one of the leaders of the expedition, to cable him as soon as

---

[38] If we take into account that, as we shall see, Whitehead adopted the action-at-a-distance approach for his alternative theories of special and general relativity, then it is somewhat surprising that he urged Russell to turn towards the field approach. It would be most interesting to know the content of Whitehead's 1884 Trinity fellowship dissertation; unfortunately, Trinity College did not begin to preserve fellowship dissertations until 1896, and, as his biographer remarked (Lowe 1985: 106), "Whitehead did not keep anything for long. He does not refer to the dissertation in any of his writings, though he often refers to Maxwell's work. If in the fame of his old age some young scholar asked him about this, his answer would surely have been that Maxwell's work was of the highest importance, and his own dissertation of no importance. He had no autobiographical inclinations, and was the sort of man who has no place in his mind for work that he did when young and passed beyond."

a preliminary study of the data indicated the likely outcome. Littlewood then cabled Russell: "Einstein's theory is completely confirmed. The predicted displacement was 1".72 and the observed 1".75 ± .06." That cable aroused great excitement in Russell's party.[39] Eight days after the public announcement of the results (6 November, 1919), Russell (1919) published his first essay on relativity theory. Many others were to follow.

"In none of these papers, or in his two books, *The ABC of Atoms* and *The ABC of Relativity*," we read in the "Introduction" (Slater 1988: xviii) to one of the volumes that assemble his writings, "does Russell claim to be presenting original views of his own. He regards himself merely as an expounder, both to the world at large and to the narrower world of philosophers, of a very difficult set of theories. His intention is to increase awareness of the new physics—philosophers in particular had not, he thought, seen just how important it was in relation to some of their traditional concerns—and at the same time to remind his readers that, because the method of tensors does not lend itself to popular exposition, what they are getting is not the sort of understanding that he enjoys, since he knows the mathematics involved, but only a second best. His younger readers, he hoped, would be motivated to undertake the study of mathematics."

Anyhow, he too used relativity to support some aspects of his philosophical system. So it happened at least when Charles Augustus Strong (1922) objected to his theory of the external world; more concretely, to Russell's theory of perception, put forward in his books *The External World* and *The Analysis of Mind* (1921), which, according to Russell, amounted to his "theory of physics."[40]

Strong was surprised to find Russell supposing particulars that were members of different pieces of matter to exist all at the same place, in case the place is one reached by light also from different pieces of matter. He also could not understand how objects could be "apparently everywhere except in the place where we see and feel them," and that "a multitude of events, all happening after 12 o'clock, should be the constituents of an event happening at 12 o'clock." To answer these, and other, queries that Russell (1922: 478) conceded, "are somewhat curious," but of curiousness due to "modern physics, for which I am not responsible (I wish I were)," he recalled the basic points of the general theory of the gravitational field as well as the Maxwellian synthesis of the electromagnetic field, pointing out that maybe these two fields could be reduced to one "if Weyl is right." Due to the existence of those two fields, "there are," he stated, "a number of things happening everywhere always";

---

[39] In his autobiography, Russell (1968: 97) wrote about this: "The general theory of relativity was in those days rather new, and Littlewood and I used to discuss it endlessly. We used to debate whether the distance from us to the post-office was or was not the same as the distance from the post-office to us, though on this matter we never reached a conclusion. The eclipse expedition which confirmed Einstein's predictions as to the bending of light occurred during this time, and Littlewood got a telegram from Eddington telling him that the result was what Einstein said it should be."

[40] Strong (1862–1940) was a "critical realist," who in a long series of books, which began in 1903 with *Why the Mind has a Body* and terminated, after many shifts on points of detail, with *A Creed for Skeptics* (1936), tried to construct a "pan-psychist" ontology.

moreover, "what we call one element of matter—say an electron—is represented by a certain selection of the things that happen throughout space-time, or at any rate throughout a large region," and "we cannot speak in any accurate sense of 'history' of a piece of matter, because the time-order of events is to a certain extent arbitrary and dependent upon the reference-body." Such facts were perhaps strange ones, but so it was, and therefore Russell felt obliged to conclude that "it is into a physical world of this description that we have to fit our theory of perception"; in particular, he wished "to include nature and mind into one single system, in a science which will be very like modern physics, though not at all like the materialistic billiard-ball physics of the past." Thus, it seems that modern physics in general, and relativity, in parti- cular, played an important role—still to be studied in detail—in the configuration of Russell's theory of perception. Also, it must be stressed that his role in the early reception of relativity in philosophical Britain was not as great as his later activity in that field may suggest.

5.1.8 Alfred Arthur Robb, a Physicist "in a Wrong Milieu"

Now it is time to consider the contributions of Alfred Arthur Robb (1873–1936), F.R.S., a curious character of whom not much is known. Joseph Larmor, the famous Cambridge Lucasian Professor, assumed the task of writing Robb's obituary for the Royal Society, and one could have expected to find there significant personal as well as scientific information. However, when reading its seven pages, we find that most of them are dedicated to discussing questions—dear to Larmor's heart—related to the Earth's motion with respect to a universal ether. Consequently, and coherently with Larmor's interests, it is not Robb's name that fills these pages but those of, among others, Kelvin, Rayleigh, Maxwell, Michelson, FitzGerald, Einstein, Minkowski, and Poincaré. Only at the end of the obituary did Larmor (1938: 320–321) say something about Robb:

> The main characteristic of A. A. Robb (who was born in Belfast on 18 January, 1873, was educated at the Royal Academic Institution and at Queen' College, Belfast, and was later of St. John's College and Emmanuel College, Cambridge) was a keen, rather rugged, independence of thought, even from early schooldays. He was thus not a very tractable pupil, and his examinations records and achievements at Belfast and later at St. John's College, Cambridge, were not remarkable. On taking his degree at Cambridge he went on to Göttingen, where he embarked on the study of the Zeeman magnetic effect, already prominent and partially understood but, as experimental developments soon showed, far from completely. A memoir prepared there under the influence of Woldemar Voigt passed on from dynamical to more symbolic treatment, on lines not dissimilar to those now rather tentatively followed, and was approved as a thesis for a degree of Ph.D. He was also Sc.D. of Cambridge and of Belfast, and a Fellow of the Royal Society, and held the French *Croix de Guerre* for Red Cross ser- vice. He became a popular figure in the Cavendish Laboratory in the prime

of J. J. Thomson's control, though not much of an experimenter himself. He was prominent as the poet laureate of that cosmopolitan community, and his topical verses describing the activities became, in various collected editions, the subject of song at the convivial dinner celebrations. Later he became rather more isolated, though he remained much in Cambridge. His final pronouncement on space and time, a revision of the early systematic logical treatise, rendered more lucid by a general introduction, was published two years ago.

We are fortunate of having a document written by the always perceptive Bertrand Russell, who in a letter to Lady Ottoline, dated 22 January, 1914, gave the best personal notice I know of Robb:[41]

> Yesterday I had a visit from a man named Robb—a wild Irishman, who wrote a book called *A theory of time and space*, not yet published, which I read for the University Press and thought highly of. It comes out of reflections on physics, but all the physicists think him mad, and he has never had any success of any sort or kind. I think my praise was the first he ever had. He is a man of about 35. He was terribly shy and nervous, and was not made any happier by Littlewood's treatment of his hat. I put a reference to his unpublished book in my paper on Sense-Data and Physics; it was this that led me to write to him and so to his coming. I sketched his theory to Littlewood, who agreed that it seemed admirable. Physicists are wrong *milieu* for him—they have no philosophy and no care for logic—one might as well expect people in the French Revolution to have a passion for balanced and judicious statements. People who fail to get appreciation through being in the wrong surroundings are pathetic. Robb is full of bad ideas as well as good ones—he might easily spend his life pursuing a will-of-the-wisp. He is fat and absurd to look at.

"Physicists are wrong *milieu* for him—they have no philosophy and no care for logic," wrote Russell, and he was right because Robb's ideas were, in more than one sense, philosophical, or at least had a deep philosophical meaning. We can even say that they participated in the kind of abstract approach that Russell himself had tried, together with Whitehead, in *Principia Mathematica*, or even David Hilbert's axiomatization of geometry.[42] Thus, although a physicist, it is my contention that it makes sense to include him in a discussion of the reception of relativity among British philosophers.[43]

That Robb himself was aware of the philosophical dimension of his work is made clear in the "Introduction" to one of his books, *A Theory of Time and Space*. There, he wrote (Robb 1914: 1):

---

[41] Quoted in Russell (1986: 334).

[42] Whitehead and Russell (1910–1913), Hilbert (1899).

[43] As to Robb's relevance to relativity, let me recall what, not in vain, though with some exaggeration, Larmor (1938: 315) wrote: "With the death of Dr. A. A. Robb…one of the main protagonists in the scientific domain now known as relativity has passed away."

In the following pages the writer proposes to give an account of an investigation of the relations of Time and Space in connection with the physical phenomena of Optics.
The subject is thus in part philosophical, in part mathematical, and in part physical [italics added].
Under the name of 'The Theory of Relativity' this subject has been much under discussion, but it is still in a condition of considerable obscurity.
Although generally associated with the names of Einstein and Minkowski, the really essential physical considerations underlying the theories are due to Larmor and Lorentz.

In that book, Robb (1914: 2–3) also explained some of his motivation in entering the relativity field:

It was to preserve symmetry that Einstein made the suggestion that events might be simultaneous to one observer, but not simultaneous to another. This remarkable suggestion was at once seized upon, without it apparently being noticed that it struck at the very foundation of Logic. That 'a thing cannot both be and not be *at the same time*' has long been accepted as one of the first principles of reasoning, but here it appeared for the first time in science to be definitely laid aside, and although many of those who accepted Einstein's view saw that there was something which was psychologically very strange about it, yet this was allowed to pass in view of the beauty and symmetry which seemed, in this way, to be brought about in the mutual relations of material systems. To others, however, this view of Einstein's appeared too difficult to grasp or analyze, and to this group the writer must confess to belong...
In 1911, the writer published a short tract entitled 'Optical Geometry of Motion, a New View of the Theory of Relativity,' in which was put forward an outline of a method of treatment in which he avoided any attempt to identify instants of time at different places. The view was advanced that the axioms of Geometry might be regarded mostly as the formal expressions of certain optical facts.

Optical Geometry of Motion, a New View of the Theory of Relativity (Robb 1911), to which he referred in the previous quotation, was neither much known (it was privately printed) nor gave a complete logical analysis of its author's ideas, though it contained some of the germs of Robb's future works. *A Theory of Time and Space* (Robb 1914), a 371-page monograph published by Cambridge University Press, was different, although Robb (1921: vi) considered it "a short preliminary account." Indeed, in some way it was so; only in 1936 did he publish a more complete version: *Geometry of Time and Space* (Robb 1936).

It would take too much space to analyze here Robb's contributions; it suffices to say that his was a sort of "synthetic" method, or "optical geometry," essentially equivalent to constructing special relativity on the basis of relations between different points established by light signals; that is—as I will comment, using Robb's words, later on—the kind of construction based on light cones and the relations "before"–"after" that would become familiar in the 1960s and 1970s.[44] What interested me specially is that Robb participated too in the new social climate that arose after the 1919 announcement of the eclipse expedition results. Thus, in 1921 he decided to publish a sort of abbreviated version (80 pages) of his 1914 monograph: *The Absolute Relations of Time and Space* (Robb 1921). Why, if not for this reason, had the reluctant-to-publish Robb decided to give to the press a not truly original work?

In spite of its brevity, *The Absolute Relations of Time and Space* is a valuable piece not only for the effort at a synthesis made by the author, but also for its "Preface." In it we find valuable information about the origin of Robb's interests. I will quote extensively from it (Robb 1921: v):

> At the meeting of the British Association in 1902, Lord Rayleigh gave a paper entitled 'Does motion through the ether cause double refraction?' in which he described some experiments which seemed to indicate that the answer was in the negative. I recollect that on this occasion Professor Larmor was asked whether he would expect any such effect and he replied that he did not expect any.
>
> In the discussion which followed reference was made to the null results of all attempts to detect uniform motion through the aether and to the way in which things seemed to conspire together to give these null results.
>
> The impression made on me by this discussion was: that in order properly to understand what happened, it would be necessary to be quite clear as to what we mean by equality of lengths, etc., and I decided that I should try at some future time to carry out an analysis of this subject.
>
> I am not certain that I had not some idea of doing this even before the British Association meeting, but in any case, the inspiration came from Sir Joseph Larmor, either at this meeting or on some previous occasion while attending his lectures.
>
> Some years later I proceeded to try to carry out this idea, and while engaged in endeavoring to solve the problem, I heard for the first time of Einstein's work.

---

[44] See, for instance, Kronheimer and Penrose (1967) and Penrose (1972). A pioneer in following Robb's approach was the Irish physicist and mathematician John Lighton Synge (1897–1995), who in his classic *Relativity: the Special Theory* wrote, after having introduced Minkowskian coordinates: "But efficient as these coordinates are, there are times when even they seem an encumbrance, hiding the *geometrical* objects which are the real concern of all geometers. Can we not deal directly with these objects as Euclid dealt with points, lines and circles? This is in fact what was done by Robb [1921], [1936]... and his point of view has had a considerable influence on the present book." And he immediately added something which can help us to understand some of the difficulties that Robb's approach had in being followed by others: "But the truth is that such 'synthetic' methods are too hard. Formulae have a way of thinking for themselves, and we cannot do without them. But let them be servants, not masters." Synge (1955, 1972: 58). Aspects of Robb's contributions have been studied in Briginshaw (1979) and Walker (1999: 111–112).

From the first I felt that Einstein's standpoint and method of treatment were unsatisfactory, though his mathematical transformations might be sound enough, and I decided to proceed in my own way in search of a suitable basis for a theory.

In particular I felt strongly repelled by the idea that events could be simultaneous to one person and not simultaneous to another; which was one of Einstein's chief contentions.

This seemed to destroy all sense of the reality of the external world and to leave the physical universe no better than a dream, or rather, a nightmare.

At this point he went on to consider the case of two physicists, A and B, who agree to discuss a physical experiment. If they agree, Robb pointed out, that agreement implies, "that they admit, in some sense, a common world in which the experiment is supposed to take place"; that is, that there is "some common sub-stratum," and this led him to try to find an objective standpoint. *Optical Geometry of Motion* was the result of such considerations. "This paper," Robb went on, "contained some of the germs of my later work and, in particular, it avoided any attempt to identify instants at different places." Then, in 1914, at the outbreak of World War I, *A Theory of Space and Time* followed. However, according to Robb, "Unhappily at that period people were concerning themselves rather with trying to sever one another's connections with Time and Space altogether, than with any attempt to understand such things; so that it was hardly an ideal occasion to bring out a book on the subject." Besides, "the subject was not an easy one, and I have been told more than once that my book is difficult reading." He had "intended making further developments of this theory, but the outbreak of the war caused an interruption" of his work.

As to the differences between his approach and Einstein's, Robb stated:

I succeeded in developing a theory of Time and Space in terms of the relations of *before* and *after*, but in which these relations are regarded as absolute and not dependent on the particular observer.

In fact it is not a 'theory of relativity' at all in Einstein's sense, although it certainly does involve relations.

These relations of *before* and *after*, serving, as they do, as a physical basis for the mathematical theory, were quite ignored in Einstein's treatment; with the result that the absolute features were lost sight of.

Even now, some six years from the date of publication of my book, comparatively few of Einstein's followers appear to realize the extreme importance of these relations, or to recognize how they alter the entire aspect of the subject.

The theory, in so far as its postulates have an interpretation, becomes a physical theory in the ordinary sense, but these postulates are used to build up a pure mathematical structure...

So far as I can at present judge, the situation is this: once coordinates have been introduced, the theory here developed gives rise to the same analysis as Einstein's so-called 'restricted relativity' and this latter cannot be regarded as satisfactory apart from my work, or some equivalent.

Einstein's more recent work is extremely analytical in character.
The *before* and *after* relations have not been employed at all in its foundation, although it is evident that, if these relations are a sufficient basis for the simple theory, they must play an equally important part in any generalization. Moreover these relations most certainly have a physical significance whatever theory be the correct one.

Robb's theories might have been—or been considered—difficult, if not obscure, but the explanation he gave in this "Preface" was luminous and, if his message had not passed unnoticed, the development of "intrinsic relativity"—that is, a geometry of space-time free of the problems related to coordinates—would have been substantially speeded, avoiding a delay of several decades.

5.1.9 The Case of Alfred North Whitehead

No analysis of the reception of relativity among British philosophers would be complete without including the case of Alfred North Whitehead (1861–1947), the never-to-be-forgotten companion to Russell in their masterful though finally frustrated *Principia Mathematica* (1910, 1912, 1913).[45] He was probably the most original contributor to the relativity world in British philosophy (and perhaps in any philosophical landscape).

What he called his "philosophical relativity" is contained in three books, *An Enquiry Concerning the Principles of Natural Knowledge* (1919), *The Concept of Nature* (1920), and *The Principle of Relativity* (1922), as well as in a not too great number of articles.[46]

That Whitehead's ideas owed much to Einstein's theories of relativity is obvious just by reading the Preface to *An Enquiry Concerning the Principles of Natural Knowledge*. There he wrote (Whitehead 1919: v–vi):

> Modern speculative physics with its revolutionary theories concerning the natures of matter and of electricity has made urgent the question, What are the ultimate data of science?... The critical studies of the nineteenth century and after have thrown light on the nature of mathematics and in particular on the foundations of geometry. We now know many alternative sets of axioms from which geometry can be deduced by the strictest deductive reasoning. But these investigations concern geometry as an abstract science deduced

---

[45] For Whitehead's biography and contributions, see Lowe (1985, 1990) and, restricted only to his philosophy of science, Palter (1960). For the topics treated in the present study, see especially Lowe (1990: chapter VI).

[46] The first of these articles is the one that Whitehead (1915–1916) published in the *Proceedings of the Aristotelian Society*; entitled "Space, time, and relativity," Einstein's name was mentioned just one time, and this within the "Supplementary notes on the above paper" section. See also Whitehead (1920b, 1921–1922), as well as his previously-discussed contribution to the February 1922 debate at the Aristotelian Society. Valuable information can also be obtained from Whitehead (1926: chapter 7).

from hypothetical premises. In this enquiry we are concerned with geometry as a physical science. How is space rooted in experience?

The modern theory of relativity has opened the possibility of a new answer to this question. The successive labors of Larmor, Lorentz, Einstein, and Minkowski have opened a new world of thought as to the relations of space and time to the ultimate data of perceptual knowledge. The present work is largely concerned with providing a physical basis for the more modern views which have thus emerged. The whole investigation is based on the principle that the scientific concepts of space and time are the first outcome of the simplest generalizations from experience, and that they are not to be looked for at the tail end of a welter of differential equations. This position does not mean that Einstein's recent theory of general relativity is to be rejected. The divergence is purely a question of interpretation. Our time and space measurements may in practice result in elaborate combinations of the primary methods of measurement which are explained in this work.

It is to Whitehead's credit that he wrote this book before the announcement of the results of the eclipse expedition: The preface is dated April 29, 1919, and contains the following phrase: "at the date of writing the evidence for some of the consequences of Einstein's theory is ambiguous and even adverse." That is, he was genuinely interested in some philosophical issues that general relativity made prominent, mainly the nature of space: According to him, space must be uniform, a point that he stressed in the Preface to *The Concept of Nature* (Whitehead 1920a: vii): "Einstein's method of using the theory of tensors is adopted, but the application is worked out on different lines and from different assumptions. Those of his results which have been verified by experience are obtained also by my methods. The divergence chiefly arises from the fact that I do not accept his theory of non-uniform space or his assumption as to the peculiar fundamental character of light-signals. I would not however be misunderstood to be lacking in appreciation of the value of his recent work on general relativity which has the merit of first disclosing the way in which mathematical physics should proceed in the light of the principle of relativity. But in my judgment he has cramped the development of his brilliant mathematical method in the narrow bounds of a very doubtful philosophy."

Whitehead's ideas culminated in *The Principle of Relativity* in which he tried to obtain a relativistic gravitational theory from the principles of natural philosophy. He put forward a physical and mathematical theory of gravitation, maintaining the old division between physics and geometry and not abandoning Minkowski's pseudo-Euclidean four-dimensional space-time (that is, he agreed with Einstein in the belief that the fundamental relations in Nature are not spatial or temporal but spatio-temporal, and that space and time are two abstractions from space-time). I will not enter into the rather well-known fact that Whitehead's gravitational theory also led to the three classical tests of general relativity, but also offered other possibilities;

I will only remark that Whitehead continued to use and defend the role of philosophy in the genesis of his formulation.[47] He wrote in this spirit (Whitehead 1922: 4–5):

> To expect to reorganize our ideas of Time, Space, and Measurement without some discussion which must be ranked as philosophical is to neglect the teaching of history and the inherent probabilities of the subject. On the other hand no reorganization of these ideas can command confidence unless it supplies science with added power in the analysis of phenomena...
>
> At the same time it is well to understand the limitations to the meaning of 'philosophy' in this connection. It has nothing to do with ethics or theology or the theory of aesthetics. It is solely engaged in determining the most general conceptions which apply to things observed by the senses. Accordingly it is not even metaphysics: it should be called pan-physics. Its task is to formulate those principles of science which are employed equally in every branch of natural sciences.

Certainly, this is what he tried to do, although his approach was a difficult one. Consequently, it is not strange that Whitehead is usually considered obscure as a philosopher. Thus, in an obituary he wrote, C. D. Broad (1948: 144) said that "he was an abominably obscure and careless writer."[48] However, not all agreed with such a statement; in fact, his ideas were carefully considered and appreciated by some philosophers as well as by a few physicists. Reviewing his *The Concept of Nature*, A. E. Taylor (1921: 77) stated: "so far as I can judge, Dr Whitehead is fully justified in his contention that his version of [Einstein's] theory is far more consistent and philosophical than any which the physicists *pur sang* have produced."

Although more critical than Taylor and despite what he wrote in Whitehead's obituary, Broad (1923b: 218) could not avoid, in his review of *The Principle of Relativity*, remarking that "What seems to me certain is that Whitehead has produced important arguments which should make us pause before deserting the traditional views so far as to make space-time non-homaloidal. In addition to this he seems to me to have shown quite conclusively that there is nothing to *force* us to a non-homaloidal theory. He has succeeded in giving a modified law of gravitation which will do all that is needed of it, and which requires only the homaloidal Space-Time of the Special Theory of Relativity."

---

[47] From the purely observational standpoint, Whitehead's theory only was "refuted" in the 1970s.

[48] More equilibrated is the judgment of Lowe (1990: 124) when he wrote: "*The Principle of Relativity* is virtually unintelligible apart from the system of natural knowledge he worked out in the *Enquiry* [*An Enquiry Concerning the Principles of Natural Knowledge*]." Also worth quoting is what John L. Synge, who during some time in the 1950s tried to developed Whitehead's relativity (see Synge 1951), wrote to Victor Lowe (1990: 127): "The further you go with Einstein, the richer the view, but with Whitehead the reverse is true."

Looking now at the works of other British philosophers, we see that White-head's ideas concerning relativity were not ignored at all by them.[49] His books, for example, were reviewed in detail, especially by Broad (1920, 1923b). Also, in *The Reign of Relativity*, a chapter of which was dedicated to "Relativity in an English form," Haldane did not forget to discuss, Whitehead's ideas, in great detail. The same happened in Nunn's *Relativity & Gravitation*. Indeed, in the Preface Nunn (1923: 7) pointed out that while writing it he had benefited from the study of Einstein's own papers, "helped out by Professor Eddington's well-known Report, Professor Jean Becquerel's lucid French treatise, and the recently published *Theory of Relativity* of Professor Whitehead."[50] Of Whitehead's book, he added: "From Professor Whitehead's book I have borrowed anything that would fit into my scheme, and I regret that I could not take more. I have been compelled to refer to it in the text only very briefly, but have ventured to express the opinion that it is a work of high moment and that its appearance raises issues of critical importance in the mathematico-philosophical discussions which the genius of Einstein and Minkowski set moving."

Later in the book, Nunn (1923: 35–36) had more to say about Whitehead's ideas:

> We have now carried the description of Einstein's main ideas as far as it is profitable to go without the aid of mathematics. But before beginning to fill in some details of the sketch we must refer briefly to a theory of relativity different in important respects from the one expounded in these pages. In a triad of very notable books [*The Principles of Natural Know-ledge* (1919), *The Concept of Nature* (1920), and *The Principle of Relativity* (1922)] Professor A. N. Whitehead has analyzed the fundamental notions of time, space and matter with unprecedented care and profundity, and, while making full use of the 'magnificent stroke of genius' by which Einstein and Minkowski transformed the old conceptions of space and time, has found himself compelled to take up a critical attitude towards some of Einstein's methods and conclusions. The most salient difference between them is that Whitehead refuses to follow Einstein in attributing physical properties, and therefore heterogeneity, to space. It is a cardinal article of his philosophic faith that temporal and spatial relations must be uniform in character, and that if we assume the contrary we surrender the basis which is essential for the knowledge of nature as a coherent system. But uniformity is not the same thing as uniqueness; there are endlessly numerous time-orders depend-ing on differences in the circumstances of motion of the observer, and there is for each time-order a corresponding space. Logically, time-order is prior

---

[49] It might be also interesting to study the reception of Whitehead's relativity theory among non-British philosophers. Although this subject lies beyond my purpose here, I refer to an early example of such a reaction: the article published in *The Monist* by the Yale University philosopher F. S. C. Northrop (1925: 6–7).

[50] Becquerel (1922). Jean Becquerel (1978–1953), son of the discoverer of radioactivity and an expert in optics, was professor at the Muséum d'Histoire Naturelle, and also director of the Laboratoire de Physique Générale of the École des Hautes Études.

to space-order; for space-order is merely the reflection into the space of one time-system of the time-orders of alternative time-systems [*The Principle of Relativity*, p. 8. Alexander (*Space, Time, and Deity*, I, pp. 50–58) has much the same idea]. The older physics was right, then, in treating physical phenomena as 'contingencies' superimposed upon the uniformity of time and space. Nevertheless, Einstein is right in contending that laws expressing their character and connection cannot be true unless they preserve the same mathematical form in all time and space systems.

Nunn recognized that Whitehead's theory "leads to exactly the same predictions, so that experience has produced, so far, no criterion by which the claims of the rival theories may be decided." His book, however, being dedicated to Einstein's theory, Nunn simply added (p. 37) that "there can be no question that Whitehead, in his wonderfully acute and convincing analysis of the fundamental presuppositions of physics—a work that will ever redound to the credit of British thought—and in the theory of relativity he has based on it, has formulated a body of doctrine with which the orthodox relativists must somehow come to terms."

Although it is true, as Broad (1948–1949: 288) wrote in one of the obituaries he dedicated to Whitehead, that his relativity theory was "never been favorably received by mathematical physicists generally," a few physicists and mathematicians showed their knowledge of and interest in Whitehead's alternative theory of relativity. George Temple (1926), for instance, wrote a paper—"A theory of relativity in which the dynamical manifold can be conformally represented upon the metrical manifold"—based on Whitehead's ideas: "The guiding principles of the theory here exposed appear to be in accordance with the views of Prof. Whitehead, as expressed in his series of three classical works on the principle of relativity." Years later, Temple would return to this theory.

Indeed, *The Principle of Relativity* was also reviewed in the *Philosophical Magazine*, something which indicates that this theory did not pass unnoticed by physicists.[51] Eddington, who had attended Whitehead's lectures when he was a student in Cambridge (Douglas 1956: 10), also showed some sensitivity towards Whitehead's ideas. Thus, in a paper he wrote in 1921—that is, before the appearance of *The Principle of Relativity*—and referring to the possibility that the curvature of space is everywhere homogeneous and isotropic, Eddington (1921: 803) wrote: "This may appear to have some connexion with the view of Dr. Whitehead that space may be Euclidean or non-Euclidean but must be homogeneous throughout." But he added immediately: "But uniform and isotropic curvature is by no means a sufficient condition for complete homogeneity of space, and it leaves room for the full range of variation of geometry from point to point required by Einstein's theory."[52]

---

[51] The review there began with the sentence (Review Phil. Mag. 1923: 1103): "Professor Whitehead in this volume has put an alternative rendering of the Theory of Relativity which calls for careful and through examination."

[52] One decade later, in his Presidential Address to the Physical Society, delivered on November 6, 1931, and while commenting on the necessity or not of the cosmological

In spite of all his possible difficulty or obscurity, it can be safely said that Alfred North Whitehead was, as I already mentioned, the most original thinker and philosopher in the history of the early—and not so early—reception of relativity in Great Britain. Besides, those who wish to understand Whitehead's philosophy must not forget that, as Lowe (1990: 123) pointed out, "The theory of relativity exercises an enormous influence on Whitehead's thinking... its main effect on him was that it accelerated the application of his logical and epistemological studies on a grand scale. A comparison of *The Principles of Natural Knowledge* with his earlier writings suggests that among specific ideas, thinking about the idea of time was what the physical theory most sharply stimulated in him."

5.2 Conclusions

Throughout this study we have seen that British philosophers reacted quickly and in general appreciatively to Einstein's relativity theories. In fact, most of them argued that Einsteinian relativity—both the special and the general theories—was compatible with their philosophical outlooks. In this sense, it did little service to philosophy, in as much as, using once more Russell's appropriate phrase, that could not "be true in all cases; and it may be hoped it is true in none." Nevertheless, that reaction was diverse and interesting, and helps us better understand the British philosophical landscape, as well as, in a few cases, the interplay between physics and philosophy in that country. In contrast this certainly was quite different from the German case, of which we know more: German philosophers like Schlick, Reichenbach, or Carnap, who gave preeminence to the lessons derived from physics (Einstein's relativity in this case) when producing their philosophical works, the British usually proceeded otherwise, giving preference to their philosophical stances, as is particularly evident in the cases of Haldane and Whitehead.

These two philosophers also illustrate another aspect of the reception of Einstein's theories in Britain: that relativity served as a stimulus to philosophical thought. Remember what I just said concerning Whitehead: "The theory of relativity exercises an enormous influence on Whitehead's thinking... its main effect on him was that it accelerated the application of his logical and epistemological studies on a grand scale."

Finally, let me point out that it might not be irrelevant to try to see if the conclusions reached in the present paper also apply to other philosophical communities that were not so prominent, nor so directly connected with the physics world in which relativity arose, as in the German one.

---

term, λ, Eddington (1932) insisted in his critic to Whitehead's theory: "The ratio of the metre to the radius of curvature is determined by λ. If λ is zero the ratio is zero and the connection breaks down. We are left with a space which does not fulfill the first conditions of a medium of measurement; and the relativity theory is laid open to criticisms such as have been brought forward by Prof. Whitehead (mistakenly, I think, as regards the existing theory) as failing to provide a 'basis of uniformity'." Quoted in Stachel (1986, 2002: 471).

References

Abir-Am, Pnina G. (1987), "Synergy or clash: disciplinary and marital strategies in the career of mathematical biologist Dorothy Wrinch," in Abir-am and Outram, eds. (1987), pp. 239–280.

Abir-Am, Pnina G. and Outram, Dorinda, eds. (1987), *Uneasy Careers and Intimate Lives. Women in Science, 1789–1979* (Rutgers University Press, New Brunswick).

Abraham, Max (1913), "Eine neue Gravitationstheorie," *Archiv der Mathematik und Physik 20*, 193–209. English translation in Renn, ed. (2007), pp. 347–362.

Alexander, S. (1920), *Space, Time, and Deity* (Macmillan, London).

Becquerel, Jean (1922), *Le principle de relativité et la théorie de la gravitation* (Gauthier-Villars, Paris).

Bensaude-Vincent, B. (1988), "When a physicist turns on philosophy: Paul Langevin (1911–39)," *Journal of the History of Ideas 49*, 319–338.

Bergson, Henri (1922), *Durée et simultanéité. A propos de la théorie d'Einstein* (Librairie Félix Alcan, París).

Biezunski, Michel (1981), *La diffusion de la théorie de la relativité en France.* Doctoral dissertation, University of Paris.

Biezunski, Michel, ed. (1989), *Albert Einstein. Oeuvres choisies*, vol. 4, *Correspondances françaises* (Éditions du Seuil, Paris).

Bridgman, Percy W. (1927), *The Logic of Modern Physics* (Macmillan, New York).

Briginshaw, A. J. (1979), "The axiomatic geometry of Space-Time: an assessment of the work of A. A. Robb," *Centaurus 22*, 315–323.

Broad, Charles D. (1915), "What do we mean by the question: is our space Euclidean?" *Mind 24*, 464–480.

—— (1916), "The nature and geometry of space," *Mind 25*, 522–524.

—— (1920), *"The Principles of Natural Philosophy.* By A. N. Whitehead," *Mind 29*, 216–231.

—— (1921a), "Review of E. Freundlich, *The Foundations of Einstein's Theory of Gravitation*," *Mind 30*, 101–102.

—— (1921b), "Review of M. Schlick, *Space and Time in Contemporary Physics*," *Mind 30*, 245.

—— (1921c), "Review of E. Cunningham, *Relativity, the Electron Theory and Gravitation*," *Mind 30*, 490.

—— (1921d), "Review of A. A. Robb, *The Absolute Relations of Time and Space*," *Mind 30*, 490.

—— (1923a), *Scientific Thought* (Kegan Paul, London).

—— (1923b), *"The Principle of Relativity, with Applications to Physical Science.* By A. N. Whitehead," *Mind 32*, 211–219.

—— (1925), *"La déduction relativiste.* By Émile Meyerson," *Mind 34*, 504–505.

—— (1948), "Alfred North Whitehead (1861–1947)," *Mind 57*, 139–145.

—— (1949–1949), "Alfred North Whitehead, 1861–1947," *Obituary Notices of Fellows of the Royal Society 6*, 283–296.

Carr, H. Wildon (1920), *The General Principle of Relativity in its Philosophical and Historical Aspect* (Macmillan and Co., London).

——— (1921), "Review of *The Reign of Relativity*," *Mind 30*, 462–467.

——— (1922), "Einstein's theory and philosophy," *Mind 31*, 169–177.

Carr, H. Wildon, Nunn, T. P., Whitehead, Alfred N. and Wrinch, Dorothy (1922), "Discussion: The idealistic interpretation of Einstein's theory," *Proceedings of the Aristotelian Society 22*, 123–138.

Clark, Ronald W. (1972), *Einstein. The Life and Times* (Avon Books, New York).

Douglas, A. Vibert (1956), *The Life of Arthur Stanley Eddington* (Thomas Nelson & Sons, London).

Eddington, Arthur S. (1919a), *Report on the Relativity Theory of Gravitation* (The Physical Society of London).

——— (1919b), "Einstein's theory of space and time," *The Contemporary Review 116*, No. 648 (December), pp. 639–643.

——— (1920a), *Space, Time and Gravitation: An Outline of the General Theory of Relativity* (Cambridge University Press, Cambridge).

——— (1920b), "The meaning of matter and the laws of nature according to the theory of relativity," *Mind 29*, 145–158.

——— (1921), "The relativity field and matter," *Philosophical Magazine 42*, 800–806.

——— (1923), *The Mathematical Theory of Relativity*. Cambridge: Cambridge University Press.

——— (1932), "The expanding universe," *Proceedings of the Physical Society 44*, 1–16.

Eddington, A. S., Ross, W. D., Broad, C. D. and Lindemann, F. A. (1920), "The philosophical aspect of the theory of relativity," *Mind 29*, 415–445.

Einstein, Albert (1917), *Über die Spezielle und die Allgemeine Relativitästheorie, Gemeinverständlich* (Vieweg, Braunschweig).

——— (1920), *Relativity, the Special and the General Theory: A Popular Exposition* (Methuen, London).

Eisenstaedt, J. and Kox, A. J., eds. (1992), *Studies in the History of General Relativity* (Birkhäuser, Boston).

Fokker, A. D. (1915), "A summary of Einstein and Grossmann's theory of gravitation," *Philosophical Magazine 29*, 77–96.

Glick, Thomas F., ed. (1987), *The Comparative Reception of Relativity* (Reidel, Dordrecht).

Graham, Loren R. (1972), *Science and Philosophy in the Soviet Union* (Alfred A. Knopf, New York).

Gray, Jeremy J., ed. (1999), *The Symbolic Universe. Geometry and Physics, 1890–1930* (Oxford University Press, Oxford).

Greenwood, Thomas (1920), "Einstein and idealism," *Mind 31*, 205–207.

Haldane, Viscount (1921), *The Reign of Relativity* (John Murray, London).

——— (1929), *An Autobiography* (Hodder and Stoughton Limited, London).

Hentschel, Ann, transl. (2006), *The Collected Papers of Albert Einstein*, vol. 10 (*The Berlin Years: Correspondence, May–December 1920, and Supplementary Correspondence, 1909–1920* (Princeton University Press, Princeton).

Hentschel, Klaus (1990), *Interpretationen und Fehlinterpretationen der speziellen und der allgemeinen Relativitätstheorie durch Zeitgenossen Albert Einsteins* (Birkhäuser, Basel).

Hilbert, David (1899), *Grundlagen der Geometrie* (Teubner, Leipzig).

Holton, Gerald and Elkana, Yehuda, eds. (1982), *Albert Einstein. Historical and Cultural Perspectives* (Princeton University Press, Princeton).

Jeffreys, Harold (1923), "*Space, Time, Matter*. By Hermann Weyl," *Mind 30*, 103–105.

Kormos Buchwald, Diana, Sauer, Tilman, Rosenkranz, Ze'ev, Illy Imes, József and Holmes, Virginia Iris, eds. (2006), *The Collected Papers of Albert Einstein*, vol. 10 (*The Berlin Years: Correspondence, May–December 19 and Supplementary Correspondence, 1919–1920* (Princeton University Press, Princeton).

Kronheimer, E. H. and Penrose, R. (1967), "On the structure of causal spaces," *Cambridge Philosophical Society 63*, 481–501.

Langevin, Paul (1911), "L'évolution de l'espace et du temps," *Scientia 10*, 31–54.

—— (1923), *La physique depuis vingt ans* (Librairie Octave Doin, Gaston Doin Éditeur, Paris).

Larmor, Joseph (1938), "Alfred Arthur Robb," *Obituary Notices of Fellows of the Royal Society 2*, 315–321.

Lindemann, A. F. and Lindemann, F. A. (1916–1917), "Daylight photography of stars as a means of testing the equivalence postulate in the theory of relativity," *Monthly Notices of the Royal Astronomical Society 77*, 140–151.

Lowe, Victor (1985), *Alfred North Whitehead: The Man and His Work*, vol. 1 (*1861–1910*) (Johns Hopkins University Press, Baltimore).

—— (1990), *Alfred North Whitehead: The Man and His Work*, vol. 2 (*1910–1941*) (Johns Hopkins University Press, Baltimore).

Maiocchi, Roberto (1985), *Einstein in Italia. La scienza e la filosofia italiane di fronte alla teoria della relatività* (Franco Angeli, Milano).

Maurice, Frederick (1937), *Haldane, 1856–1915. The Life of Viscount Haldane of Cloan, K.T., O.M.* (Faber and Faber Limited, London).

—— (1939), *Haldane, 1915–1928. The Life of Viscount Haldane of Cloan, K.T., O.M.* (Faber and Faber Limited, London).

Meyerson, Émile (1925), *La deduction relativiste* (Payot, Paris).

Northrop, F. S. C. (1925), "Relativity and the relation of science to philosophy," *The Monist 35*, No. 1 (January), pp. 1–26.

Nunn, Percy (1923), *Relativity & Gravitation* (University of London Press, London).

Oakeley, Hilda D. (1930–1931), "In memoriam: Herbert Wildon Carr," *Proceedings of the Aristotelian Society 31*, 285–298.

Palter, R. M. (1960), *Whitehead's Philosophy of Science* (University of Chicago Press, Chicago).

Passmore, John (1966), *A Hundred Years of Philosophy* (Duckworth, London; first edition of 1957).

Paty, Michel (1987), "The scientific reception of relativity in France," in Glick, ed. (1987), pp. 113–167.

Penrose, Roger (1972), *Techniques of Differential Topology in Relativity* (SIAM, Philadelphia).

Reeves, Barbara J. (1987), "Einstein politicized: the early reception of relativity in Italy," in Glick (1987), pp. 189–229.

Renn, Jürgen, ed. (2007), *The Genesis of General Relativity*, vol. 3. J. Renn and Matthias Schemmel, eds., *Gravitation in the Twilight of Classical Physics* (Springer, Dordrecht).

Review Phil. Mag. (1923), "Whitehead's *The Principle of Relativity*," *Philosophical Magazine 45*, 1103.

Robb, Alfred A. (1911), *Optical Geometry on Motion, a New View of the Theory of Relativity* (W. Heffer, Cambridge).

—— (1914), *A Theory of Time and Space* (Cambridge University Press, Cambridge).

—— (1921), *The Absolute Relations of Time and Space* (Cambridge University Press, Cambridge).

—— (1936), *Geometry of Time and Space* (Cambridge University Press, Cambridge).

Ross, W. D. (1921), "Book reviews," *Mind 30*, 232–233.

Russell, Bertrand (1897), *An Essay on the Foundations of Geometry* (Cambridge University Press, Cambridge).

—— (1914a), *Our Knowledge of the External World as a Field for Scientific Method in Philosophy* (Open Court, Chicago).

—— (1914b), "The relation of sense-data to physics," *Scientia 16* (July), 1–27 (English version), 3–34 of supplement (French version). Reprinted in Russell (1986), pp. 5–26.

—— (1917), "Review of May Sinclair's *Defence of Idealism: Some Questions and Conclusions*," *The Nation 21* (8 September), pp. 588, 590. Reprinted in Russell (1986), pp. 106–110.

—— (1919), "Einstein's theory of relativity," *The Atheneum*, no. 4,672 (14 November), 1,189. Reprinted in Russell (1988), pp. 207–209.

—— (1922), "Physics and perception," *Mind 31*, 478–485.

—— (1926), "Relativity: philosophical consequences," *Encyclopedia Britannica*, 13th edition (London), pp. 331–332. Reprinted in Russell (1988), pp. 228–234.

—— (1959), *My Philosophical Development* (Allen & Unwin, London).

—— (1968), *The Autobiography of Bertrand Russell*, vol. II ("1914–1944") (Allen & Unwin, London).

—— (1975), *My Philosophical Development* (Allen & Unwin, London).

—— (1986), *The Philosophy of Logical Atomism and Other Essays, 1914–19* (Unwin Hyman, London).

—— (1988), *Essays on Language, Mind and Matter, 1919–26* (Unwin Hyman, London).

Ryckman, Thomas (2005), *The Reign of Relativity. Philosophy in Physics, 1915–1925* (Oxford University Press, Oxford).

Schilpp, Paul Arthur, ed. (1959), *The Philosophy of C. D. Broad* (Open Court, La Salle, Ill.).

Schlick, Moritz (1922), "Die Relativitätstheorie in der Philosophie," *Verhandlungen der Gesellschaft Deutscher Naturforscher und Ärzte. 87. Versammlung, Hundertjahrfeier* (Leipzig), pp. 58–69. English translation Schlick (1979b).
—— (1979a), *Moritz Schlick: Philosophical Papers*, H. L. Mulder and B. van de Velde-Schlick, eds., vol. 1 (Reidel, Dordrecht).
—— (1979b), "The theory of relativity in philosophy," in Schlick (1979a), pp. 343–353.
Slater, John G. (1988), "Introduction," in Russell (1988), pp. xiii–xxv.
Spirito, Ugo (1921), "Le interpetrazioni [*sic*] idealistiche delle teorie di Einstein," *Giornale critico della filosofia italiana 2*, no. 2, 63–75.
Stachel, John (1986), "Eddington and Einstein," in *The Prism of Science*, E. Ullmann-Margalit, ed. (Reidel, Dordrecht), pp. 225–250. Reprinted in Stachel (2002), pp. 453–475.
—— (2002), *Einstein from 'B' to 'Z'* (Birkhäuser, Boston).
Strong, C. A. (1922), "The meaning of meaning," *Mind 31*, 69–71.
Synge, J. L. (1951), *The Relativity Theory of A. N. Whitehead*, Lecture Series 5 (Institute for Fluid Dynamics and Applied Mathematics, University of Maryland).
—— (1955), *Relativity: the Special Theory* (North-Holland, Amsterdam).
—— (1972), *Relativity: the Special Theory*, second edition (North-Holland, Amsterdam).
Taylor, A. E. (1921), "*Relativity, the Special and General Theory: A Popular Exposition*. By Albert Einstein; *Space, Time and Gravitation: An Outline of the General Theory of Relativity*. By A. S. Eddington; *The Concept of Nature*: Tarner Lectures delivered in Trinity College, November, 1919. By A. N. Whitehead," *Mind 30*, 76–83.
Temple, George (1926), "A theory of relativity in which the dynamical manifold can be conformally represented upon the metrical manifold," *Proceedings of the London Mathematical Society 25*, 401–416.
Thomson, Joseph John (1936), *Recollections and Reflections* (G. Bell and Sons, London).
Turner, J. E. (1916), "The nature and geometry of space," *Mind 25*, 223–228.
—— (1922a), "Dr. Wildon Carr and Lord Haldane on scientific relativity," *Mind 31*, 40–42.
—— (1922b), "Relativity, scientific and philosophical," *Mind 31*, 337–342.
Vucinich, Alexander (2001), *Einstein and Soviet Ideology* (Stanford University Press, Stanford).
Walker, Scott (1999), "The non-Euclidean style of Minkowskian relativity," in Gray, ed. (1999), pp. 91–127.
Whitehead, Alfred N. (1915–1916), "Space, Time and Relativity," *Proceedings of the Aristotelian Society*, 104–129.
—— (1919), *An Enquiry Concerning the Principles of Natural Knowledge* (Cambridge University Press, Cambridge).
—— (1920a), *The Concept of Nature* (Cambridge University Press, Cambridge).

—— (1920b), "Einstein's theory: an alternative suggestion," *Times* (London), *Educational Supplement*, February, 13, p. 83.

—— (1921–1922), "The philosophical aspects of the Principle of Relativity," *Proceedings of the Aristotelian Society 22*, 215–223.

—— (1922), *The Principle of Relativity with Applications to Physical Science* (Cambridge University Press, Cambridge).

—— (1926), *Science and the Modern World* (Cambridge University Press, Cambridge).

Whitehead, Alfred N. and Russell, Bertrand (1910–1913), *Principia Mathematica*, 3 vols. (Cambridge University Press, Cambridge).

Wrinch, Dorothy (1922), "On certain methodological aspects of the theory of relativity," *Mind 31*, 200–204.

—— (1924), "*Scientific Thought*. By C. D. Broad," *Mind 33*, 184–192.

—— (1927), "The relations of science and philosophy," *Journal of Philosophical Studies 2*, 153–166.

—— (1929–1930), "Scientific method in some embryonic sciences," *Proceedings of the Aristotelian Society*, 229–242.

—— (1937), "The cyclol hypothesis and the 'globular proteins,'" *Proceedings of the Royal Society of London A 161*, 505–524.

Wrinch, Dorothy and Jeffreys, H. (1921), "The relation between geometry and Einstein's theory of gravitation," *Nature 106*, 806–809.

6

Science and Ideology in Einstein's Visit to South America in 1925

Alfredo Tiomno Tolmasquim

Museu de Astronomia e Ciências Afins, Brazil

Einstein visited South America early in 1925, spending one month in Argentina and one week in both Uruguay and Rio de Janeiro. The visit was one of many trips he made in the first half of the 1920s around Europe and to other continents. From the end of 1919, as his fame spread beyond the scientific milieu, Einstein suddenly found himself in the public eye. He was known and sought out the world over and could give voice to his scientific, political, or ideological points of view. The voyages he made at this time are of precisely this nature, in addition to his interest in visiting different places and experiencing other customs. His journey to the USA in 1921 with Chaim Weizmann, a biochemist who was President of the World Zionist Federation, had the dual aim of giving lectures at American universities and raising funds for the construction of the Hebrew University of Jerusalem. Likewise, his trips to the UK that same year and to France the following year were underpinned by a desire to reestablish relations between the scientific community in Germany and its counterparts in these countries, which had been severed during the war. In 1922, Einstein went to Japan, which provided him with the chance not only to discuss scientific issues, but also to learn about an unfamiliar and entirely different culture (Sujimoto 1989). He then went to Palestine, where the Jewish state would be created in later years. Likewise, his trip to South America brought together scientific, ideological, and even tourist interests.

At the time, Argentina, the country that issued the original invitation, was regarded as one of the leading centers of physics in the southern hemisphere (Pyenson 1985). They were investing heavily, albeit temporarily, in attracting foreign scientists, many of them Germans, to help establish a local scientific environment. The leading institution in physics was the Institute of Physics at the Universidad de La Plata, set up by Emil Bose (Pyenson 1985; Bibiloni 2001), a former director of the Danzig Institute of Technology in Denmark. It was he who was responsible for attracting to La Plata Konrad Simon, professor of electrotechnics also at Danzig, and Johann Jacob Laub, one of the first physicists to seek out Einstein after the publication of his 1905 papers and with whom he published some articles in 1908 and 1909. In 1925, the director of the Institute of Physics was Richard Gans, previously a professor at the University of Strasbourg, and author of important works on magnetism.[1] Also in Argentina were Jorge Duclout, a former student of the Zurich Polytechnic who moved to Argentina in 1884, and the physiologist Georg Nicolai, who together with Einstein wrote the "Manifesto to Europeans," also known as "Anti-Manifesto," which argued against the engagement of scientists in the war.

There were also some initiatives being taken in the country that were linked specifically with the Theory of Relativity. In 1911, Laub gave a course at the Universidad de La Plata on the Theory of Special Relativity, and in 1919 he published, an article in *Revista de Filosofía* about Einstein's ideas concerning space and time. In 1923, Argentine professors Teófilo Isnardi, José Collo, and Félix Aguilar published a paper on the implications of the Theory of Relativity for astronomy. Jorge Duclout and the Spanish mathematician Julio Rey Pastore also gave a number of lectures on relativity. After the assassination of the German foreign minister Walther Rathenau in July 1922, when it was feared that Einstein's life was also in danger, Argentine writer Leopoldo Lugones, together with the Centro de Estudiantes de Ingenería and the Instituto Nacional del Professorado Secundário, put forward the suggestion that a fund be set up to finance a trip by Einstein to Argentina and to create a chair for him at the university which should serve as a refuge (Ortiz 1995).

Einstein had already received some show of interest in his work from the country's scientific community. In April 1922, *Revista del Centro de Estudiantes de Ingenería* requested his authorization for the first-ever translation and publication in Spanish of his article "Die Grundlage der allgemeinen Relativitätstheorie," which had been published in 1916 in *Annalen der Physik*.[2] In August of the same year, the Universidad de Buenos Aires, through an initiative of Jorge Duclout's, who was at the time a member of its board representing the Faculty of Exact, Physical and Natural Sciences, granted Einstein the honorary title of *Doctor Honoris Causa* in physics and mathematics.[3]

Yet despite the considerable presence of foreign scientists in Argentina, especially Germans, and even some interest in his work, the country lacked an academic output of any international weight. Just a few days before he set off, Einstein confided to Ernesto Gaviola, an Argentine physicist who was taking a course at the Berlin University at the time, that he held out little hope for the development of a scientific culture in tropical countries (Gaviola 1952). Later, in a press release published during his stay in Argentina, he expressed the opinion that a better option than bringing foreigners to Argentine universities, who often did not assimilate into the community,

---

[1] Gans returned to Germany in 1925 shortly after Einstein's visit, having been hired by the University of Königsberg. After the end of World War II, Gans returned to Argentina, where he died in 1954.

[2] Einstein agreed to the publication on the condition that a copyright fee should be paid if it appeared in a single publication. *Revista del Centro de Estudiantes de Ingenería* to Einstein, Buenos Aires, April 5, 1922 (EA 44–740) and Einstein's answer, Berlin, May 31, 1922 (EA 44–741).

[3] *Revista de la Universidad de Buenos Aires*, año XIX, tomo I, 1922, pp. 605, 680, 808, and 871. See also Rectorado de la Universidad de Buenos Aires to Einstein, Buenos Aires, October 13, 1922, EA 30–165.

would be to educate Argentines abroad, who would then return to build up a scientific community in their homeland.[4]

The formal invitation to Einstein to visit Argentina was issued by the Universidad de Buenos Aires on December 31, 1923, but the machinations that preceded it reveal the involvement of groups with interests other than the purely scientific. In September 1922, the Institución Cultural Argentino-Germana had been set up in Argentina with the purpose of strengthening scientific and cultural relations between Argentina and Germany. It comprised academics from Argentina and Germans living in the country. The first group saw the institution as a forum for scientific and cultural activities devoid of any political leanings. Meanwhile, the German ambassador to Argentina, A. Pauli, viewed it as an opportunity to counterbalance French cultural influences, with the added advantage that it did not bear the stigma of official propaganda. At the very first meeting, a suggestion was made to invite Einstein, which received support from the scholars but was rejected by the German community, whose discourse in Latin America mirrored that of the right-wing groups in Germany, which saw Einstein as a traitor to the fatherland. It was this difference of opinion that prevented Einstein from being invited by the institution.[5]

In the following year, there was a shift in power within the Institución Cultural Argentino-Germana, and pressure applied by the academically-inclined group led to an invitation finally being given. When they received no reply, they made another attempt, this time appealing to the personal standing of the editor of the *Revista Mensual de Medicina y Cirurgia Vox Medica,* Dr. Stutzin,[6] who then lived in Berlin. In the end, Einstein turned down the invitation with the explanation that he was short of time.[7] When the first meetings of the Cultural Institution had been held, the academic faction, dissatisfied with the German members' refusal, had taken their idea of the invitation to the Universidad de Buenos Aires' board, which approved it and decided that the university should raise the necessary funds. However, no concrete action was taken at this point.[8]

Parallel to this, a Jewish group based at the Asociación Hebraica was planning to bring Jews of scientific and artistic renown to Argentina to show the nation what contribution these new immigrants could make. Having heard about the other group's frustrated attempts, they contacted Max Strauss, a friend of Einstein's who was director of the Carl Lindström Aktiengesellschaft and lived in Berlin. They asked him to pass on the association's invitation, with guarantees of the sum of four thousand

---

[4] *Mundo Israelita*, May 2, 1925, p. 5.

[5] A. Pauli, Deutsche Gesandtschaft to Auswärtige Amt, Buenos Aires, Report 415, September 22, 1922; Report 432, October 10, 1922, Report 181 May, 14, 1923. Politisches Archiv des auswärtigen Amtes, Bonn (hereafter PadAA) – R64677. Some transcriptions may be found in Kirsten and Treder 1979, vol. 1, n. 152; vol. 2, n. 720, 721, 723.

[6] Stutzin to Einstein, Berlin, 30 October 1923. EA 43–088.

[7] For more on Einstein's rejection of the invitation, see *Mundo Israelita*, March 7, 1925, p. 1.

[8] Session of the Superior Council of the Buenos Aires University of October 30, 1922. Exp. 2374/1922, *Revista de la Universidad de Buenos Aires*, año XIX, tomo L, 1922, p. 888; and Resolution n. 130, October 30, 1922, idem, p. 825.

dollars plus first-class return tickets.[9] In lieu of Einstein, who was away in Leiden at the invitation of Paul Ehrenfest, Strauss was received by Einstein's wife Elsa. She explained that because of previous problems, Einstein had decided that he would only accept official invitations from scientific institutions so as to prevent his name being inappropriately used for the propaganda purposes of one group or another. The Asociación Hebraica then contacted the Universidad de Buenos Aires, telling them of Einstein's willingness to visit South America provided it was on the invitation of an academic institution, and offering to provide financial backing for the trip.[10] As the university was already considering such an invitation, it needed no further encouragement to take the next steps.

Einstein did, however, know about the Jewish community's intercession in the invitation and even set up an unofficial correspondence directly with the Asociación Hebraica to discuss preparations for his trip. Ten days after the university sent Einstein their invitation, which spent an inordinate length of time going through official channels, the Asociación Hebraica itself gave Einstein notice that it was on its way.[11] He replied that he could not go to Argentina that year, but that he would do so as soon as possible, giving the following justifications:

Berlin, March 8, 1924

Asociación Hebraica de Buenos Aires,

I gratefully acknowledge receipt of your letter dated 9th January.

Your invitation touched me so that my immediate desire was to accept it. However, after considering the matter I am convinced that in 1924 I will not be able to go to South America. First of all, because I am here very busy with different scientific projects that I cannot leave incomplete. Secondly, I have spent so much time away from Berlin over the last few years that I would not dare request leave of absence from the authorities once again.

In thanking you for your kind invitation, I ask you to keep it open until I am able to accept it. In exchange I promise that I shall not accept any other before having visited you.

I remain most faithfully yours,

Albert Einstein[12]

The university's board transferred the invitation to the following year and eventually, in July 1924, Einstein agreed to the trip, setting the date for March 1925.[13]

---

[9] Max Strauss to Einstein, Berlin, November 5, 1923, EA 45–084.

[10] *Mundo Israelita*, March 28, 1925, p. 4. According to Ortiz, the original letter may be found in the Archive of the Buenos Aires University, Expediente 3373, 1923. (Ortiz 1995, p. 83, note 54.)

[11] Asociación Hebraica to Einstein, Buenos Aires, January 9, 1924, EA 43–089.

[12] Translated from a Spanish version, printed in *Mundo Israelita*, March 28, 1925.

[13] Legación de la Republica Argentina to Einstein, Buenos Aires, January 9, 1924, EA 43–089.

The Hebrew Association placed 4,660 pesos (some US $1,500) at the disposal of the university, and the university pledged a further 7,700 pesos ($2,500).[14] The Institución Cultural Argentino-Germana contributed 1,500 pesos, and the universities of Córdoba, La Plata, and Tucumán added a further 4,500 pesos, bringing the total amount raised to almost US $2,000 more than originally planned.

If in Argentina the Jewish community's backstage involvement was evident, in Brazil it was entirely explicit. The Asociación Hebraica contacted Rabbi Isaiah Raffalovich, leader of the Jewish community in Rio de Janeiro, about Einstein's passage through this city, suggesting that he set up contact with a Brazilian university, which was duly done (Raffalovich 1952) See Fig. 6.1. However, in Brazil the invitation did not come from the university but from the rabbi himself in his own name and in that of Paulo de Frontin and Aloysio de Castro, the directors of the faculties of engineering and medicine, respectively.[15]

At the same time, the Hebrew Association in Buenos Aires also contacted the Universidad de Montevideo in Uruguay, including it in the Argentine invitation. The university pledged 1,000 pesos and also undertook to obtain an invitation from Santiago University in Chile, in case Einstein should wish to return via the Pacific.

In Brazil and Uruguay, the two other countries that Einstein agreed to visit and give talks in, the scientific communities were at a much more incipient stage than in Argentina. Brazil only had one university, the Universidade do Rio de Janeiro, which had been set up in 1920 and was actually an artificial union of the traditional Polytechnic School, the Medical School, and the Law Faculty. There was the Academia Brasileira de Ciências, founded in 1919, whose activities were concentrated in Rio de Janeiro and which, according to some authors, was much more a place for the advancement of science than a strictly scientific institute. The only real interest in the Theory of Relativity was expressed by a small group of professors at the Escola Politécnica do Rio de Janeiro, from among whom the names of Manoel Amoroso Costa and Roberto Marinho de Azevedo merit special mention. Costa gave the first talk on the subject at the Polytechnic and in 1922 published the first book in Brazil on relativity, entitled *Introdução à Teoria da Relatividade* (Amoroso Costa 1995). In 1919, before the results of the 1919 eclipse were published, which confirmed the predictions made by the Theory of Relativity, Azevedo published an article on the topic in *Revista de Ciências* (the journal of the Sociedade Brasileira de Ciências, forerunner of the Academia Brasileira de Ciências).[16] In Uruguay, studies in physics were also limited, and basically only took place at the Universidad de Montevideo's School of Engineering and at the Asociación Politécnica del Uruguay.

On accepting the invitation to South America, Einstein turned down another invitation extended by Robert Millikan to spend some months at Mount Wilson,

---

[14] Rectorado of the Universidad de Buenos Aires to Einstein, Buenos Aires, May 16, 1924, EA 43–096; and June 16, 1924, EA 43–097. Expediente 3.373/1923, May 26, 1924 Session of the Superior Council. *Revista da Universidad de Buenos Aires*, Seccion I, tomo II, 1924, p. 38.

[15] Raffalovich to Einstein, Rio de Janeiro, January 27, 1925, EA 44–010.

[16] The list of papers published on the Theory of Relativity in Brazil in the period 1919 to 1935 may be found in (Moreira 1995).

JEWISH COLONIZATION ASSOCIATION

Cable adress Raffaelli

CAIXA POSTAL N..........

c/o London & Brazilian Bank Ltd.,

Candelaria, 19,

RIO DE JANEIRO, 27. 1. 1925.

My Dear Sir,

At the request of Dr. Aloysio de Castro, Director of the Faculty of Medicine in our University, (whom you met at the League of Nations Meeting at Geneva,) and Dr. Paulo de Frontin, Director of the Rio de Janeiro Polytechnic School and President of the Engineering Club, I had the honour of cabling you last week extending to you a cordial invitation on behalf of the above-mentioned Institutions to honour our Capital with your presence on your return from Argentine. The Jewish Community, of which I have the honour to be the spiritual head, anticipate with pride and pleasure the honour which your presence will reflect upon them. We are looking forward to hear that you have accppted the invitation and I pray you to honour me with your kind reply at your convenience in order to enable the Institutions to make the necessary arrangements for your reception.

I have the honour, My Dear Sir, to remain,

Yours Faithfully,

Dr. I. Raffalovich

Grande Rabbino.

Professor

Dr. Albert Einstein,

Berlin.

Fig. 6.1. Rabbi Raffalovich's letter inviting Einstein to visit Rio de Janeiro. (Courtesy of Albert Einstein Archives, Hebrew University of Jerusalem, Jerusalem.)

which would surely have been much more interesting for his scientific work. Caltech was already one of the leading research centers in astrophysics, and Millikan ran a program in which he invited renowned physicists to spend some time at Mount Wilson. Two of these visiting physicists were Lorenz and Ehrenfest, the latter of whom had actually transmitted the invitation to Einstein.[17]

[17] Ehrenfest to Einstein, London, May 25, 1924, EA 10–084; Einstein to Ehrenfest, Berlin, May 31, 1924 (EA 10–088) and July 12, 1924 (EA 10–090). Correspondence between Einstein and Millikan, EA 17–291 to 17.299, and EA 17–357.

If Einstein officially required that the invitation originate from an academic institution, in practice his journey from the very outset bore a clear commitment to the local Jewish communities. However, it does not seem that his journey had the aim of raising funds for the Hebrew University, as did his earlier visit to the United States. Since the South American Jewish communities were not as wealthy as their American counterparts, his correspondence with Weizmann before the trip centered on the opening of the Hebrew University that would take place at the beginning of 1925, rather than his trip to South America.

It is not clear what really led Einstein to undertake a three-month journey on his own to a group of countries that had so little to add to his scientific work. There are many theories (Ortiz 1995; Fölsing 1998), but the most likely is that it was a set of factors. He was certainly keen to interact with scientists all over the world to help build up a truly international scientific community, and to disseminate the concepts of relativity and the most current issues in physics. However, aside from scientific concerns, there was the desire to visit a different continent and experience the customs of its people since he had been so fascinated by his trips to Japan and Palestine. His travel journal on his South American trip is a mixture of attempts to record ethnographic information with comments about people and places. In it, he noted how overawed he was by the wildlife, he made observations about the nature of the towns and other places he visited, and he noted his interest in indigenous cultures and even the influence of the climate on people's behavior. On reaching South America, Einstein observed everything with great interest, much like an anthropologist first coming into contact with a new world. In his journal he wrote about the impact of his first contact with Rio de Janeiro's landscape at the beginning of the trip: "Botanic Gardens and the flora in general surpass the dreams of the 1001 nights. Everything lives and grows before one's eyes, so to speak. There is a delightful ethnic mix to be seen. Portuguese-Indian-black in all possible combinations. Spontaneous like plants, subjugated by the heat. Fantastic experience. An indescribable abundance of impressions in a few hours."[18]

The facilities and resources placed at his disposal by the Argentinian institutions played a role in his decision to make the journey and gave no room for excuses of a technical nature. In addition, he felt the responsibility of the commitment he had made to the Argentines. In the letter to Ehrenfest, in which he turned down the invitation sent by Millikan, Einstein said that he had already committed himself to visit South America and that his next foray outside Europe would be there. Later, at the insistence of Ehrenfest, who may not have understood his decision to go to South America rather than Mount Wilson, Einstein justified his decision, saying that "I should go to South America next June, they (the South Americans) won't stop bothering me."[19]

Einstein left the port of Hamburg on the morning of March 5, 1925, on board the Cap Polônio. It reached Rio de Janeiro on March 21, where it made a one-day maintenance stop before proceeding to Buenos Aires, which it reached on March 25.

---

[18] Einstein's travel journal for Argentina, Uruguay, and Brazil. EA 29–133.

[19] Einstein to Ehrenfest, Berlin, July 12, 1924, EA 10–690.

Einstein spent most of his time in the capital city, though he did visit Cordoba and La Plata to give some talks, and spent some time reposing at the country home of the Wasserman family in La Vajole. On the night of April 23 he set off on the steamship Ciudad de Montevideo to the Uruguayan capital, which he reached the following day. Finally, on May 1, he embarked on the French vessel Valdivia for Rio de Janeiro, which he reached on the afternoon of May 4. After a week in the Brazilian capital, Einstein set off on his journey back to Germany on board the Cap Norte.

Einstein's visit to the three South American countries comprised something of a mix of elements, rather like the motives behind his trip. He was greatly involved with the local Jewish communities, which paid for his board and lodging in all three countries. In Buenos Aires he was put up at banker Bruno John Wasserman's mansion, and in Montevideo he stayed at the residence of the Russian Jew Naum Rosenblatt. In Rio de Janeiro Einstein stayed at a hotel (Gloria Hotel), but was chaperoned at almost all his events by Izidoro Kohn, a Jew of Austrian birth. He also paid visits to a number of Jewish organizations. In Buenos Aires, these included a Yiddish newspaper called *Das Volk*, the Hospital Israelita, the Federación Sionista Argentina, a girls' orphanage (Asilo Argentino de las Órfãs Israelitas), and a Sephardic synagogue; while in Rio de Janeiro he went to the Biblioteca Scholem Aleichem and the Federação Sionista do Brasil. He also took part in receptions hosted by the local Jewish communities. In Argentina, one such reception was held at the Savoy Hotel, and at the Capitol Theatre he also gave a lecture organized by the Asociación Hebraica called "A few thoughts about the situation of the Jews," and took part in a large event at the Coliseo, the largest theater in the city, organized by the Zionist Federation to celebrate the creation of the Hebrew University on the same date. In Rio de Janeiro, he attended a large reception at the Automóvel Clube do Brasil, at which almost all the Jews living in the city were present.

In Argentina, Einstein's visit provided a backdrop for the expression of disagreements within the Jewish community. The main aim of the small group of businessmen who had helped finance his visit was to enhance the image of Jews in the country, strengthening ties with the local intelligentsia and politicians. The idea was to establish a good neighbor policy that would prevent expressions of anti-Semitism, which was already prevalent in the country, as well as to move a few rungs up Argentina's social ladder. Additionally, the Hebrew Association could take on a leading role within the Jewish community with a demonstration of strength and standing in the Argentine academic world. However, the pro-Zionist groups took advantage of Einstein's presence to publicize their ideas, and gained much ground during his visit by taking part in the reception committee, his visit to the Zionist Federation, and the great ceremony in honor of the University of Jerusalem. Not only did they steal the spotlight from the Asociación Hebraica, but they jeopardized the overriding purpose of the trip as they put their efforts into promoting the Zionist cause.[20] The biggest fear of those connected with the Hebraica was that Einstein's visit might fail to highlight the key role played by the Jewish community in Argentina, and would actually associate Jews ever more strongly with the idea of Zionism.

---

[20] *Mundo Israelita*, April 25, 1925, p. 2.

Fig. 6.2. Einstein and some members of the Argentine Jewish Community, April 1925. (Courtesy of Archivo General de la Nacion, Buenos Aires.)

At the time Einstein was actually very engaged with the Zionist cause and the creation of the University of Jerusalem, about which he made no bones. He thought that the Jews in South America should be encouraged to contribute to the building of the university and to help the needy Jewry of Europe, opinions he repeated at virtually all the events he took part in with the Jewish communities in the countries he visited. Yet his vision of Zionism, as he tried to explain in his talks, was not that all Jews should automatically emigrate to Israel. He thought of Israel not as a physical center to which all Jews must return, but as a center for the combined cultural expression of a people spread across the world.

Another social group involved in Einstein's visit to South America was that of German immigrants and their descendants. The German-speaking community in Argentina was the most numerous in South America, adding up to around 30,000 people. It was also highly organized, and had a number of newspapers, schools, a hospital, institutions to provide assistance for immigrants, and a series of associations of a social, political, and economic nature (Newton 1977). It was also a significant player in the academic world, and was responsible for setting up different technical and scientific entities, such as the Asociación Argentina de Ingenieiros Alemanes, the Asociación Argentina de Químicos Alemanes, the Asociación Argentino-Germana de Médicos, the Asociación Científica Alemana, and the Instituición Cultural Argentino-Germana.

Although it did not represent a single ideological viewpoint, this community did have a predominantly right-wing attitude that defended German nationalism, denounced the Versailles Treaty, supported the restoration of the monarchy, and reviled those they deemed left-wingers and pacifists as traitors. On the day that

Einstein reached Buenos Aires, the newspaper *La Prensa* published his article "Pan-Europa," in which he highlighted the difference between the European cultural unity existing since ancient times, and temporary political divisions, which "through an idiotic small-minded patriotism tried to limit intellectual persons to the political borders of their states" (Einstein 1925). This article further reinforced the arguments propounded by those who had been against Einstein's visit. A second article by Einstein published in *La Plata Zeitung* was a further attempt to clarify his ideas, but it failed to change the mood of the local German community. No reception or welcome was given by the Sociedad Científica Alemana or even by the Institución Cultural Argentino-Germana, which had provided funds for his trip. The whole community's attention was ostensibly and deliberately turned to the German admiral Paul Behnke, who was visiting the country at the same time.

In Montevideo and Rio de Janeiro, the circumstances were different. Both German communities organized receptions for Einstein at the cities' respective German social clubs. The events were agreeable and almost no mention was made of politics. The German consuls in all three South American countries also held receptions to welcome their illustrious visitor. Einstein wrote about them in his travel journal: "A funny lot, the Germans. I am to them a putrid flower that they keep putting back in their buttonhole."[21] The German consuls saw Einstein's visit as a chance to offset the influence of French culture and science. As the consul in Argentina, Karl Gneist, put it: "Unfortunately, the influence of German culture in Argentina and indeed in the whole of South America is still infinitely behind the French," adding that "for the first time, a German academic of global repute has come, whose simple, friendly and maybe slightly transcendent manner has been remarkably in tune with the nation here. One could not imagine finding a better man to counteract the hostile propaganda and shatter the myth of the German barbarian."[22]

In scientific terms, the most significant contribution was in Argentina, where Einstein gave a complete course on the Theory of Relativity involving eight lectures. The lectures were taken down in shorthand by engineer Teófilo Isnardi and published in the newspaper *La Nacion* and later in the *Revista de Filosofia*. Einstein also gave a talk at the Faculty of Philosophy and Language of the Universidad de Buenos Aires, called "Positivism and idealism: geometry and the finite and unbounded space of the general theory" (Alberini 1925), and took part in a question and answer session with a small group of scientists at the Academia Nacional de las Ciencias Exactas, Fisicas y Naturales. He wrote about this event in his travel journal: "In the afternoon, meeting at the Academia. I was made a foreign member. They asked me silly scientific questions, to the point that I had trouble keeping a straight face."[23] In Uruguay, the second port of call, the course on the Theory of Relativity was slimmed down to three lectures, and in Rio de Janeiro, his last destination, it comprised only two lectures.

---

[21] Einstein's travel journal to Argentina, Uruguay, and Brazil. EA 29–133.

[22] Karl Gneist, Deutsche Gesandtschaft to Auswärtige Amt, Buenos Aires, April 4, 1925, PadAA – R64678, (Kirsten and Treder, 1979, v. 2, n. 725.)

[23] Einstein's travel journal to Argentina, Uruguay, and Brazil. EA 29–133.

Fig. 6.3. Einstein and some members of the German Club in Rio de Janeiro, May 1925. (Courtesy of Archive of the German Club, Rio de Janeiro.)

In Rio, however, he did have one significant engagement, which was a lecture at the Academia Brasileira de Ciências. Though this was his final port of call and he was exhausted by the lengthy journey, Einstein wrote an article on the current discussions that were taking place about the nature of light, which he entitled "Bemerkungen zu der gegenwärtigen Lage der Theorie des Lichtes." In the paper, Einstein gave a short explanation of the differences between the electromagnetic wave theory and the quantum theory of light, and talked about the BKS theory (by Bohr, Kramers, and Slater), which attempted to provide a theoretical explanation for the energy proper- ties of light without using the hypothesis of quanta as particles. He then commented on the experiment being developed by Berlin-based physicists Geiger and Bothe to test the BKS theory, and finished by saying that, "When I left Europe, the experi- ments had not been finished. However, the results so far obtained seem to show the existence of such a correlation. If this correlation is actually shown to exist, it is a valuable new argument in favor of the reality of light quanta" However, he avoided giving an opinion on the results of the experiment. Actually, the preliminary results were published in April 1925 when Einstein was in South America, and the definitive findings came out in June of the same year, reinforcing the quantum concept of light. There were apparently no comments or questions during the conference or even in the newspapers at the time, which published numerous articles by scientists, showing that the concerns of the moment in Brazil were more focused on the theory of rela- tivity and not on the constitution of light (Tolmasquim and Moreira 2002; Moreira 2005).

Nevertheless, despite the different states of development of science in each coun- try, they did share some common ground as regards their relationship with relativity.

The first aspect was the significant French influence in both scientific and cultural terms. Even Argentina, which attracted many visiting German scientists, was under strong French influence. In Brazil the relationship was even more marked, and appropriately illustrated by the fact that the local academics had heard about the theory of relativity by means of French scientists, despite the considerable resistance to relativity in France. The first foreign scientist to lecture on relativity in Brazil was Emile Borel, in 1922, and Costa's book was written after he had spent two years in France, in 1920 and 1921. The influence was also clear in Einstein's decision to

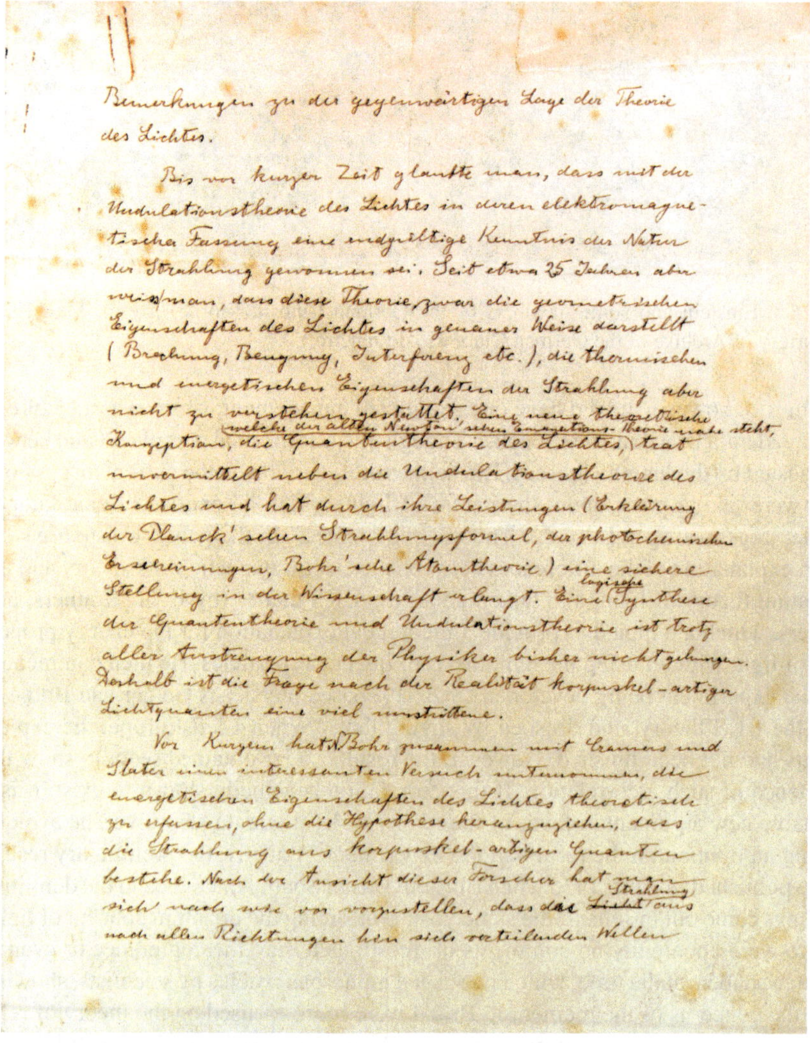

Bemerkungen zu der gegenwärtigen Lage der Theorie des Lichtes.

Bis vor kurzer Zeit glaubte man, dass mit der Undulationstheorie des Lichtes in deren elektromagnetischer Fassung eine endgültige Kenntnis der Natur der Strahlung gewonnen sei. Seit etwa 25 Jahren aber weiss man, dass diese Theorie zwar die geometrischen Eigenschaften des Lichtes in genauer Weise darstellt (Brechung, Beugung, Interferenz etc.), die thermischen und energetischen Eigenschaften der Strahlung aber nicht zu verstehen gestattet. Eine neue theoretische Konzeption, die Quantentheorie des Lichtes, welche der alten Newton'schen Emanations-Theorie nahe steht, trat unvermittelt neben die Undulationstheorie des Lichtes und hat durch ihre Leistungen (Erklärung der Planck'schen Strahlungsformel, der photochemischen Erscheinungen, Bohr'sche Atomtheorie) eine sichere Stellung in der Wissenschaft erlangt. Eine logische Synthese der Quantentheorie und Undulationstheorie ist trotz aller Anstrengung der Physiker bisher nicht gelungen. Deshalb ist die Frage nach der Realität korpuskel-artiger Lichtquanten eine viel umstrittene.

Vor Kurzem hat N. Bohr zusammen mit Cramers und Slater einen interessanten Versuch unternommen, die energetischen Eigenschaften des Lichtes theoretisch zu erfassen, ohne die Hypothese heranzuziehen, dass die Strahlung aus korpuskel-artigen Quanten bestehe. Nach der Ansicht dieser Forscher hat man sich nach wie vor vorzustellen, dass die ~~Licht~~ Strahlung aus nach allen Richtungen hin sich verteilenden Wellen

Fig. 6.4. Einstein's paper on the constitution of light addressed to the Brazilian Science Academy. (Archive of Getulio das Neves' family.)

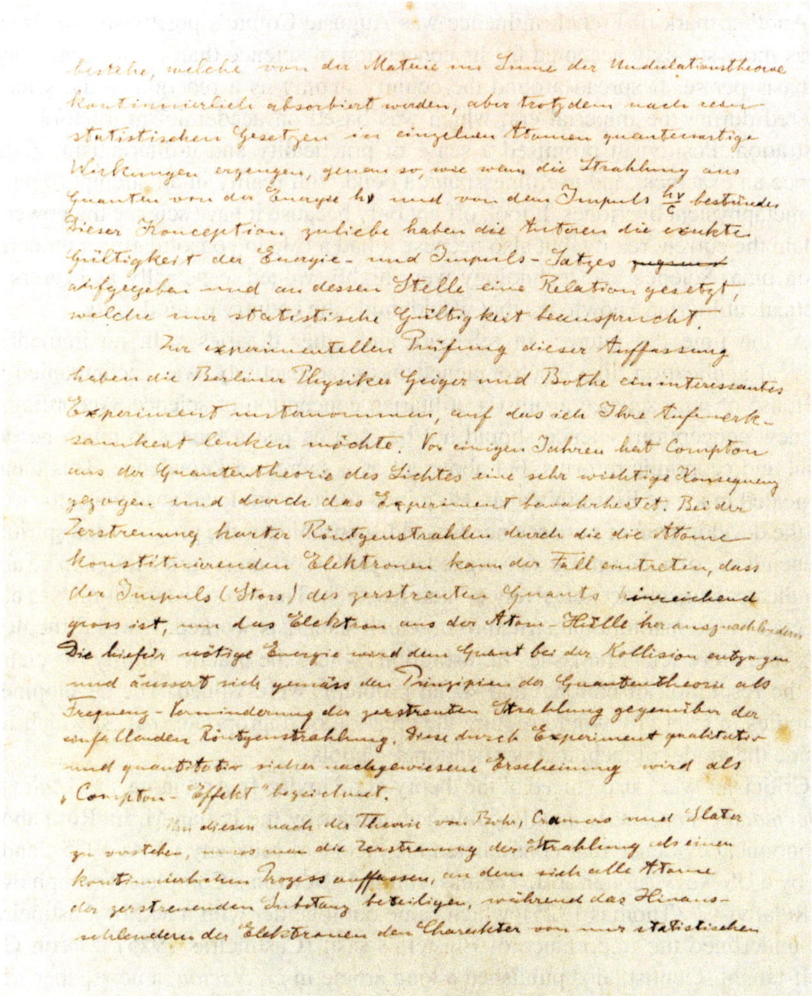

Fig. 6.4. (Continued).

speak in French in the three countries he visited, a fact that was noted by the German embassies and included in their reports.[24]

---

[24] Karl Gneist, Deutsche Gesandtschaft in Buenos Aires to Auswärtige Amt, Buenos Aires, April 4, 1925, PadAA – R64678; Arthur Schmidt-Elskop, Deutsche Gesandtschaft in Montevideo to Auswärtige Amt, Montevideo, July 4, 1925, PadAA – R64678; Hubert Knipping, Deutsche Gesandschaft in Rio de Janeiro to Auswärtige Amt, Rio de Janeiro, May 20, 1925, PadAA – R64678; Kirsten and Treder 1979, v. 1, n. 155 and 156; v. 2, n. 725, 727, 728.

Another mark of French influence was Auguste Comte's positivism. In Brazil, it was more strongly accepted for its conception of science than for its mathematical basis per se. It spread around the country mainly as a reaction to the science fostered during the imperial era, which was based on academicism, rhetoric, and illustration. Positivism promised a sense of practicality and utilitarianism, giving science an exactness and usefulness, and a bond with reality in an attempt to purge any metaphysical overtones. It took off not only because it gave science the power to explain the current reality, but also because it had a role in consolidating a modernization plan. Science and technology were highly valued, especially as bearers of practical, utilitarian knowledge that would guide the country to modernity.

At the time the interest in relativity and other theories with no immediate practical application, like electromagnetism or radioactivity, was accompanied by a defense of pure science against a utilitarian conception of science. According to this new conception, science should not be seen as just a means to much needed social and economic reforms, but above all as a source of knowledge. This theme permeated much of Einstein's visit. He joined forces with local academics to stress that the development of pure science was indispensable for the progress and spiritual independence of a nation. He also argued that professors must earn enough to be able to dedicate themselves exclusively to their research. The positivist tradition was also reflected in the institutional structure in which scientists worked. It was difficult to work on purely scientific issues in institutions where the practical utility of science and the resolution of national and social problems were valued. The development of studies in relativity would require the creation of appropriate spaces, which fell outside the model of technical or vocational schools.

Criticisms were also voiced of the theory of relativity. In Argentina, the *Anales de la Sociedad Científica Argentina* published articles by the Italian M. La Rosa about astronomical evidence that contradicted the theory of relativity (Rosa 1925), and a text by a US Navy captain, J. J. Thomas, entitled "Newton's Complete Triumph over the Relativists" (Thomas 1925), which came out together with a text by Castiñeiras that underlined the importance of Einstein's visit (Castiñeiras 1925). Martin Gil, a self-taught scientist, also published a long article in *La Nación*, a newspaper with a large readership, in which he said he was unable to accept that an "imaginary" mathematical formula could shatter the concept of absolute time.[25] In Brazil, criticisms were made by Justice Pontes de Miranda, author of a weighty volume called *Systema de Sciencia Positivista do Direito*, who sought to build up legal science based on the positivist ideas of the national hero and aviator, Gago Coutinho; and by Licínio Cardoso, a professor at the Escola Politécnica and member of the Academia Brasileira de Ciências, who wrote a text entitled "A Relatividade Imaginária." The articles were published in *O Jornal*, a newspaper with a large circulation, and the latter was also presented at a session of the Academia Brasileira de Ciências, to be followed up by responses and discussions at a number of subsequent meetings (Pontes de Miranda 1925; Coutinho 1925; Cardoso 1925).

---

[25] *La Nación*, May 11, 1925, p. 4 and May 12, 1925, p. 4.

Overall, the criticisms of the Theory of Relativity were of a scientific nature and did not get into personal or political issues, as was often the case elsewhere. They were generally made within a framework of different philosophical conceptions. In the eyes of some, Einstein was the forefather of a new science or religious doctrine which stood against the gospel of positivism. In many cases, the critics of the theory of relativity mentioned its excessive abstractness and the uncomfortable distance it had from the physical world.

Einstein's visit to South America was not, nor could it have been, enough to cause a broad acceptance of relativity. Nevertheless, it helped spread the research that was being done in the area and encourage scientists to become more actively involved. As Castiñeiras saw it, "the visit by doctor Einstein was extremely useful as it awakened interest among scholars to deepen their study of his theory, and of all the physical and mathematical questions related to it" (Castiñeiras 1925). Einstein's presence also provided support for those people who called for specific spaces for pure science, which ultimately had different consequences in each country. However, the absorption of relativity and the production of knowledge in this field was a gradual process as new generations replaced older ones, and as specific institutions were set up to develop it.

To Einstein, the journey for the most part was burdensome. He suffered from the heat and the many social engagements. In every city he visited he was hounded by the press and had to take part in one event after another hosted by the countries' presidents, the deans and other authorities from academic institutions, ambassadors, and others. Some of them were agreeable, like the reception given by Argentine students before he left, and the weekends spent at the home of Bruno Wasserman in La Vajole. But overall, they were tiring appointments full of speeches, and pomp and circumstance. He particularly noticed the local concern with dress and formalities, once describing his hosts as "lacquered Indians."[26] His general mood also changed during the trip. If at the beginning, it was all new and exciting, as time went by his appointments became a burden. He wrote in his journal on the way from Montevideo to Rio de Janeiro, the last leg of his trip: "My nerves are on edge. I would give anything not to have to go up on the tightrope one more time in Rio, but I must hold firm." One good point, as Einstein saw it, was his contact with the Jewish communities. In a letter to Michele Besso a few days after his return, he wrote, "everywhere when I arrived there was a hearty welcome from the Jews because I am a symbol of unity for them. This gave me great joy and I hope that this expectation bears good fruit."[27] He returned convinced that he would no longer take on voyages that did not have a strictly scientific purpose. Yet he kept to this decision only in part. The next time he traveled beyond Europe was in 1931, when he eventually visited Caltech at Millikan's invitation, but he did not manage to avoid the official ceremonies held during his time in New York and in Havana, Cuba.

---

[26] Einstein's travel journal to Argentina, Uruguay, and Brazil. EA 29–133.

[27] Einstein to Michele Besso, Berlin, July 5, 1925, EA-At 7.352-2.

References

Academia Brasileira de Ciências (1925). "Professor Albert Einstein: sua visita ao Brasil e homenagens recebidas durante sua estada no Rio de Janeiro". Rio de Janeiro: Typ. Jornal do Commercio.

Academia Nacional de Ciencias Exactas, Fisicas y Naturales (1995). "A 70 años de la visita de Albert Einstein: Preguntas y respuestas sobre aspectos entonces desconocidos de la Fisica Moderna". Buenos Aires: Academia Nacional de Ciencias Exactas, Fisicas y Naturales.

Agulla, Juan Carlos (1988). "Einstein en la Argentina". *Todo es Historia*, 247, 38–49.

Alberini, Coriolano (1925). "La Reforma Espistemológica de Einstein". *Revista de la Universidad de Buenos Aires*, 2ª série, sección II, tomo II, March–May 1925, 7–16.

Amoroso Costa, Manoel (1995). *Introdução à Teoria da Relatividade*. Rio de Janeiro: Ed. UFRJ. First published in 1922, Rio de Janeiro: Livraria Scientifica Brasileira.

Azevedo, Roberto M. (1920). "O Princípio da Relatividade". *Revista de Sciencias*, (parte I) v. 4, n°1 and (parte II) v. 4, n°2.

—— (1920). "A Teoria da Relatividade de Einstein". *Revista Brasileira de Engenharia*, v. 2, n°1.

Bibiloni, Anibal G. (2001). "Emil Hermann Bose y Margrete Elisabet Heiberg-Bose, pioneros de la investigación en física en la Argentina". In *Encontro de História da Ciência*. Videira, Antonio Augusto P. and Bibiloni, Aníbal G., eds. Rio de Janeiro: CBPF, 20–60.

Caffarelli, Roberto Vergara (1979). "Einstein e o Brasil". *Ciência e Cultura*, vol. 31, n. 12.

Cardoso, Licinio (1925). "Relatividade Imaginária". *O Jornal*, May 16.

Castiñeiras, J. R. (1925). "La visita del Professor Alberto Einstein". *Anales de la Sociedad Científica Argentina*, XCIX, 1–14.

Clark, Ronald W. (1973). *Einstein: the life and times*. London: Hodder & Stoughton.

Coutinho, Gago (1925). "Palestras sobre a Teoria da Relatividade". *O Jornal*, May 6.

Einstein, Albert (1925). "Pan Europa". *La Prensa*, March 24. Reproduced in Moreira, Ildeu C. and Videira, Antonio Augusto, eds. (1995). *Einstein e o Brasil*. Rio de Janeiro: UFRJ, 71–76.

—— (1926). "Observação sobre a situação actual da Theoria da Luz". *Revista da Academia Brasileira de Sciencias*, no. 1, 1–3. Reproduced in Moreira, Ildeu C. and Videira, Antonio Augusto, eds. (1995). *Einstein e o Brasil*. Rio de Janeiro: UFRJ, 61–64.

Fölsing, Albricht (1998). *Albert Einstein: a biography*. New York: Viking.

Gaviola, Enrique (1952). "Alberto Einstein (Premio Nobel de Física, 1921)". *Ciência y Investigación*, May, 08, 234–238.

Kirsten, Christa; Treder, Hans-Jürgen, *et al.* (1979). *Albert Einstein in Berlin 1913–1933*. Vol. I: Darstellung und Dokumente. Vol II: Spezialinventar. Berlin: Akademie-Verlag.

Moreira, Ildeu C. (1995). "A recepção das idéias da relatividade no Brasil". In *Einstein e o Brasil*. Moreira, Ildeu C. and Videria, Antonio Augusto P., eds. Rio de Janeiro: UFRJ, 177–206.

Newton, Ronald C. (1977). *German Buenos Aires, 1900–1933: Social Change and Cultural Crisis*. Austin: University of Texas Press.

Ortiz, Eduardo L. (1995). "A Convergence of Interests: Einstein's visit to Argentina in 1925". *Ibero-Amerikanisches Archiv*. 21, n° 1/2, 67–126.

Pais, Abraham (1982). *Subtle Is the Lord: The Science and the Life of Albert Einstein*. Oxford: Oxford University Press.

Pereira, Francisco Lafayette Rodrigues (1926). "Recepção de Einstein". *Revista da Academia Brasileira de Sciencias*, n. 1, 77–9.

Pontes de Miranda, Francisco (1925). "Espaço-Tempo-Matéria: um dos problemas philosóphicos da Theoria da Relatividade Generalizada". *O Jornal*, May 6.

Pyenson, Lewis (1985). *Cultural Imperialism and Exact Sciences: German Expansion Overseas 1900–1930*. Frankfurt am Main: Lang.

Raffalovich, Isaiah (1952). *Tziunim ve Tmurim be shiv'im shanot nedudim: otobayografi* (in Hebrew). Tel Aviv: Defus Shoshani.

Revista do Clube de Engenharia (1925). "Alberto Einstein". Rio de Janeiro: Revista do Clube de Engenharia.

Rosa, M. La (1925). Prove Astronomiche contrarie allá "teoria della relativitá". *Anales de la Sociedad Científica Argentina*, tomo C, Jul./Dec., p. 85.

Schwartzman, Simon (1979). *Formação da Comunidade Científica no Brasil*. São Paulo: Nacional/Rio de Janeiro: FINEP.

Sujimoto, Kenji (1989). *Albert Einstein: a photographic biography*. New York: Schoken Books.

Thomas, J. J. (1925). "Newton's Complete Triumph over the Relativists". *Anales de la Sociedad Científica Argentina*, tomo C, Jul./Dec. p. 133–140.

Tolmasquim, Alfredo T. (1996). "Constituição e diferenciação do meio científico brasileiro no contexto da visita de Einstein em 1925". *Estudios Interdisciplinarios de America Latina y el Caribe*, 7, no. 2, 25–44.

—— (2003). *Einstein: o viajante da relatividade na América do Sul*. Rio de Janeiro: Vieira & Lent.

Tolmasquim, Alfredo T.; Moreira, Ildeu C. (2002). "Einstein in Brazil: The communication to the Brazilian Academy of Science on the constitution of light". In *History of Modern Physics—Proceedings of the XXth International Congress of History of Science* (Liège, 20–26 July 1997). Kragh, Helge; Vanpaemel, Geert; Marage, Pierre, eds. Brussels: Brepols, 229–242.

7

The Reception of Einstein's Relativity Theories in Literature and the Arts (1920–1950)

Hubert F. Goenner

University of Göttingen – Institut für Theoretische Physik, Germany

7.1 Introduction

Among the many statements about Einstein during the year 2005, the suggestion that he had helped to bridge the gap between Snow's "two cultures," the exact sciences and the humanities (Snow 1959) can also be found: He not only stood for excellent physics but also had been an "engaged cultural scientist who even stepped into political debates" (Lossau 2005). As to the "cultural scientist," he may have been confused with his unrelated namesake, the art critic, writer, and dramatist *Carl* Einstein. This other Einstein looked out for new structural models for literature and the arts with an eye for precise experimental systems (Gnam 2003, 200). In the following, I shall try to convince you that, from the point of view of a physicist, *Albert* Einstein's new ideas about space, time, and relativity, while stimulating for some artists and writers, were neither fully grasped by them nor had a readily noticeable influence on the development of their work.

Obviously, notions like *space* and *time*, and the relations between them have been a topic of literature and art ever since there were writers, painters, and sculptors. Since *Gotthold Lessing*, poetry, plays, and music were given the predicate of "Zeitkünste" (arts of time); painting, sculpture, and architecture then were "arts of space" (Lessing 1766, 1954). Also, the connections between *subject* and *object* have always intrigued philosophers as well as painters and writers. With the advent of Einstein's relativity theories,[1] and later with quantum mechanics, all of these concepts had to be considered afresh. Thus, it cannot be ruled out that, after 1905, Special Relativity, after 1908, Minkowski's space-time picture, after 1915, General Relativity, and after 1917, Relativistic Cosmology may have influenced other cultural fields including the arts and literature. The fact that Einstein's new theories, in part apparently in conflict with common sense, were considered important in physics, can have helped painters, writers, architects, and composers to more courageously follow their own startling ideas. Yet the mere chronological coincidence of the appearance of creative or "revolutionary" works in different cultural fields, by itself, cannot

---

[1] By "theories of relativity," both Special and General relativity are subsumed.

establish a *causal* link among them. I do not belong to those admirers of Albert Einstein who seem to let history begin with him, e.g., the history of non-Euclidean geometry, or whose first mention of the concept of four-dimensional space is his use of it. Moreover, the same word, e.g., *space*, means different things in different fields. How shall a physicist, in this context, deal with the following remark by Arthur Köstler:

> Einstein's space is no closer to reality than van Gogh's sky. The splendour of the exact sciences originates in no deeper truth than that of Bach and Tolstoj; it starts from the very act of creation. The discoveries of a scientist imprint his own order on the chaos, and depend on the system of reference of the observer; it changes from century to century like a nude by Rembrandt and one by Manet. (Köstler 1970, 253)[2]

I will mainly look at the reception of the content of Einstein's theories during his lifetime, in German- and English-speaking countries. The focus will lie on literature and painting. My ordering principle will be to follow the work of a few selected personalities one by one. In contrast, Calcraft (1980) classifies the relevant literature into texts giving *explicit* reference to Einstein or his relativity theories, and into those exhibiting "a more subtle if not less pervasive influence." In a subclassification, he then distinguishes texts in which the *philosophical* implications, *moral* issues, or *technical* aspects of Einstein's work are in the foreground, and those which comprise a mere *metaphorical* usage[3] of relativity and its language. In the following, science fiction with its time travel stories is *not* dealt with although it might be intriguing to discuss fiction writers anticipating part of the space-time picture (cf. Wells 1895).

As an appetizer, three literary examples from more recent times are now listed briefly. The first, Friedrich Dürrenmatt's play *The Physicists*, may be known to a broader public (Dürrenmatt 1962). In it, in a mental home, Einstein and his fellow patients Newton and Möbius are leading a murderous hunt for a secret, direful world-formula. The gist of the play does not consist in a discussion of Einstein's theories but in the author's pointing to the responsibility of physicists, or rather their irresponsible behaviour, with regard to the welfare of the world.

In his story *Einstein crosses the River Elbe near Hamburg*, the writer Siegfried Lenz has Einstein ride on a ferry boat and, at some point, makes time stand still. Somehow, Lenz must have been occupied with the fact, described by Special Relativity, that measured time intervals may be longer or shorter, depending on the relative speed of observers, in this case of the boat with regard to the river bank. For clocks to "stand still," however, the relative motion would have to be with the velocity of light. This is possible only for some elementary particles—not for a macroscopic body as

---

[2] This is a back-translation of the English original, which was unavailable to me at the time of writing. Unless stated otherwise, all translations from German into English have been made by the author.

[3] By *metaphorical* usage I understand the use of physical concepts in a nonphysical context. As an example, take Sartre's remark that "the theory of relativity applies in full to the universe of fiction." His "universe of fiction" makes sense only metaphorically. (Sartre 1955, 23, quoted from Holton 1996, 132.)

a boat. When the time freeze ends, and the passengers leave the ferry, an old man (presumably Einstein) leaves first. This causes Lenz to conclude:

> [..] does not this one leave who himself determines what a fact is? (Lenz 1975, 139)

Possibly, he has misunderstood Einstein's remark that a physical theory is not determined by facts of nature, but are a free invention of the human mind. Einstein did not mean that facts are laid down by theory, but rather that these facts can be interpreted or explained differently according to the specific theoretical modeling.

The last modern example is formed by a poem by the playwright Rolf Hochhuth describing an aspect of Einstein's character[4]:

> Einstein whom a man
> had left waiting at a bridge
> answered, as the man apologized
> "No hurry – what I must do
> can be done everywhere."

7.2 Understanding and Misunderstanding

Sometimes, the opinion is voiced that the new observer-dependent picture of the world by which Einstein's relativity theories replaced the Newtonian–Kantian notion of *absolute* space and *absolute* time "had strong resonances with the contemporary revolutions occuring in art, music, theater, and literature" (Friedman and Donley 1985, 20). This *cannot* mean that these other "revolutions" have been generated by Einstein's theories. Most of these "revolutions" had occurred *before* Einstein's name became familiar to everybody on the street, including artists in their ateliers and cafés, i.e., after November 1919. If we take painting as an example, many of the new styles had surfaced before this date, such as cubism; futurism; expressionism, represented by groups such as *Der blaue Reiter* ("Blue rider": F. Marc, W. Kandinsky, and P. Klee) and geometric abstraction; surrealism; and constructivism, represented by *Die Brücke ("The bridge": E. L. Kirchner, E. Heckel, K. Schmidt-Rottluff, E. Nolde, M. Pechstein, etc.).*[5] For the evolution of literature and the arts, the technical-industrial development (automats, airplanes, radio, cinema, and telegraph), the

---

[4] Poem taken from the preface of (Hartmann 2004). The German original reads:

> Einstein, den ein Mann
> an einer Brücke hatte warten lassen,
> gab, als der sich entschuldigte,
> zur Antwort: "Keine Eile -
> was ich zu tun habe, kann überall geschehen."

[5] From the annotations to the contemporary edition of *Carl* Einstein's influential book (Einstein 1926, 1996), we may date *analytical* cubism to Braque's exhibition in 1908, to Picasso's *Demoiselles d'Avignon* from 1907, and *synthetical* cubism as starting in 1912/13.

increased tempo of life, social unrest (Marxism-Leninism), and the First World War played a more decisive role than any influence that might have resulted from progress in the exact sciences—including Einstein's theories of 1905 and 1915. In literature, a similar situation applies. I am surprised by the claim that the use of notions like "point of view" or "the inner positioning with regard to things," which are considered important in Robert Musil's novel *Die Verwirrungen des Zöglings Törleß* (The confusion of the pupil Törless), are interpreted as a sign of the influence of Einstein's relativity theory on Musil (Roth 1972, 49): the novel appeared in 1906 when Special Relativity was virtually unknown, even to the ordinary physicist. I am similarly reserved with regard to a statement linking Marinetti's demand for "destroying the 'I' in literature" with Einstein's (Special) Relativity; both are said to mean "the disintegration of the traditional web of understanding in a world without a fixed reference point" (Kraft 1972, 13).

Some of those who view all parts of cultural life as being bound together by a dense network of mutual relationships do not hesitate to argue by help of the catchword *autonomy*. Although usually reflecting isolation more than proximity, the gain of autonomy by different cultural developments is seen as a sign of togetherness:

> The works of Braque and Picasso from the years 1910–1911 signify [.] the sanction of autonomous painting which now marches out for the conquest of the real, the modern world. [.] Exactly, I repeat, as Einstein does with the experience of common sense, who breaks with the three dimensions of perceivable phenomena in order to confirm the autonomy of physics. (Daix 1973, 90)

Apart from the fact that this is a misinterpretation of Einstein's famous papers of 1905 and 1915, if there is autonomy in physics, then it is reflected by the laws of nature, not by the intentions of physicists.

7.2.1 The Fourth Dimension

The notion of "space-time," a four-dimensional "world," today dominates both the popular presentations and the teaching of Einstein's relativity theories. A specialist on Picasso's work writes:

> We find ourselves in the era of debates among scientists, who more and more take leave of their senses [.] Hertz animated the air with invisible waves. In 1900, Planck assured us that energy and matter have a discontinuous structure. Einstein introduced the fourth dimension [.]. (Daix 1973, 66)

Futurism begins with Marinetti's *Manifeste du Futurisme* of 1909; the expressionist group *Die Brücke* was active from 1905 to 1913 while *Der blaue Reiter* established itself in 1912; constructivism and suprematism (Malevich) started in 1914 and 1916, respectively. Apollinaire introduced the concept of 'surrealism' in 1917.

The fourth dimension did not originate with Albert Einstein, nor did Einstein himself readily welcome this picture. It was the Göttingen mathematician Hermann Minkowski who invented this elegant mathematical representation. He dramatically overstressed it:

> From this hour on, space by itself and time in itself must become complete shadows; only some kind of union of both will retain its independence. (Minkowski 1908)

Since then, the new $(1+3)$-dimensionality (time and 3 dimensions of space) has often been misunderstood to be the same as a structure formed by four *spatial* dimensions. Four- (and higher-) dimensional space had been a theme in mathematics during the nineteenth century for Gauss to Riemann, for Cayley to Helmholtz. In the second half of the century, the concept of *four*-dimensional space gained in popularity outside of pure science—not least through a union with spiritualism. Thus, the astrophysicist Johann K.F. Zöllner of Leipzig terrified his colleagues with spiritualist seances in which, through a medium, a link with a fourth spatial dimension was said to have taken place (Zöllner 1880). In principle, knots could be untied without cutting them, and a closed box could be entered from the fourth dimension without the slightest damage being done to it. This fourth *spatial* dimension *and* non-Euclidean geometry had been the subject of theoretical discussions among painters well before 1905. Even at the beginning of the twentieth century, English physicists like Oliver Lodge and Sir William Crookes, as well as French astronomer Camille Flammarion were deeply interested in the implications of "ether vibrations" for psychic phenomena, as, e.g., for the transmission of thought (Henderson 2002, 131–132).

More seriously, the famous French mathematician Henri Poincaré made a comparison with 2-dimensional projections of a 3-dimensional body in order to "visualize" a *four*-dimensional body: its projections from different directions would then be 3-dimensional bodies (Poincaré 1916, 88–91).[6] Conceptionally, Minkowski's 4-dimensional *space-time* must be strictly separated from this 4-dimensional pure *space*. Although time and space are closely interwoven in the space-time representation of the Lorentz group, they can be easily distinguished within both, special and general relativity.[7] The use of the space-time picture is not imperative in Special Relativity, but in General Relativity, Einstein's theory of gravitation. "All attempts to visualize a fourth [spatial] dimension are futile. It must be connected to a time experience in three space" (Hinton 1904, 207).[8] The Belgian writer and symbolist Maurice Maeterlinck did publish a book on this, but somehow he could not get along without the vision of a "spatialization" of time, and interprets "time, for want of a better explanation, [.] as movement of space, and space as the rest of time." His conclusion

---

[6] A very readable introduction into the nineteenth century background to the popularity of a fourth *spatial* dimension, acceptable to physicists, is given by Henderson (1983, 2003).

[7] This distinction is cursorily blurred, again, when in a certain kind of research in relativistic quantum field theory, time t is continued into the complex domain. Then, t is replaced by $i\tau$ with $i = \sqrt{-1}$. τ and the three space variables in fact form a 4-dimensional Euclidean space.

[8] From Henderson (1983, 324).

is "that, caught between space and time, we end in some kind of cosmic cul-de-sac" (Maeterlinck 1929, 97–98). In the meantime, unlike Maeterlinck, we became accustomed to distinguishing between what is called "reality" and a mathematical model for describing it. In a section of her book, Henderson convincingly shows how the long tradition of *pre*-Einstein discourse among painters on 4-dimensional space and non-Euclidean geometry, *after* the appearance of Einstein's theories, was rewritten by some by naively replacing the fourth *spatial* dimension by time (Henderson 1983, 342–352).[9] This included the Russian suprematist El Lissitzky (1890–1941).

7.2.2 $E = mc^2$ and the Atom Bomb

The best-known formula of the twentieth century, Einstein's relationship between energy E and inertial mass m i.e.:

$$E = mc^2,$$

has led to another misunderstanding.[10] The formula says that to any form of energy, be it the energy of electromagnetic radiation or of mechanical motion, a certain inertial mass is assigned. This also holds for gravitational mass due to its equality with inertial mass. One result is that *light* responds to gravitational attraction. Conversely, any mass contains an immense amount of energy, calculable through Einstein's formula. An inadmissible interpretation of this formula by writers and journalists, even by popularizing presentations of some physicists, is that matter is 'de-materialized'; everything that exists would be energy taken as a 'spiritual' entity. A further misunderstanding takes Einstein's formula as the key to the building of an atom bomb. In order for this to take place however, research areas like nuclear physics and nuclear chemistry first had to come into existence. Also, the technical problems of controlling the ignition of an uncontrolled nuclear chain reaction had to be solved. After Hiroshima, Einstein expressed himself as follows:

> I do not see myself as the father of the release of atomic energy. My contribution to this was very indirect. Really, I would not have believed that nuclear fission could be realized during my lifetime. [.] (Einstein, *Atlantic Monthly*, November 1945, quoted from Calaprice 1999, 117)

On the other hand, the photographer Philippe Halsman reports that Einstein told him in 1947, during a photo session how depressed he had been that his formula $E = mc^2$ and his letter to President Roosevelt had made the atom bomb possible (French 1979, 28).

In our context, these remarks become significant if seen against the background of a critical claim with regard to the exact sciences by certain literary and philosophical circles: the exact sciences, apart from their technical applications, are held to

---

[9] Henderson's book is a beautifully written and almost inexhaustible source of everything you ever wanted to know about art and a fourth (spatial) dimension. Cf. also the review by S. Sigurdsson (1989).

[10] Here, c is the speed of light in vacuum.

be irrelevant for human thinking. The philosopher and social scientist J. Habermas presents this opinion:

> The insights of atomic physics remain—taken by themselves—without consequence for the interpretation of our way of life ("Lebenswelt"). [.] Only when nuclear fission is performed, by help of physical theories, only when informations are used for generating productive or destructive forces, their revolutionizing *practical* consequences may enter literary consciousness [.]. Poems are written by looking at Hiroshima, not by working through hypotheses about the transformation of mass into energy. (Habermas 1968, 82)[11]

Correspondingly, there seem to be any number of writers who take no notice of developments in the exact sciences. On the other hand, even some of those quite removed from physics, such as the essayist, critic, and writer Franz Blei (1871–1942) could not resist displaying their knowledge of new developments. In connection with the lecture of an economics professor, he wrote:

> His dislike of final theories possessed something of the precaution of the modern physicist who abstains from calculating the future history of an atom from its present condition [.], because the physical world is not as strongly deterministic as was believed in the past two hundred years. (Blei 1930, 202)

7.3 On the Reception of the Theories of Relativity in Literature (ca. 1915–1955)

Einstein is seen as a role model for researchers in the exact sciences. After having attained world fame, he entered public discourse in a way that reached beyond his research in theoretical physics. Unlike most of his collegues in physics, he received some satisfaction, even some enjoyment, from his interaction with the media. This gave him the possibility to spread his ethical and political views. His statements focussed on the prevention of war, disarmament, international understanding, Zionism, and he sympathized with the socially and politically less well off. With an astonishing amount of energy and time, he seemingly tried to answer all questions directed to him. No reader of a daily, no openminded contemporary could fail to notice his presence in the media. Thus, Einstein certainly exerted some influence in cultural sectors other than physics. During the Einstein Year 2005, some have given Einstein an even larger weight.

As an example from literature, his alleged influence on "driving back the absolute" is claimed, or more concretely, an influence on the shift of perspective in the narrative from a single, omniscient, distanced subject as a storyteller to a presentation of different persons with different perspectives. Alternatively, a single narrator may now

---

[11] This is nothing more than a description of the status quo, i.e., the lack of education in the exact sciences of poets—not to speak of social scientists.

occur moving in a number of external and inner 'spaces,' with the latter being represented as dreams, hallucinations, or mental disorders. However, all these techniques
of writing were already in use before Einstein and his theories.

A number of the writers and artists who tried to make their works reflect
Einstein's relativity theories had encountered physics or mathematics during their
university studies; this should have been easier for them than for their colleagues
without an education in the exact sciences. Such authors included the physicians
Gottfried Benn, Alfred Döblin, and the writers Hermann Broch and Robert Musil.

7.3.1 German-Speaking Countries

Thomas Mann (1875–1955)

We start with a Nobel Prize-winning author *without* training in science: Thomas
Mann. He gained information concerning Einstein's relativity theories mainly from
the press and from second-hand essays (Weishaupt 1994, S. 36). Although acquainted
with Einstein and associated with him in many a political action, he obviously
avoided being informed by him *directly*, supposedly because of his ambiguous stance
concerning Einstein's research.

Of all themes, in an essay entitled "Okkulte Erlebnisse" (Occult Experiences)
Mann takes on Albert Einstein. He speaks about the relationship between metaphysics and natural science:

> The fact that I know and understand not much about the teachings of
> Mr. Einstein (except, e.g., that things yet own a 'fourth dimension,' i.e.,
> time), do hinder me as little as other intelligent laymen to notice that in this
> teaching the borderline between mathematical physics and metaphysics has
> become blurred. Is it still 'physics,' or what is it really when it is claimed
> (and today it is said so) that matter is only a manifestation of energy, and its
> 'smallest' parts, but who already are neither small nor large, and surrounded
> in fact by statio-temporal force-fields; however, they themselves are said to
> be *outside of time and space*? (Mann 1993, 183)

Possibly, Thomas Mann refers here to *popularizing* interpretations of physics by
physicists, whose reading, occasionally, may be disconcerting. For example, the
astronomer and mathematical physicist Arthur S. Eddington held that: "It is perti-
nent to remember that the concept of substance has disappeared from fundamental
physics: what we ultimately come down to is form [.]. In modern physics, form, particularly waveform, is at the root of everything" (quoted from Friedman and Donley
1985, 146). Yet neither 'substance' nor 'form' in this generality are concepts native
to physics.

Thomas Mann was not always as emotionally uninvolved. In his story "Sea
Voyage with Don Quijote," he describes his attitude to Einstein as follows:

> What kind of thoughts of a pupil! But, is it not so that something puerile
> adheres to the cosmological world view, in comparison with its opposite, the

psychological? In this, I am reminded of the blank rounded childlike eyes of Albert Einstein. I cannot help it, but to me the immersion into human life is of a riper, more adult character than the speculation about the Galaxy. ("Meerfahrt mit Don Quijote," in Mann 1935, 209–270, here page 234)

The influence of relativity theory on Thomas Mann's work has been investigated particularly through his novels *Der Zauberberg* (The Magic Mountain) (Mann 1924) and *Felix Krull* (Mann 1985). In the former, it is the comparison between the flow of *inner* time of the novel's hero, in a sanatorium in Davos, and the happenings in the *external* world of the "lowlands" which plays a role. It remains open whether Mann wanted to hint at the relativity of simultaneity and its consequence, i.e., time dilation. We do have a cryptic note by the author:

It was in 1924, after endless intermissions and difficulties, that there finally appeared the book which, all in all, had had me in its power not seven but twelve years. [.] It is certain that ten years earlier the book would not have found readers—nor could it have been written.[12]

That the book could not have been written in 1914, does this mean that Mann needed the acclaim Einstein received from 1919 on in order to learn about relativity theory? Perhaps, at the time he read a popularization of this theory containing a statement similar to that by theoretical physicist Steven Hawking:

A single absolute time does not exist in relativity theory. According to it, each person has his own measure of time depending on where he is, and with what speed he moves. (Hawking 1988, 51)

Hawking's statement certainly is correct. However, the fact that he does not give the slightest hint that it is without meaning for a human perception of time, instigates the uninitiated into believing either that Einstein's theories are abstruse, or, as Mann might have tried, to use the remark as an explanation of the perception of *psychological* time. In the novel *Felix Krull*, modern instruments like X-ray machines, gramophones, and stereoscopes play a bigger role than physical theories.[13] Thomas Mann also endeavoured, unconvincingly, to imbed the idea of 4-dimensional space-time into his writing (see Weishaupt 1994, 37).

Robert Musil (1880–1942)

In 1908, the writer Robert Musil obtained his doctoral degree from the philosophical faculty of Berlin University with the thesis "Contributions to a critical examination of Mach's teachings."[14] In developing Special Relativity, Albert Einstein had also been

---

[12] From http://ebooks.unibuc.ro/lls/AncaPeiu-STEVENS/chronology.htm, accessed April, 2011.

[13] Cf. Danius (2002), and its review in Misa (2005).

[14] This refers to Ernst Mach (1838–1916), physicist and natural philosopher. Musil's examiners had been philosophers Carl Stumpf (1848–1936) and Alois Riehl (1844–1924), mathematician Hermann Schwarz (1843–1921), and experimental physicist Heinrich Rubens (1865–1922).

strongly influenced by Mach's empiro-critical analysis of the foundations of physics. Musil, who thus had some education in mathematics and the natural sciences, had noticed Einstein as early as 1918 when writing:

> Intellect fills mankind with new happiness and brings progress, descends from abacus to infinite series, and from Thales to Professor Einstein. (Musil 1978, 1023; Roth 1972, 141)

In his "Profile of a Program" Musil tries to link the arts and natural science:

> Today all soulful courage lies in the exact sciences. We will not learn from Goethe, Hebbel, Hölderlin, but from Mach, Lorentz, Einstein, Minkowski, Couturat, Russell, Peano... [15] And in the program of this art, a single work of art can have this reason: mathematical courage, the dissolving of souls into elements, unlimited permutation of such elements, all is connected with all, and may be built up from such. The building-up does not prove: it's made out of such, but: it is linked with such. (Roth 1972, 365)

In her book, *The Mastering of Speed*, A. Gnam investigates how Musil and W. Benjamin included the nonliterary knowledge of their time into their writings. An important topic is "the metaphorical link of the ever rising 'tempo' in everybody's life with 'speed' as defined by physics." Supposedly backed by the new scientific insights, the authors would project into relativity theory "a spiritually understood 'exit from the space-time-relation'." According to Gnam, "such imaginations have almost nothing to do with Einstein's research." The new effects of relativity theory are based on velocities close to the speed of light and cannot be transferred—without change—to velocities of everyday life (Gnam 1999; cf. also Gnam 2004). Nevertheless, in his essay "Der mathematische Mensch" (The Mathematical Man), Musil demands a transfer of mathematical virtues in thinking to the realm of feelings (Gnam 2001, 132).

Alfred Döblin (1878–1957)

The physician and writer Alfred Döblin is the author of an important novel, *Berlin Alexanderplatz*, which describes the life of a man in the big city (Döblin 1930). Döblin did not hold the press responsible, but rather the scientists themselves when asking who would force upon the public "the importance of Einstein's teachings. The hierarchy of scientists, the secret association, the conspiracy, and free masonry of mathematicians [.]?" (*Berliner Tageblatt*, 24 Nov. 1923, evening edition, page 2).[16] Döblin's article is a reaction to Einstein's popular book of 1917 (Einstein 1917); he continues:

---

[15] The last three names belong to mathematicians and logicians (philosophers).

[16] The article's title is: "Die abscheuliche Relativitätslehre" (The Abominable Relativity Theory). Döblin expressly forbade reprints.

This little book did not stimulate me, yet it aroused a lot of anger and confusion. It began seemingly popular; after a few pages, the formulas broke loose, the infamous cabalistic symbols of mathematics. [.] Mathematics is the foe of nature and of insight into nature. A human being who possesses mathematical knowledge, the lingo of formulas, and with it approaches nature, must be like a woman who, with soaped hands, wants to grab a fish: surely, she will not seize it. Yet it is an unprecedented arrogance of mathematicians to pose in front of the world and of nature, and to say that they only had eyes for things. [.] Millions of well-educated persons either do not understand relativity theory, or they do not know what they are supposed to do with it.

Readers did react to Döblin's article; after a while, the editors of *Berliner Tageblatt* (in the evening edition of 13 Dec. 1923, page 2) left the final word to Döblin but not without providing an introduction:

[.] that the ultimate final word must be left to a certain Immanuel Kant (whom also Dr. med. Döblin will acknowledge), and who says in the preface to his 'Metaphysische Anfangsgründe der Naturwissenschaft' that in each particular teaching about nature only so much genuine science is to be found… as there is mathematics contained in it.

In another essay "Insight into Nature, not Natural Science," Döblin repeats his reproach toward mathematics:

Mathematics never comes close to things. The mathematical method in the natural sciences leads in the wrong direction. Mathematics is a small, often comfortable, important in practice, and clean auxiliary discipline. Yet with it, it's like with the French in the Ruhr Basin: the French are good, but in the Ruhr Basin they are bad.[17]

Max Brod (1884–1968)

In his best-selling novel *Tycho Brahes Weg zu Gott* (1916) (*The Redemption of Tycho Brahe*, 1928), the writer and administrator of Kafka's papers (Kafka archive), Max Brod, models his main character Kepler on Einstein, as an invulnerable, persevering man who is withdrawing himself from all external influence.

Indeed, [Einstein] seemed to enjoy going through all possibilities of dealing with a scientific topic with indefatigable courage. He never committed himself; jokingly and with virtuosity he never shunned any multifariousness yet kept taking hold in a confident and creative manner. In Kepler's figure of my Tycho-Brahe-novel, I molded his characteristic scientific courage and readiness to always start afresh [.]. I never noticed such an egoistic trait

---

[17] Döblin refers to the French occupation of the Ruhr Basin after the First World War in order to enforce reparation payments from Germany.

of Einstein; on the contrary, I always found him kindhearted, helpful, and astonishingly open-minded. (Brod 1969, 201–202)

Hermann Broch (1886–1951)

After training as a textile engineer, the Austrian writer Hermann Broch began to study mathematics, philosophy, and psychology at university; thus he became familiar with the mathematical language used in physics. In 1928, he published an essay concerning "fundamental philosophical problems of an empirical science" in which, however, he only treated mathematics (Broch 1977, 131–146). As a writer, he did not directly put Einstein's relativity theories to use but turned to Einstein's personality. In his novel *Die unbekannte Größe* (The Unknown Quantity) he introduced the mathematician Richard Hieck as an intellectual striving solely for recognition. For Broch, mathematics was "the type of tautological knowledge bound clearly to itself." The question he then posed was how such a man could come "to the solution of the rationally intractable but still missing insight into the great questions about death, love, and the fellow man." In 1935, Broch also delivered a film script "Das unbekannte X" (The Unknown X); in the movie, the mathematician was given the name "Professor Weitprecht," and endowed with traits of Albert Einstein. In an introduction to the script, Broch explained:

> Neither Einstein's nor the theory of relativity's names are mentioned explicitly [.] but everything is presented such that the 'poetic reality' is actually correct, and it provides an, if somewhat compressed but essentially correct, image of the events. (Broch 1977, 251)

For whatever reasons, the film did not materialize. Broch thought it

> [.] awful to do anything making relativity theory, hence also Einstein, look banal. It pays to be finicky when it can be seen that [Einstein] still is attacked rudely [.]. (H. Broch to Ruth Norden, 16 Jan. 1936, in Broch 1981a, 384)

I do not know whether Einstein read this novel of Broch; however, he enthusiastically welcomed another novel *Der Tod des Vergil* (The Death of Vergil) and wrote to Broch (around 1945):

> The book makes very clear to me what induced me to flee when I assigned myself to science with all my bones. I had been conscious of it previously, but not as distinctly *); what you have said about the intuitive in your letter, is as spoken by my soul. Logical form does exhaust the essence of perception as little as measure of verse does with regard to the essence of poetry. [.] The essential remains mysterious and will remain so forever; it may be felt but cannot be grasped. [.] *) Flight from the I and We to the It. (Broch 1981c, 18–19)

Einstein's letter made Broch feel ill at ease; he admired him very much: "indeed, he is the most magnificent human being I ever encountered. As Polgar put it, the second best happiness to [us] Earthly children lies in the personality of fellow man" (H. Broch to Elke Spitz, 24 Jan. 1945; in Broch 1981c, 437).

Albert Einstein, who loved to smoke his pipe, became a witness when Broch obtained his US citizenship in 1944. This event led Broch to write the following verse[18]:

> If at the bowl-brim it is knocked
> the pipe most often remains clogged.
> You would not solve this genuine riddle
> (- forgive my carrying owls to Athens -)
> than by knocking it with quiet hand
> not at the edge, but in the middle:
> so says, in allegory, the pipe's bowl
> wishing health, strength and happiness
> to my citizenship's grandiose witness.
> (Broch 1980, 126–127)

With Broch, we find here the same sort of less than poetic rhyming also done by Einstein on many occasions—not to forget other famous physicists.

Gottfried Benn (1886–1956)

While Gottfried Benn wrote an essay "Goethe and the Natural Sciences," we do not know of a corresponding one "Einstein and the Humanities"—in all likelihood because of the missing points of contact between Einstein and literary culture. Nevertheless, in Benn's work, occasional traces can be found that show his awareness of Einstein and his theories. One example is the sequel of scenes "Drei alte Männer" (Three Old Men) in which he has the host say:

> We shall spend this evening wholly immersed in the objective world. The paper, in which Robert Mayer set up the law of the conservation of energy, is said to be four pages long - it became the measure of the century. Relativity theory in its first draft presented itself as three formulas of a hand's breadth on one page. The relation of quantity to quality obviously is rather special.

---

[18] My free and unrhymed translation. The original is:

> Wird sie am Schalenrand geklopft,
> so bleibt die Pfeif' zumeist verstopft.
> Nie löst sich anders ein Problem
> (- verzeiht die Eule in Athen -)
> als wenn beklopft mit sicherer Hand
> in seiner Mitten statt am Rand:
> Dies sagt die Schale gleichnishaft
> und wünscht Gesundheit, Glück und Kraft
> dem großen Zeugen meiner Bürgerschaft.

To bring our existence into one short formula - how would this look? (Benn
2003, 113)

Possibly, with "the three formulas," he meant the special Lorentz transformation,
linking two systems of reference moving with constant relative velocity. Note,
though, that the *content* of Special Relativity is *not* used by Benn. The same author
becomes very poetic in a piece describing the thoughts of a Berlin lecturer in philosophy; only Einstein's name is inserted:

Oh, postman among values of hebephrenia![19] Respectable bag at the belly
strap, content postage to the point and rubber cell rolled into the full beard;
still gaze, stiff out of an animal's face, towards the edge of being, from
Ephesos to Einstein this weighty antinomy, arch of the yoke like Spinoza:
determinatio est negatio - yet motive of sunflower, there is transition, there is
gentle song, ionic tragedy, at the crust of abyss warbling of butterflies: olive
gentle, agave crusty, [.]. (From "Der Garten von Arles" (Garden of Arles)
1920, quoted from Benn 1987, 118)

Generally, Benn was dissatisfied with the never-ending replacement of one "truth" in
science by another one:

Space and Time! Just now, geometry was an axiomatic system, in Euclidean
formulas nature surrendered, and Kant, once and forever completed decisively ideas deliberated through centuries. This truth held 30 years. Human
spirit, never at rest, particularly the progressive 19th-century, discovered
that to a straight line several parallels are possible, non-Euclidean geometry, spherical geometry. This truth lasted for four decades. Since then,
it is spoken only of relational definitions, space and time are reference
systems among rigid bodies, the famous and truly grand theory. But already
it appears illegal to introduce space into the physically small, the atom, for
quantum mechanics the causal theory of space does not hold. (Benn 1987,
232–247, esp. 235 as quoted by Emter 1995, 149)

Carl Einstein (1885–1940)

At the end of our line of German writers stands the one closest to the arts, Carl
Einstein.[20] Central concerns in his work, particularly in his theory of visual art,
may be seen in his criticism of the concept of causality, of the duality between subject and object, and of his attempt to reintroduce the "absolute" into the arts and
which, according to him, had vanished from philosophy. Carl Einstein had no education in any of the exact sciences, nevertheless he tried to acquire some knowledge of
non-Euclidean geometry and of modern physics. Perhaps he was influenced more by

---

[19] Hebephrenia is a particular kind of schizophrenia.

[20] No kinship with Albert Einstein in the sense that Carl is not among the many cousins—as
much removed they might have been—mentioned by Albert Einstein.

Bergson's *Duration and Simultaneity* than by Albert Einstein.[21] In modern art, e.g., cubism, he saw a new kind of perception of reality.

> Thanks to technical science it has been understood that reality can be changed and enhanced in an unforeseen measure. Physics provides the intellectual basis for such technical courage. Riemann had established that geometry works with conventions, not at all with unambiguous realities [.].
> (Einstein 1996, 157–158)

In his criticism of constructivism he argued that "the wish is to create, consciously, new objects, a new space; he is groping between Riemann not grasped and Lobachevski not understood [.]." The fundaments of Carl Einstein's theory of visual art were formed when relativity theory had become the subject of small talk but prior to Heisenberg's breakthrough in quantum mechanics (Emter 1995, 161).

Franz Blei immortalized Carl Einstein in his *Grand Bestiarium* depicting literary notables:

> DER EINSTEIN. This is a cometary matter as far as the said Einstein is a comet or wandering star of the metaphysical heavens; therefrom, once in a while, he deviates into the earthly atmosphere, inexplicably, because his path cannot be calculated. There he will glow, and flash, and spit. His appearance on Earth is catastrophic for middle-class brains, whose pappy substance will furiously boil upon Einstein's closest approach to Earth. Thereupon, the said Einstein will again continue his metaphysical path whose future direction his most meticulous observer, Rowohlt, does not know. (Blei 1924, S. 28)[22]

Occasionally, an addition to this entry is quoted ("Das kleine Einstein" (The Little Einstein)) which is said to refer to *Albert* Einstein and which sounds as a later contrivance:

> THE LITTLE EINSTEIN. The little Einstein is an animal like the big one - only in relativity. Yet it differs from the great Einstein by living relatively in relativity while the great [animal] lives in it absolutely. Once in a while, the little [animal] establishes relations and tries to make them absolute with only relative success. The little Einstein has its feet at its large head, and moves in a circle. (Blei 1920, 20; quoted from Kraft 1972, 50)

In his comparison of Carl and Albert Einstein, K. H. Kiefer asks whether "sciences and arts etc. can illuminate, influence or interpenetrate each other." For the period around 1915, he favours "a relative unity. In the case of Carl Einstein's theory of art and literary work this means: the close analogy of the theory of relativity founded by Albert Einstein starts to replace the tradition of Bergson and Ernst Mach [.]" (Kiefer 1982, 190; his English summary). In view of Kiefer's following remark

---

[21] He referred to Bergson twice in connection with futurism but did not mention *Albert* Einstein in his book on twentieth century painting (C. Einstein 1996, 172, 176).

[22] In a reprint of 1995, the editor erroneously identifies this Einstein with Albert (Blei 1995, Register). Rowohlt is the editor of Carl Einstein.

in his article, physicists might not be strongly convinced of his understanding of physics: "It seems to me that, already in 1915, Carl Einstein takes account of this epistemological state of affairs by occupying the 'right place' which Albert Einstein assigns to the observer. As cannot be otherwise, this place is located in non-Euclidean [geometry], it implies Heisenberg's uncertainty relations" (Kiefer 1982, 189).

7.3.2 English-Speaking Countries

Again, only some prominent poets and writers are chosen. For a larger list, cf. (Friedman and Donley 1985).

William Carlos Williams (1883–1963)

The enthusiasm about Einstein in the United States reverberated also in the work of poets and writers. One example is given by the pediatrician in Rutherford, New Jersey, and recognized American poet, William Carlos Williams. After Einstein's first visit to the United States, he succeeded in writing a hymnic poem "St. Francis Einstein of the Daffodils":

In March's black boat
Einstein and April
have come at the time in fashion
up out of the sea
through the rippling daffodils
in the foreyard of
the dead Statue of Liberty
whose stonearms
are powerless against them
the Venusremembering wavelets
breaking into laughter -

Sweet Land of Liberty,
at last, in the end of time,
Einstein has come by force of
complicated mathematics
among the tormented fruit trees
to buy freedom
for the daffodils
till the unchained orchards
shake their tufted flowers -
Yiddishe springtime!

At the time in fashion
Einstein has come
bringing April in his head

up from the sea
in Thomas March Jefferson's
black boat bringing
freedom under the dead
Statue of Liberty
to free the daffodils in
the water which sing: Einstein has remembered us
Savior of the daffodils!

Einstein is characterized as a rebellious, easygoing, great mathematician, with his Jewish identity being accented: "[.] Sing of Einstein's Yiddishe peachtrees, sing of sleep among the cherryblossoms.[.]" Concepts from his Relativity Theory are introduced only metaphorically: "[..] Einstein [..] has come [..] shouting that flowers and men were created relatively equal [.]," or, " [.] The shell of the world is split and from under the sea Einstein has emerged [.]." After seven verses the poem ends with:

It is Einstein
out of complicated mathematics
among the daffodils -
spring winds blowing
four ways, hot and cold,
shaking the flowers!

(Friedman and Donley 1985, Appendix 195–198.) Williams did regard his experiments with rhyme and meter as being directly related to Einstein's theories:

How can we accept Einstein's theory of relativity, affecting our very conception of the heavens above us of which the poet writes so much, without incorporating its essential fact—the relativity of measurements—into our own category of activity? Do we think we stand outside the universe?[. . .] Relativity applies to everything.

Poems being measures of space and time, for Williams:

[..] poems cannot any longer be made following a Euclidean measure, beautiful as that may make them. The very grounds for our beliefs have altered[. . .] Relativity gives us the cue. We have to do with the poetic, as always, but with a relatively stable foot, not a rigid one. (From Donley 1981, 217)

It is left to the readers to discover Einstein's theories incorporated into "rhyme and meter" of this poem.

Archibald MacLeish (1892–1982)

Another one intrigued by Einstein's personality and scientific achievements was the American lawyer, writer, poet, and civil servant in the US government[23] Archibald

---

[23] Among other posts held, he was librarian of Congress and assistant head of the US delegation to UNESCO.

MacLeish. In 1926, he published a long poem "Einstein" considering in it Einstein and the philosophical implications of his work. The poem begins:

Standing between the sun and moon preserves
A certain secrecy. Or seems to keep
Something inviolate if only that
His father was an ape. Sweet music makes
All of his walls sound hollow and he hears
Sighs in the paneling and underfoot
[...]
The mirror moon can penetrate his bones
With cold deflection. He is small and tight
And solidly contracted into space
Opaque and perpendicular which blots
Earth with its shadow. And he terminates
In shoes which bearing up against the sphere
Attract his concentration,
[...]

The eleven parts of the poem are accompanied by ten brief descriptions by the author written on the margin like:

> Einstein upon a public
> bench Wednesday the
> ninth contemplates finity

MacLeish's comment for the third section of the poem is: "Einstein descends the Hartmannsweilerstrasse." I assume that, in the author's memory, Einstein's address in Berlin, "Haberlandstrasse," and the name "Hartmannsweiler (Kopf)," a mountain in Alsace, one of the most fervently fought-for places in The Great War,[24] became merged.

Decorticate
The petals of the enfolding world and leave
A world in reason which is in himself
And has his own dimension. Here do trees
adorn the hillside and hillsides enrich
The hazy marches of the sky and skies
[...].

These lines are taken from the fourth section: "Einstein provisionally before a mirror accepts the hypothesis of subjective reality." What a beautiful description of the transformation by Einstein of living nature into a rational cosmological model! The following entry of MacLeish then points to the change in Einstein's philosophical attitude: "[...] rejects it." Of course, violins are not missing in the poem; the

---

[24] There, from 1914 to 1918, 60 000 men on both sides were slaughtered, together with their commanding officers. The bodies found are resting now in a big cemetery at the slope of the mountain.

seventh section is accompanied by: "Einstein dissolved in violins invades the molecular structure of F. P. Paepkes Sommergarten. Is repulsed," and opens with

> But violins
> Split out of trees and strung to tone can sing
> Strange nameless words that image to the ear
> What has no waiting image in the brain.
> [...].

I have not been able to find out what "Paepke's Sommergarten" refers to. A concrete hint at (Newtonian) physics is seen in a section headed "Einstein hearing behind the wall of the Grand Hôtel du Nord the stars discover the Back Stair":

> [..] He lies upon his bed
> Exerting on Arcturus and the moon
> Forces proportional inversely to
> The squares of their remoteness and conceives
> The universe.
> Atomic.
>
> He can count
> Oceans in atoms and weigh out the air
> in multiples of one and subdivide
> Light to its numbers.
> If they will not speak
> Let them be silent in their particles.
> Let them be dead and he will lie among
> Their dust and cipher them - undo the signs
> Of their unreal identities and free -
> The pure and single factor of all sums -
> Solve them to unity. (MacLeish 1934, 67–75)

As rhythmically beautiful as the poem may be, for me it remains difficult to find a successful symbiosis of poetic language and Einstein's ideas concerning relativity. Einstein's solitary inner life amidst all the outside clamour is wonderfully described; in fact, MacLeish sees him as isolated from if not repulsed by a wordless nature, as standing outside the universe. Even so, Einstein ends "no farther, as, beyond extensively the universe itself."

Like quite a few writers on both sides of the Atlantic Ocean, MacLeish also expressed his feelings on the haste of modern life. In his poem "Signature for Tempo" he deals with past and future, the relative speed that joins them, and some of the effects to be read into motion:

> [.] Think that this silver snail the moon will climb
> All night upon time's curving stalk
> ....
> These live people,

These more
Than three dimensional
By time protracted edgewise into heretofore
People,
How shall we bury all
These queer-shaped people,
In graves that have no more
Than three dimensions?
Can we dig
With such sidlings and declensions
As to coffin bodies big
With memory?
(MacLeish 1933, 151–152)

It nevertheless is startling how unhurried the poem's lines flow.

E. E. Cummings (1894–1962)

Some of the poems of the American writer Edward Estlin Cummings are "classics" in spite of excentric typography and his "syntactically and semantically anomalous" texts (E. Hesse in Cummings 1994, 114). Cummings was no friend of fixed rules of thinking; this left him reserved with regard to the exact sciences. His poem "Space being(don't forget to remember)Curved" reflects his attitude. While taking note of Einstein's general relativity and the ensuing "grandiose" picture of the universe, he remained sceptical about human beings:

Space being(don't forget to remember)Curved
(and that reminds me who said o yes Frost
Something there is which isn't fond of walls)
an electromagnetic(now I've lost
the)Einstein expanded Newton's law preserved
conTinuum(but we read that beFore)
of Course life being just a Reflex you
know since Everything is Relative or
to sum it All Up god being Dead(not to
mention) inTerred
                LONG LIVE that Upwardlooking
Serene Illustrious and Beatific
Lord of Creation, MAN:
                        at a least crooking
of Whose compassionate digit, earth's most terrific
quadruped swoons into billiardBalls !

"Frost" refers to the American poet Robert Frost who himself was ambivalent toward science. The last lines point to a curved finger at the trigger of a rifle; from the killed elephant's tusks billiard balls are made. For Cummings, man, notwithstanding his endeavour to explain the universe, fails to be a wise "god" for our Earth. The elder Einstein was not far from this pessimistic view; five years before his death he told Hermann Broch:

> Humanity stays as stupid as it always was, and we need not pity it [in case of its doom]. And yet I will be full of regret, because nobody will play then Bach and Mozart anymore. (From Steiner 2004, 22)

James Joyce (1882–1941)

It has been claimed repeatedly that ideas from the exact sciences, e.g., Einstein's theories, notably, the relativity of simultaneity did influence the work of James Joyce (Friedman and Donley 1985, 102–108; Duszenko 1997). In particular, *Finnegans Wake* is referred to. The name of a pub owner "H. C. Earwicker," via "ear" (hearing) $\sim$ time and through "wick" $\sim$ space, is linked with space-time; correspondingly, his sons "Shem" and "Shaun" are seen to stand for time and space. Joyce is said to have been impressed by Einstein's [or rather Minkowski's] unification of space and time into space-time, and by his cosmology. Duszenko interprets the dependency of time- and space-measurements on the observer as the possibility of influencing *subjectively* situations in space-time. He also maintains, erroneously, that in Einstein's relativity theory no borderline between past and future exists. Of course, similar claims go back as far as Plato for whom, so it is said, among *was*, *is*, and *will* only the *is* made sense (cf. Calcraft 1980, 173). In the end, in *Finnegans Wake*, only puns using Einstein's name and words from relativity such as "Eins within a space" (Calcraft 1980, 178, note 19) may be found. In another case, Einstein's name is played with: "On the hike from Elmstree to Stene" (Elm = Ulme (in German) → Ulm (Einstein's birthplace); Stene → Stein (German for stone) → Einstein. Newton's and Einstein's theories of gravitation are referred to in the expression: "Let's hear what science has to say, pundit-the-next-best-king. Splanck!-Upfelbowm." Here, "Splanck" is heard as the sound of a falling apple with "Upfelbowm" being the appletree (in German: Apfelbaum). On the other hand," Splanck!" also might refer to (Max) Planck (Friedman and Donley 1985, 106–107). While I admire the performance as detectives of the colleagues in English Departments, I agree with *the first sentence* in the following judgment by H. Broch, even when it is based on a lack of physical understanding:

> *Ulysses* has nothing to do with relativity theory which is not even mentioned in it. Nevertheless, one may reasonably state that the epistemological essence of relativity theory is established [there] by the discovery of the logical medium within the sphere of physical observation.

Somehow, when focussing on the observer, Broch did mix up relativity and quantum theory, and misrepresented both:

> Relativity theory has discovered that [.] a principal source of error exists, the act of seeing itself, observing by itself [.], and in order to overcome this

source of errors [.] a theoretical unity of physical object and physical subject must be created. (Broch 1975, 63–94; here 77)[25]

Lawrence Durrell (1912–1990)

His best-known work, *The Alexandria Quartet*, he considered "an investigation of modern love"; in it Durrell aimed at fusing Western perceptions of space and time with the metaphysics of the East. However, the quadruple set of novels is also an endeavour to replace *formulas* representing geometry by the form and structure of a text. The "Quartet" comprises the novels *Justine* (1957) (made into a film), *Balthazar* (1958), *Mountolive* (1958), and *Clea* (1960). According to the author:

> Modern literature offers us no Unities, so I have turned to science and am trying to complete a four-decker novel whose form is based on the relativity proposition.
> Three sides of space and one time constitute the soup-mix recipe of a continuum. The four novels follow this pattern.
> The first three parts, however, are to be deployed spatially [.] and are not linked in a serial form. They interlap, interweave, in a purely spatial relation. Time is stayed. The fourth part alone will represent time and be a true sequel. (*Balthazar*, New York 1958, 5; quoted from Friedman and Donley 1985, 86)

However, as Calcraft, in his extended discussion, tells us:

> it is clear that Durrell was not concerned with a precise representation of relativity or quantum theory in his novel; rather, he sought a fairly loose translation of aspects of these scientific theories into the human idiom. Indeed one of the chief functions of relativity theory in the 'The Alexandria Quartet' is as a source of literary metaphor for the treatment of interpersonal relationships and human perception. (Calcraft 1980, 176)

Thus, also in this case, Einstein's relativity theories form only a source of *metaphors* for literary purposes. But can we expect from writers to do more than that? I remind you of the texts of some contemporary French philosophers who also tried to incorporate physics into their essays, unsuccessfully though, and who were then ridiculed by Sokal's hoax (Sokal 1996, 1998; Sokal and Bricmont 1997). Such a hoax had been anticipated in a way, in literature, in John Abulafia's story "Foolscap" (Abulafia 1977; cf. also Calcraft 1980, 170).

---

[25] For Broch's ignorance vis-à-vis the difference between relativity theory and quantum mechanics cf. (Emter 1995, 133, note 325).

7.4 On the Reception of the Theories of Relativity in the Visual Arts (ca. 1915–1955)

Some writers with a background in the exact sciences claim that twentieth century science had a noticeable influence on twentieth century painting, in particular, within the first half of this period (cf. Waddington 1970). Mostly, little proof is given. In painting and sculpture, it seems more difficult than in literature to discover an interaction with the relativity theories because the artists were even less inclined to theorize than their colleagues, the writers. But art historians were no less skilled in discovering evidence invisible to the uninitiated eye than professors in language departments. One of many examples to be found in Linda Henderson's book is "Time" of 1919 by John Covert, a mixture of collage and painting (Henderson 1983, 228–229; figure 68). Apart from geometrical figures like cones, parts of circles, and straight lines as well as compasses, all known from elementary school, the number "4" and a "faulty formula" $x = \frac{c}{t}$ on the object, nothing points to a connection with physics—not to mention relativity theory. But now, the number 4 is interpreted to refer to the fourth dimension, and the garbled formula to allude to relativity.

7.4.1 Einstein and the Fine Arts

When it comes to Einstein's assumed influence, he himself had no particular interest in the fine arts; it was mainly music that he loved. In his apartment, he did not surround himself with notable paintings or other works of art. Even when visiting galleries and exhibitions with his step-daughter Margot, with regard to modern art he always remained aloof (cf. also Goenner 2005). He would not have recognized his belief in the harmony, beauty, and truth reflected by the laws of nature in any of the directions painting and sculpture took during his lifetime.

Einstein Portrayed

Any number of painters and sculptors drew, sketched, portrayed, and formed his face, bust, or figure. Among them were the Berlin impressionists Max Liebermann, and Emil Orlik, teacher of George Grosz and Hannah Höch. None of the painters occupied with Einstein's likeness belonged to "modern art," not to mention the avant-garde, regardless of whether expressionism, futurism, Dada, constructivism, or "Neue Sachlichkeit" (new objectiveness, Berlin realism) is considered. Among sculptors, there is the exception of Jacob Epstein, who, as a follower of Rodin and Brancusi and inspired by primitive art, reached a form of modern realism. Busts of Einstein were made also by less-known sculptors such as Harald Isenstein and Sinowij M. Wilenski. A similar situation applies to painting; we do possess wonderful drawings and portraits of Albert Einstein by Leonid Pasternak, Hermann Struck, Alexander Sander, Erich Büttner, Charlotte Behrend-Corinth, and Ben Shan—to name a few. Well-known contemporaries like Oskar Kokoschka, Otto Dix, the Berlin painters George Grosz, Christian Schad, and Rudolf Schlichter did not approach Einstein. A mild deviation from the bourgeois stylistic attitude of Albert and Elsa Einstein as, e.g., seen by them in his portrait by Josef Scharl led to this rejection.

Einstein's Reception in the Visual Arts

It seems obvious that Einstein's universal fame, his charisma, and his ethically founded commitment to the socially neglected attracted artists more than his physical theories did. With Einstein's public presence in the media, in the 1920s and 1930s, even one or the other avant-garde artists took notice of him. Hence, Einstein's head appears in a photomontage by Hannah Höch "Cut with the Dada Kitchen Knife through the Last Weimar Beer-Belly Cultural Epoch in Germany" (1919/20, National Gallery, Berlin).

He is just one figure among many persons and objects, arranged into four themes: science, the military, industry, and Dada "revolutionaries." Directly under his head we may read a slogan of the movement:

> Hey, hey, young man
> Dada isn't an artistic style.

The detailed and interesting interpretation of Höch's collage by Anke te Heesen (Heesen 2005, 43–48) lets the artist "[.] incorporate Einstein and relativity theory into the sphere of Dada. The physicist forms the superstructure, the theoretical guarantor, for an anti-bourgois, partly anarchistic, in any case an artistic movement. [.]" It is open whether Hannah Höch was thinking along such lines; if she did she could not have been further away from Einstein's bourgeois attitude toward the visual arts, including Dada, and—not to forget—toward women. Being a revolutionary thinker in physics did not automatically make him anti-bourgeois in other ways of life.

The art historian Bettina Schaschke, in her book *Dadaistische Verwandlungs-kunst*, presents a collage "The Human Eye and a Fish, the Latter Petrified" by a member of the Cologne Dada group, Johannes Theodor Baargeld (1892–1927). This collage contains illustrative elements both from Minkowski's space-time interpretation of Special Relativity, i.e., the *lightcone*, and from General Relativity: the relativistic path of a planet on a spiraling curve (perihelion motion, in the lower left corner) (Schaschke 2004, 191 ff. and 212 ff.).

Probably, the connection between Baargeld and Einstein's theory came about through the painter Otto Freundlich (1878–1943), brother of the astrophysicist Erwin Finlay Freundlich. Otto Freundlich, together with Baargeld and Max Ernst, set up the first Dada exhibition in Cologne in 1919. In Germany, Erwin Freundlich was one of the first who tried to do empirical astronomical checks on Einstein's theory of gravitation. He became the driving force for the construction of the laboratory and observation station, the "Einstein tower," near Potsdam, built by the architect Erich Mendelsohn. As Otto was closely connected with Berlin leftist literary circles, e.g., Pfemfert's expressionist magazine *The Action*, he might have helped to place his brother Erwin's article on Relativity Theory in the literature and poetry monthly *Die weißen Blätter* (The White Leaves) (Freundlich 1920).

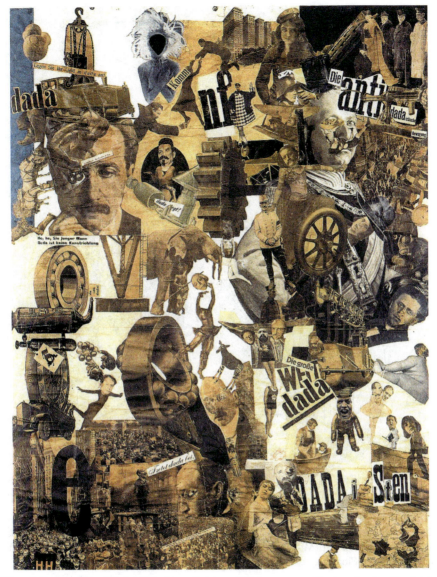

Fig. 7.1. *Hannah Höch: Schnitt mit dem Küchenmesser Dada durch die letzte Weimarer Bierbauchkulturepoche Deutschlands.* 1920. Berlin, Neue Nationalgalerie. © VG Bild-Kunst, Bonn. 2009.

Paul Klee (1879–1940)

Another painter who, in the context of his investigations concerning the "Metamorphosis of Form," was brought into the neighborhood of the changing images of space in the natural sciences, is Paul Klee. It is possible that Helmholtz's and

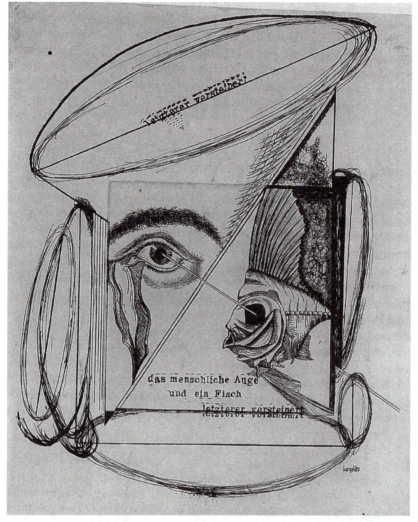

Fig. 7.2. Johannes Baargeld: *The Human Eye and a Fish, The Latter Petrified*, 1920. New York, Museum of Modern Art (MoMA). © 2005. Digital image, The Museum of Modern Art, New York/Scala, Florence.

Mach's psychology of perception influenced Klee's imagery (Schuster 1998, 20). Klee wrote:

> Also in art, there is enough room for precise research. [.] For this, mathematics and physics provide the handle in the form of rules for the preservation and the variation. The necessity to first give attention to the functions, and not yet to the completed form is beneficial. Algebraic, geometric, and mechanical excercises provide moments of training toward a direction to

the essential, functional, vis-à-vis the impressive. (From Paul Klee: *Das bildnerische Denken*, quoted from Ferrier 2001, 202)

Klee's paintings with titles like *Nichtcomponiertes im Raum* (Uncomposed in Space), *Spiel der Kräfte einer Lechlandschaft* (Playful Balance of Forces in a Landscape of the River Lech), *Kräfte des Paukers* (Forces of the Timpanist), and *Gleich Unendlich* (Equals Infinity) attest to this view. However, I am unable to relate the first title listed above to a response of Klee to "the great esthetic challenge of relativity theory," a claim voiced during the Einstein Year (Grothues 2005). Among the connections Klee included into his courses "Beiträge zur bildnerischen Formlehre" (contributions to the teaching of form as captured in pictures) were "Relations between music and painting. Rythmic exercises." This fits well to the constructivist Naum Gabo's: "Space cannot be seen by the eye [.]. Sculpture is to be heard like music" (quoted from Courtenay 1980, 155).

Further known painters have been made witnesses for an influence of Einstein's relativity theories on their work, i.e., El Lissitzky, Theo van Doesburg (1883–1931), and Piet Mondrian (1872–1944). Lissitzky's conception of space:

We see that suprematism has swept away from the surface the illusions of planimetric 2-dimensional space, the illusions of 3-dimensional perspectivistic space, and has created the last illusion of *irrational* space with infinite stretchability in depth and foreground. (Shadova 1978, 48)

seems quite unrelated to the notion of physical space. Also, Mondrian always "maintained his commitment to the absolute, above and beyond the relativization caused by Einstein" (Henderson 1983, 318). Finally, van Doesburg found curvature too "baroque"; he used the expression "non-Euclian" alternatively for "not 3-dimensional" and "not static" (Henderson 1983, 326–327).

Picasso, Cubism, and Einstein

In the context of discussing Picasso's cubist painting *Les Demoiselles d'Avignon* of 1907, P. Daix wrote:

I do not want to construe neither relations nor causal links between totally different kinds of research. Yet the fact remains that our picture of the world was shaken *simultaneously* by physics and modern painting: at *the same time* in fields related by the necessity to doubt the representation of the external world given by our senses. [.] A physicist of 27 and a painter of 25 years have found the same courage, to oppose concepts as well grounded as to be taken as axioms or dogmas. (Daix 1973, 67)

This implies a comparison with Einstein's "discovery" of Special Relativity in 1905. But was it really "the same courage"? Daix's opinion has been taken up again, recently, i.e., "that art and science should have progressed in a parallel manner in the twentieth century" (Miller 2001, 8). In addition, Miller credits Picasso with a

re-interpretation of Poincaré's suggestion to map a fourth *spatial* dimension with projections on different two-dimensional planes: "Picasso, with his visual genius, saw
that the different perspectives could be shown in spatial simultaneity. Thus emerged
the *Demoiselles*." (Miller 2001, 105). He then compares paintings of Picasso with the
figures in a textbook on four-dimensional geometry (Jouffret 1903): "May Picasso
have taken the idea of faceting from Jouffret's projections of four-dimensional solids
onto a plane?" (Miller 2001, 108). In contrast, Picasso distrusted theory:

> Mathematics, trigonometry, chemistry, psychoanalysis, music, and whatnot
> have been related to cubism to give it an easier interpretation. All this has
> been pure literature, not to say nonsense, which has only succeeded in blind-
> ing people with theories. (quoted from Richardson 1996, 103)

Admittedly, "evidence" in art history is rather removed from the evidence of empirical data in the natural sciences. Hence, Braque may be quoted as saying that "tactile
space fascinated him 'so much, because that is what early cubist painting was, a
research into space' " (Miller 1996, 418). The same painter, on another occasion,
described his manner of dealing with space as "tactile," that it enabled him "to make
people want to touch what has been painted as well as look at it" (quoted from
Richardson 1996, 105). The artists may have drawn on the 19th-century idea of a
fourth dimension of *space* as well as on other images for the outside world. Yet
Einstein and his relativity theories had no bearing at all on cubistic painting; in this
I strongly side with Henderson (1983, 352–365). The same opinion was held by
Einstein himself. He refused to accept the idea of a relationship between his relativity
theories and cubism when, after World War II, an art critic, Paul M. Laporte, sent him
his manuscript entitled "Cubism and Relativity." According to Einstein, works of art
and results in the exact sciences would have to be judged differently; the arts were
dependent on culture, yet results in physics were not.[26] The use of notions like "the
fourth dimension" or "non-Euclidean geometry" by artists in their interpretation of
cubism and the one by physicists and mathematicians should not be equated. In his
letter to Laporte Einstein wrote:

> I find your comparison rather unsatisfactory [.] the essence of the theory
> of relativity has been incorrectly understood in it, granted that this error is
> suggested by attempts at a popularization of the theory.

In relativity theory, *one* point of view, i.e., one coordinate system, would suffice to
explain a physical situation. Einstein goes on:

> This is quite different in the case of Picasso's painting, as I do not have to
> elaborate any further. Whether, in this case, the representation is felt as an
> artistic unity depends, of course, upon the artistic precedents of the viewer.

---

[26] The cultural independence of physics is to be understood in terms of its empirically testable
results—not with regard to the kind of questions posed to nature. In comparison with the
spontaneity (of the ideas springing up) in literary and artistic works, in the development of
physics a certain rigidity applies.

Fig. 7.3. Salvador Dali: *Crucifixion (Corpus Hypercubicus)*. 1954. New York, The Metropolitan Museum of Art, Chester Dale Collection. © 2008. Image Copyright The Metropolitan Museum of Art/Art Resources/Scala, Florence.

This new artistic 'language' [i.e., cubism] has nothing in common with the Theory of Relativity. (Laporte 1966, cited in Holton 1996, 130–131 and Henderson 1983, 356.)

As an early believer in and connoisseur of cubism, Carl Einstein hoped that *in arts* cubism "would become quite a practical thing"—in the same way that non-Euclidean geometry had become for the physics of gravitation (quoted from Gnam 2003, 204). He did not mention a relationship between the development of cubism and relativity (C. Einstein 1926) although he might have selected the paintings for an exhibition "Picasso and Negro Sculptures," in 1913, in Otto Feldmann's "Neue Galerie Berlin" (Richardson 1996, 317). The opinion of Paix and followers have been reiterated, by journalists, during the Einstein Year 2005. Pablo Picasso, as an artist, is said to have tried much of the same with his cubism as did Einstein, as a scientist, with his relativity theory: to penetrate the geometry of the world in a new and deeper way. Thanks to this kind of reductionism, some may keep their illusion of an uncomplicated world.

Whereas Picasso did not care about relativity theory, Salvador Dali played with it. His "limp watches" on many paintings and water color drawings point to "time bent" and may be seen as influenced by General Relativity.

A particular example is Dali's painting *The Persistence of Time* of 1931. In fact, according to Henderson "Dali discussed the watches in the context of his comments on non-Euclidean versus Euclidean geometry and the theories of Einstein." The immediate visual source for Dali seems to have been a plate of Camembert cheese which made him describe the melted watches as "the extravagant and solitary Camembert of time and space" (Henderson 1983, 347). Another example would be Dali's *Corpus Hypercubus* (1954), showing his interest in four-dimensional *space*[27] (cf. also Henderson 1983, 349).

7.5 Summary

Since the 1920s, some writers have tried to transform the chronological *parallelism* between "revolutionary" changes in the arts and in modern physics, notably Einstein's relativity theories and, later, quantum mechanics, into a resonant bond.

A remarkable coincidence of intellectual experiences: almost simultaneously with the birth of the new expressionistic art, the new relativity theory (Einstein above all) started to seize the natural sciences. [..] Relativity, too, is freeing everything and every event from the rigidity of statics, and resolves it into a cosmic dynamics. All in motion. (Hatvani 1917, column 148–149)[28]

From the examples given, it seems to me that the developments in modern literature and the visual arts, relative to those in physics, at the beginning of the twentieth

---

[27] In his painting of the crucified Christ, though, only six of the eight cubes which can be folded so as to form the surface of a hypercube are shown. Cf. figure 5 of Henderson (1983).

[28] The expressionistic group "Die Brücke" was founded in Dresden on 7 July 1905; Einstein's famous paper on the electrodynamics of moving bodies was submitted to *Annalen der Physik* on 30 June 1905.

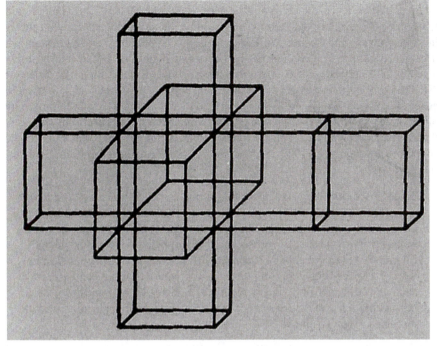

Fig. 7.4. H. P. Manning: *Eight Cubes Which Can Be Folded So As to Form a Hypercube.* In *Geometry of Four Dimensions*, [Macmillan 1914] New York: Dover 1956, p. 240. With kind permission of Dover Publications Inc.

century are little more than a coincidence—within a common cultural context already knowledgeable about higher dimensions and non-Euclidean geometry. *After* the new mathematical and physical representations of space, time, and causality had come to life, some of the writers and artists attempted to absorb them for their own purpose. Although using some scientific vocabulary, generally they did not get beyond its metaphorical and symbolical application: after all, they were not educated to understand Einstein's relativity theory in any depth.[29] As to literature, all of the German language writers discussed above were critical with respect to the methods of the exact sciences: in their view, they led to too much emphasis being given to rational thinking. Nevertheless, at least H. Broch and R. Musil hoped for "a connection between the 'two cultures'. For both [.], an essential part of [their] reflections is focussed on overcoming the antinomy between rationality and irrationality" (Emter 1995, 117). Thus, from the perspective of literature, a positive transfer from relativity to literature and the arts may have occurred.

---

[29] In the essay "A. and Pangeometry," El Lissitzky blames fellow painters: "[.] they have based their ideas, in a flexible, superficial way, on the most modern scientific theories, without acquiring any knowledge of them (multi-dimensional spaces, the theory of relativity, the universe of Minkovsky, and so on)." Cf. (Henderson 1983, 296).

In cultural science, the concept "Die Moderne" (The Modern Era) is often used to integrate all new developments in society in the first third of the twentieth century including science and the arts. While this concept is helpful in *describing* a situation, it cannot *explain* it, and should not be overstressed such as to necessarily imply a continuous interaction of the various cultural subfields. One important aspect of "Die Moderne" is the rapid growth of the extent and speed of communications, more generally, of technology and its intrusion into everyday life. Due to the proximity of technology and the exact sciences, knowledge about physics and some physical theories had spread as well. Thus, popularizations of Einstein's theories had become common knowlege since 1920. As Moholy-Nagy, a former teacher at the Bauhaus, stated in 1942:

> Since 'space-time' may be a misleading term, it especially has to be emphasi-
> zed that space-time problems in the arts are not necessarily based on
> Einstein's theory of relativity. This is not meant to discount the relevance
> of his theory to the arts. But artists and laymen seldom have the mathemati-
> cal knowledge to visualize in scientific formulae the analogies to their own
> work. [.] Space-time is not only a matter of natural science or of esthetic
> and emotional interest. It deeply modifies the character of social ends, even
> beyond the sense that pure science may lead to a better application of our
> resources. (Henderson 1983, 337–338)

Hence, also the arts reacted to Einstein and his theories, and, possibly, this led to some of the barriers between the two cultures being torn down. However, this interaction did not create any important new stylistic direction despite the theorizing about their work by some artists, among them Marcel Duchamp (1887–1968), Theo van Doesburg, and El Lissitzky. In fact, it seems safe to conclude that Einstein's *personality* rather than his scientific theories fascinated artists and writers. As a human figure he appears in novels, poems, and plays; for some poets, he even must have played the role of muse.

Acknowledgments

Part of this article was used in a lecture within the series "Natur und Geist 2005 – Albert Einstein – Physiker und Philosoph"; my thanks for the invitation go to Prof. Dr. Hans Jürgen Wendel and Dr. Olaf Engler, Max-Schlick-Forschungsstelle, University of Rostock. I also am grateful to Privatdozentin Dr. Andrea Gnam, Humboldt University, Berlin, for sending me her papers, and for encouraging remarks. Skuli Sigurdsson, Berlin and Reykjavik, gave some methodological advice and let me participate in his broad knowledge. With his critical remarks, a referee helped me to broaden my perspective. I also thank Lindy Divarci for her editing work.

References

Abulafia, John, *et al.* (1977). "Foolscap", in: *Introduction 6: Stories by New Writers.* London: Faber & Faber.

Benn, Gottfried (1987). *Gottfried Benn. Sämtliche Werke.* Band III, Prosa 1. Stuttgarter Ausgabe. Hrsg. G. Schuster in Verbindung mit Ilse Benn. Stuttgart: Klett-Cotta.

—— (2003). *Gottfried Benn. Sämtliche Werke.* Band VII/1. Stuttgarter Ausgabe. Hrsg. G. Schuster in Verbindung mit Ilse Benn. Stuttgart: Klett-Cotta.

Blei, Franz (1920). *Bestiarium literaricum das ist: genaue Beschreibung deren Tiere des literarischen Deutschland verfertigt von Dr. Peregrin Steinnhövel,* München.

—— (1924). *Das Grösse Bestiarium der Literatur.* Die moderne Literatur – eine Darstellung; Erster Band. Berlin: Ernst Rowohlt.

—— (1930). *Erzählung eines Lebens.* Leipzig: Paul List Verlag.

—— (1995). *Das Grosse Bestiarium der Literatur.* Herausgegeben und mit einem Nachwort versehen von Rolf-Peter Baacke. Hamburg: Europäische Verlagsanstalt.

Broch, Hermann (1975). "James Joyce und die Gegenwart. Rede zum fünfzigsten Geburtstag." In: *Schriften zur Literatur 1. Kritik.* Kommentierte Werkausgabe. Ed. Paul M. Lützeler. Band 9/1. Frankfurt: Suhrkamp.

—— (1977). *Hermann Broch. Die unbekannte Grösse.* Kommentierte Werkausgabe. Paul M. Lützeler (Hrsg.) Band 9/1. Frankfurt: Suhrkamp.

—— (1980). Band 8 der kommentierten Werkausgabe: Gedichte.

—— (1981a). *Hermann Broch. Briefe 1 (1913–1938)* Dokumente und Kommentare zu Leben und Werk. Frankfurt: Suhrkamp.

—— (1981b). *Hermann Broch. Kommentierte Werkausgabe.* Paul M. Lützeler (Hrsg.) Band 13/2. Frankfurt: Suhrkamp.

—— (1981c). *Hermann Broch. Briefe 3 (1945–1951)* Dokumente und Kommentare zu Leben und Werk. Frankfurt: Suhrkamp.

—— (1986). Kommentierte Werkausgabe, Bd. 2. *Die unbekannte Größe.* Suhrkamp-Taschenbuch Nr. 393. Frankfurt am Main: Suhrkamp.

—— (1987). *Hermann Broch. Kommentierte Werkausgabe.* Paul M. Lützeler (Hrsg.) Band 8. Frankfurt: Suhrkamp.

—— (1991). *Hermann Broch. Kommentierte Werkausgabe. Der Tod des Vergil: Roman.* Paul M. Lützeler (Hrsg.) Band 4. Frankfurt: Suhrkamp. (The Death of Virgil. New York: Pantheon Books 1945).

Brod, Max (1915). *Tycho Brahes Weg zu Gott.* Leipzig: Kurt Wolff Verlag. English translation: "The Redemption of Tycho Brahe." New York: Alfred Knopf 1928.

—— (1969). *Streitbares Leben 1884–1968.* München: Herbig.

Calaprice, Alice (1999). *Einstein sagt. Zitate, Einfälle, Gedanken.* Zürich: Piper.

Calcraft, Lee (1980). "Einstein and Relativity Theory in Modern Literature", in: *Einstein—The First Hundred Years.* M. Goldsmith, A. Mackay & J. Woudhuysen, eds., 163–179. Oxford: Pergamon Press.

Clarke, Bruce and Linda D. Henderson, eds. (2002). *From Energy to Information. Representation in Science and Technology, Art, and Literature*. Stanford: Stanford University Press.

Courtenay, Philip (1980). "Einstein and Art", in: *Einstein—The First Hundred Years*. M. Goldsmith, A. Mackay & J. Woudhuysen, eds., 145–157. Oxford: Pergamon Press.

Cummings, E. E. (1994). *Poems—Gedichte*. Bilingual edition, selected, translated and commented by Eva Hesse. Ebenhausen near Munich: Langewiesche-Brandt.

Daix, Pierre (1973). *Picasso: Der Mensch und sein Werk*. Aus d. Franz. übersetzt von Eva Rapsilber. (Paris: Galerie Somogy 1964). Gütersloh: Bertelsmann Lesering o.J.

Danius, Sara (2002). *The Sense of Modernism: Technology, Perception, and Aesthetics*. Ithaca: Cornell University Press.

Döblin, Alfred (1930). *Berlin Alexanderplatz: Die Geschichte vom Franz Biberkopf*. Berlin: S. Fischer. (English translation by Eugene Jolas: "Berlin Alexanderplatz: The Story of Franz Biberkopf." A Frederic Ungar Book. New York: Continuum International.

Donley, Carol C. (1981). "Einstein's Influence on Modern Poetry", in: *After Einstein. Proceedings of the Einstein Centennial Celebration at Memphis State University 1979*. P. Barker & Cecil Shugart, eds., 213–219. Memphis: State University Press.

Dürrenmatt, Friedrich (1962). *Die Physiker*. Zürich: Die Arche.

Duszenko, Andrzej (1997). "The Joyce of Science: New Physics in *Finnegans Wake*"—"The Theory of Relativity in *Finnegans Wake*." http://duszenko.northern.edu/joyce/

Einstein, Albert (1917). *Über die spezielle und allgemeine Relativitätstheorie, gemeinverständlich*. 1. Aufl. Braunschweig: Vieweg.

Einstein, Carl (1926, 1996). *Die Kunst des 20. Jahrhunderts*. Propyläen Kunstgeschichte, Bd. 16. Berlin: Propyläen 1926. Newly edited and annotated by U. Fleckner and Th. W. Gaehtgens. Berlin: Fannei & Walz.

Emter, Elisabeth (1995). *Literatur und Quantentheorie. Die Rezeption der modernen Physik in Schriften zur Literatur und Philosophie deutschsprachiger Autoren (1925–1970)*. Berlin: de Gruyter.

Ferrier, Jean-Louis (2001). *Paul Klee* Edition Terrail. Deutschsprachige Ausgabe Komet MA-Service, Frechen. Paris: Pierre Terrail.

French, Anthony Philip, ed. (1979). *Einstein, a Centenary Volume*. London: Heinemann Educational Books Ltd.

Freundlich, Erwin (1920). "Die Entwicklung des Physikalischen Weltbildes bis zur Allgemeinen Relativitätstheorie", in: *Die weißen Blätter*, 7. Jahrgang, viertes Heft. Berlin: Paul Cassirer.

Friedman, Alan J. and Donley, Carol C. (1985). *Einstein as Myth and Muse*. Cambridge: University Press.

Gnam, Andrea (1999). *Die Bewältigung der Geschwindigkeit. Robert Musils Roman 'Der Mann ohne Eigenschaften' und Walter Benjamins Spätwerk.* München: Wilhelm Fink.

—— (2001). "Zum Beispiel die Lust am Automobilfahren inmitten alter Motive. Robert Musils behutsamer Umgang mit den Naturwissenschaften." *Jahrbuch zur Kultur und Literatur der Weimarer Republik* **6**, 127–142.

—— (2002). "Schnell wie ein Überschallflugzeug oder ein virtueller Tanz? Überlegungen zur Geschwindigkeit." *Studia austriaca* **X**, Fausto Cercignani, ed., 153–159.

—— (2003). "Carl Einsteins Wahrnehmung von Raum, Körper und Geschwindigkeit im Kontext der zeitgenössischen Diskussion." In: *Die visuelle Wende der Moderne. Carl Einsteins Kunst des 20. Jahrhunderts.* Klaus H. Kiefer, Hrsg. München: Wilhelm Fink.

—— (2004). "Die Bewältigung der Geschwindigkeit." http://www.andrea-gnam.de/a.php?dir=publikationen&id=habil1

Goenner, Hubert (2005). *Einstein in Berlin.* München: C. H. Beck.

Grothues, Diana (2005). "Einstein und die Kunst." Radio-broadcast ZDF", Aspekte", Jan. 28, 2005. http://www.zdf.de/ZDFde/inhalt/12/0,1872,2248780,00.html.

Habermas, Jürgen (1968). "Technik und Wissenschaft als 'Ideologie'." In: *Technik und Wissenschaft als Ideologie.* J. Habermas, ed., 48–103. Frankfurt: Suhrkamp.

Hartmann, Karl-Heinz (2004). *Zeiten der Gelassenheit. Ein literarischer Jahresbegleiter.* München: Claudius Verlag.

Hatvani, Paul (1917). "Versuch über den Expressionismus." *Die Aktion* Nr. 11/12, column 146–150.

Hawking, Steven (1988). *Eine kurze Geschichte der Zeit.* Reinbek: Rowohlt. English original: "A Brief History of Time." London: Bantam Books.

Heesen, Anke te (2005). "Dada/Einstein. Ein Physiker in Papier." In: *Einstein on the Beach. Der Physiker als Phänomen.* M. Hagner, ed., 40–56. Frankfurt am Main: Fischer Taschenbuchverlag.

Henderson, Linda D. (1983, 2003). *The Fourth Dimension and Non-Euclidean Geometry in Modern Art.* Princeton: University Press; new ed., Cambridge: MIT Press.

—— (2002). "Vibratory Modernism: Boccioni, Kupka, and the Ether of Space." In: *From energy to information,* B. Clarke & L. D. Henderson, eds. Stanford: Stanford University Press.

Hinton, Charles H. (1904). *The Fourth Dimension.* London: Swan Sonnenschein & Co.

Holton, Gerald (1996). *Einstein, History, and Other Passions.* Reading: Addison-Wesley.

Jouffret, Esprit Pascal (1903). *Traité élémentaire de géométrie à quattre dimensions.* Paris: Gauthier-Villars.

Kiefer, Klaus H. (1982). "Einstein & Einstein. Wechselseitige Erhellung der Künste und Wissenschaften um 1915." *Komparatistische Hefte* **5/6**, 181–191.

Köstler, Arthur (1970). *The Act of Creation.* London: Pan Books.

Kraft, Herbert (1972). *Kunst und Wirklichkeit im Expressionismus*. m. e. Dokumentation zu C. Einstein. Bebenhausen: Rotsch.

Laporte, Paul M. (1966). "Cubism and Relativity, with a Letter of Albert Einstein." *Art Journal* **25**, no. 3, 246.

Lenz, Siegfried (1975). *Einstein überquert die Elbe bei Hamburg. Erzählungen.* Hamburg: Hoffmann und Campe.

Lessing, Gotthold Ephraim (1954). "Laokoon oder Über die Grenzen der Malerei und Poesie." (1766) In: Gesammelte Werke in 10 Bänden, Paul Rilla, ed. Band 5. Berlin: Aufbau Verlag.

Lossau, Norbert (2005). "Festlicher Auftakt zum Einstein-Jahr 2005 am Mittwoch. Vor 100 Jahren schuf der Physiker Albert Einstein die Spezielle Relativitätstheorie—Im Gedenkjahr werden Berlin und Potsdam die wichtigste Bühne sein." *Berliner Morgenpost* vom 16. 1. 2005.

MacLeish, Archibald (1926). *New & Collected Poems, 1917–1976*, p. 137–144. Boston: Houghton Mifflin Company.

—— (1933). *Poems, 1924–1933*. Boston: Houghton Mifflin Co. Undated reprint Cambridge (MA): The Riverside Press.

Maeterlinck, Maurice (1929). *Die Vierte Dimension*. Leipzig: Deutsche Verlags-Anstalt. French original edition: "La vie d'espace." Paris: Eugène Fasquelle 1928; New York: Dodd, Mead & Co. 1928.

Mann, Thomas (1924). *Der Zauberberg*. Berlin: S. Fischer.

—— (1935). *Leiden und Grösse der Meister*. Berlin: S. Fischer.

—— (1985). *Bekenntnisse des Hochstaplers Felix Krull*. Gesammelte Werke in Einzelbänden, Hrsg. und mit Nachbemerkungen versehen von Peter de Mendelssohn. Frankfurt am Main: S. Fischer.

—— (1993). *Essays*. Band 2: Für das neue Deutschland 1919–1925. H. Kurzke und St. Stachorski (Hrsg.). Frankfurt: S. Fischer.

Miller, Arthur I. (1996). *Insights of Genius. Imagery and Creativity in Science and Art.* New York: Springer (Copernicus).

—— (2001). *Einstein, Picasso: Space, Time, and the Beauty that Causes Havoc.* New York: Basic Books.

Minkowski, Hermann (1908). *Raum und Zeit*. Vortrag auf der 80. Naturforscher-Versammlung zu Köln am 21.9.1908. Leipzig: B. G. Teubner.

Misa, Thomas J. (2005). "A Grammophone in Every Grave. On Sara Danius The Senses of Modernism: Technology Perceptions, and Esthetics." *History and Technology* **21**, 351–355.

Musil, Robert (1978). *Prosa und Stücke. Kleine Prosa, Aphorismen, Autobiographisches, Essay und Reden, Kritik.* Gesammelte Werke, Bd. 11. Adolf Frisé (Hrsg.). Reinbek: Rowohlt.

Poincaré, Henri (1916). *La science et l'hypothèse*. Paris: Flammarion. [1. edition 1902]

Richardson, John (1996). *A Life of Picasso*. vol. II: 1907–1917. London: Jonathan Cape.

Roth, Marie-Louise (1972). *Robert Musil—Ethik und Ästhetik. Zum theoretischen Werk des Dichters*. München: Paul List.

Sartre, Jean-Paul (1955). "Francois Mauriac and Freedom", in: *Literary and Philosophical Essays*. New York: Criterion Books.

Schaschke, Bettina (2004). *Dadaistische Verwandlungskunst. Zum Verhältnis von Kritik und Selbstbehauptung*. Berlin: Gebr. Mann Verlag.

Schuster, Peter-Klaus (1998). "'Diesseits bin ich gar nicht fassbar.' Klee's Erfindungen der Wirklichkeit." In: *Klee aus New York*. Ausstellungskatalog von Sabine Bewald. Berlin: Nicolai.

Shadova, Larissa (1978). *Suche und Experiment. Geschichte der russischen und sowjetischen Kunst zwischen 1910 und 1930*. Dresden: Verlag der Kunst.

Sigurdsson, Skuli (1989). "Linda D. Henderson. The Fourth Dimension and Non-Euclidean Geometry in Modern Art." *ISIS* **80**, no. 4, 737–738.

Snow, Charles P. (1959). *The Two Cultures and the Scientific Revolution*. Rede-Lecture. Cambridge: Cambridge University Press.

Sokal, Alan D. (1996). "A physicist experiments with cultural studies." *Lingua Franca*, May/June, 62–64.

—— (1998). "What the *Social Text* affair does and does not prove." *Critical Quarterly* 40(2): 3–18.

Sokal, Alan D. and Jean Bricmont (1997). *Impostures intellectuels*. Paris: Odile Jacob.

Steiner, George (2004). *Grammatik der Schöpfung*. München: Deutscher Taschenbuch Verlag.

Waddington, C. H. (1970). *Behind Appearance: A Study of the Relations between Painting and the Natural Sciences in this Century*. Cambridge, Mass.: MIT Press.

Weishaupt, Heike (1994). *Albert Einstein und Thomas Mann*. Magisterarbeit am Historischen Institut der Universität Stuttgart, Abteilung Geschichte der Naturwissenschaften und Technik.

Wells, H. G. (1895). *The Time Machine*. London: W. Heinemann; New York: Holt & Co.

Zöllner, Johann C. F. (1880). *Transcendental Physics: An Account of Experimental Investigations from the Scientific Treatises of Johann Carl Friedrich Zöllner*. Transl. Charles Carleton Massey. London: W. H. Harrison.

Part III

The Emergence of the Relativistic Worldview

8

Hilbert's Axiomatic Method and His "Foundations of Physics": Reconciling Causality with the Axiom of General Invariance

Katherine A. Brading and Thomas A. Ryckman

University of Notre Dame and Stanford University, USA

> So it is that all human cognition begins with intuitions,
> proceeds from there to concepts, and ends with ideas.
> Kant, *Critique of Pure Reason.* (A702/B730)
> Epigram to Hilbert (1899)

8.1 Introduction

In November and December 1915, Hilbert gave two presentations to the Royal Göttingen Academy of Sciences under the common title 'The Foundations of Physics'. Distinguished as 'First Communication' (Hilbert, 1915b) and 'Second Communication' (Hilbert, 1917), the two 'notes', as they are widely known, eventually appeared in the *Nachrichten* of the Academy. The first quickly entered the canon of classical general relativity but has recently become the object of renewed scholarly scrutiny since the discovery (Corry, Renn and Stachel 1997) of a set of printer's proofs dated December 6, 1915 ((Hilbert, 1915a), henceforth 'Proofs'). Hilbert's second presentation has not received the same detailed reconsideration, with the recent exception of an extended study offered by Renn and Stachel (1999/2007). While we agree with much of their detailed technical reconstruction, we profoundly disagree with the assessment of Renn and Stachel that the second note shows that Hilbert had abandoned his own project (set out in the first note), and is working on a variety of largely unrelated problems within Einstein's. In our opinion, this assessment rests on misunderstandings concerning the aims, content, and significance of the second communication, as well as its links to the first. Our aim in this paper is to offer an alternate narrative, according to which Hilbert's second note emerges as a natural continuation of the first, containing important and interesting further developments of that project, and above all shedding needed illumination on Hilbert's assessment of the epistemological novelty posed by a generally covariant physics.

Hilbert's notes on 'Foundations of Physics' traditionally have been assessed solely in terms of the contributions they made to general relativity, as that theory is known in its completed form. From this vantage point, they present a mixed record of achievement, ranging from genuine insight (the Riemann scalar as the suitable invariant for the gravitational action) through incomprehension (Hilbert's interpretation of electromagnetism as a consequence of gravitation) to abject failure (attachment to the untenable electromagnetic theory of matter of Gustav Mie). The usual implication is that Hilbert's principal intent in November 1915 was to arrive at a theory of gravitation based on the principle of general invariance in one blinding flash, masterfully wielding an arsenal of advanced mathematics. Our main contention is that such assessments radically occlude internal motivations, which are largely logical and epistemological, and so cast them in a misleading light. In particular, the explicitly stated epistemological intent of the 'axiomatic method' is ignored, as are Hilbert's own express assertions regarding his construction as a triumph of that method. But set within the context of the 'axiomatic method', Hilbert's two notes may be seen to have the common goal of pinpointing, and then charting a path towards resolution of, the tension between causality and general covariance that, in the infamous 'hole argument', had stymied Einstein from 1913 to the autumn of 1915. Unlike Einstein's largely informal and heuristic extraction from the clutches of the hole argument, Hilbert stated the difficulty in a mathematically precise manner as an ill-posed initial value problem, and then indicated how it can be resolved. As we will show, material cut from the Proofs establishes this essential thematic linkage between the two notes and redeems Hilbert's claim that tension between causality and general covariance, precisely formulated in Theorem I of the first note, was the 'point of departure' for his axiomatic investigation.

8.2 The Essential Context: Hilbert's Axiomatic Method and Kantian Epistemology

Hilbert's first note opens with a declaration that the ensuing investigation of the foundations of physics is undertaken 'in the sense of the axiomatic method' ('*im Sinne der axiomatischen Methode*'), and it concludes with the striking claim that the results obtained redound 'certainly to the most magnificent glory of the axiomatic method.' Unless mere rhetorical embellishment, these passages establish that the 'axiomatic method' (whatever that may be) played an integral role in the enterprise at hand. Understanding the significance of Hilbert's setting his results squarely within the frame of the axiomatic method is accordingly essential.

What, then, is the axiomatic method? In the literature, it has been widely assumed that Hilbert's references to 'axiomatic method' simply signal his derivation of 14 fundamental field equations, as well as several subsidiary theorems, from two principal axioms (e.g., Guth, 1970, 84; Mehra, 1974, 26, 72 n. 145; Pais, 1983, 257). However, as can be documented in numerous lecture courses going back at least to 1905, the term not only implicates a typical mathematical concern with the rigorous explicit statement of a theory, but also connotes a specifically *logical and epistemological* method of investigation of mathematical theories (including those of physics) that Hilbert pioneered, and which he saw as closely tied to the nature of thought itself.[1]

In published articulation, the axiomatic method debuted in Hilbert's classic Gauss–Weber *Festschrift* essay, *Grundlagen der Geometrie* (1899). The essay's epigraph has been little noticed, yet is worth quoting in the original German, for it is Kant's most concise statement of how cognition requires, and results from, the distinct sources of intuition, concepts, and ideas:

> *So fängt denn alle menschliche Erkenntnis mit Anschauung an, geht von da zu Begriffen und endigt mit Ideen.* (A702/B730)

As Kant's directive prescribes, the axiomatic method is conceived as a logical analysis of cognitions that begins with certain 'facts' presented to our finite intuition or experience. Both pure mathematics and natural science alike begin with 'facts', i.e., singular judgments about 'something ... already ... given to us in representation *(in der Vorstellung)*: certain extra-logical discrete objects, that are intuitively present as an immediate experience prior to all thinking'.[2] Analysis next determines the concepts under which such given facts can be classified and arranged, and then attempts to formulate the most general logical relations among these concepts, a 'framework of concepts' *(Fachwerk von Begriffen)* crowned with the fewest possible number of principles. The axioms standing at the pinnacle of the *Fachwerk von Begriffen* are not only general but also *ideal*. They are, as far as possible, independent of the particular intuitions (and so, concrete facts) from which the process started. By virtue of their ideality, and thus their severance from experience and intuition, the self-sufficiency of the mathematical subject matter (which may then be developed autonomously), quite apart from any particular reference associated with particular terms or relations, is thereby highlighted. Axioms thus play a hypothetical or guiding role in cognition. As will be seen, Hilbert considered axioms to be 'things of thought' or indeed, 'ideas' in Kant's regulative sense, effecting a separation between logical/mathematical and intuitional/experiential thought, even as the latter has thus been arranged in deductive form. Indeed, it is just 'the service of axiomatics'

> to have stressed a separation into the things of thought *(die gedanklichen Dinge)* of the (axiomatic) framework and the real things of the actual world, and then to have carried this through.[3]

Use of the axiomatic method does not aim, at least in the first instance, at the discovery or recognition of *new* laws or principles, but at the conceptual and logical

---

[1] Hallett (1994), 162, quotes from Hilbert's 1905 Summer Semester Lectures '*Logische Prinicipien des mathematischen Denkens*', 'The general idea of [the axiomatic method] always lies behind any theoretical and practical thinking.'

[2] Hilbert (1922); Engl. trans., 1121. Of course, for Hilbert, the basic objects of number theory, the positive integers or rather the *signs* that are their symbolic counterparts, are given in a quasi-spatial, but not in a *spatial* or *temporal* intuition.

[3] Hilbert Winter Semester lectures 1922–1923 *Wissen und mathematisches Denken.* Ausgearbeitet von Wilhelm Ackermann. Mathematische Institut Göttingen. Published in a limited edition, Göttingen, 1988; as cited and translated in Hallett (1994), 167.

clarification or reconstruction of known ones.[4] Ultimately, the axiomatic method is concerned with demonstrating that the axioms selected for a theory possess the three meta-logical properties or relations of mutual consistency, independence, and completeness.[5] Combining these aspects together, successful pursuit of the axiomatic method leads to a 'deepening of the foundations' *(Teiferlegung der Fundamente)*, i.e., of the *mathematical foundations*, of any theory to which it is applied, and this, indeed, is the overall objective.[6]

8.2.1 Mie's Theory and the Axiomatic Method

We recall that the task of the axiomatization of physical theories was the sixth in the famous list of 23 mathematical problems Hilbert posed at the 1900 International Congress of Mathematicians in Paris. Inclusion of the axiomatization of physics among the other purely mathematical problems appears rather incongruous until Hilbert's lifelong interest in physics is taken into account.[7]

> The investigations on the foundations of geometry suggest the problem: To treat in the same manner, by means of axioms, those physical sciences in which mathematics plays an important part.... If geometry is to serve as a model for the treatment of physical axioms, we must, with a small number of axioms, try to include as large a class of physical phenomena as possible, and then by adjoining new axioms to arrive gradually at the more special theories.... As he has in geometry, the mathematician will not merely have to take account of those theories coming near to reality *(Wirklichkeit)*, but also of all logically possible theories. He must be always alert to obtain a complete survey of all conclusions derivable from the system of axioms assumed. Further, the mathematician has the duty to test exactly in each instance whether the new axioms are compatible with the previous ones. The physicist, as his theories develop, often finds himself forced by the results of his experiments to make new hypotheses, while he depends, with respect to the compatibility of the new hypotheses with the old axioms, solely upon these experiments or upon a certain physical intuition, a practice which is not admissible in the rigorously logical construction of

---

[4] See Majer (2001), 19.

[5] Hilbert's 1905 Summer Semester Göttingen lectures *'Logische Prinzipien des mathematischen Denkens'* already characterized the general idea of the axiomatic method as striving for the consistency, independence, and completeness of an axiom system. See Peckhaus (1990), 59.

[6] Hilbert (1918), 407 (Engl. trans., 1109): 'The procedure of the axiomatic method, as it is expressed here, amounts to a *deepening of the foundations* of the individual domains of knowledge, just as becomes necessary for every edifice that one wishes to extend and build higher while preserving its stability.'

[7] Corry (2004) amply demonstrates the extent of this interest, examining in considerable detail Hilbert's many lecture courses and seminars devoted to various physical theories or questions of current physics.

a theory. The desired proof of the compatibility of all assumptions seems to me also of importance, because the effort to obtain such proof always forces us most effectively toward an exact formulation of the axioms.[8]

Three items of interest mark this passage.

- Geometry is regarded as a model for the axiomatization of physical theories.
- In axiomatizing, the mathematician is to take account of 'all logically possible theories', not just theories 'near to reality', and so the axiomatic method is ideally suited for setting up a speculative or hypothetical theory.
- Axiomatization has the express purpose of testing the consistency of new hypotheses with previously adopted axioms and assumptions, a task that requires 'the rigorously logical construction of a theory' in place of its informal statement in experiential or intuitive terms.

Above all, we wish to stress the *hypothetical* character of Hilbert's axiomatic approach to physics. This aspect was explicitly underlined by Hilbert's former student and Göttingen physics colleague, Max Born, in a tribute on the occasion of Hilbert's 60th birthday.

[B]eing conscious of the infinite complexity he faces in every experiment [the physicist] refuses to consider any theory as final. Therefore . . . he abhors the word 'axiom' to which the sense of final truth clings in the customary mode of speech. . . . Yet the mathematician does not deal with the factual happenings, but with logical connections; and in *Hilbert's* language the axiomatic treatment of a discipline in no way signifies the final setting up of certain axioms as eternal truths, but the methodological requirement: Place your assumptions at the beginning of your considerations, stick to them and investigate whether these assumptions are not partially superfluous or even mutually inconsistent.[9]

These points are of special significance for understanding the role of the Mie theory in Hilbert's two notes on the 'Foundations of Physics'.

As both Einstein and Hilbert were aware in 1915, Einstein's gravitational theory, though in principle capable of encompassing all matter fields into space-time geometry, did not itself suppose any particular theory of matter. Hilbert knew of the Mie theory at least since the discussion of it at the Göttingen Mathematical Society in December 1912, and again in December 1913, when Born had put it into a more canonical mathematical form (Corry, 1999, 176). Certainly, the fact that Mie had sought to derive field equations of a generalized Maxwellian electrodynamics from the axiom of a Lorentz invariant 'world function' fitted very naturally into Hilbert's axiomatic approach. A central attraction of the Mie theory was that then, coupled with Einstein's theory of gravitation, it enabled a *hypothetical axiomatic completion*

---

[8] As translated in Gray (2000), 257–258.

[9] Born (1922), 90–91, our translation. Unless otherwise noted, all translations in this paper are our own.

of physics that could be studied by drawing consequences from the amalgamation of the two theories. In this regard, Hilbert's 'theory' is a canonical illustration of the mode of investigation of the 'axiomatic method', in Hilbert's own most precise characterization of that method, as the 'mapping' *(Abbildung)* of a 'domain of knowledge' *(Wissensgebiet)* onto

> a framework of concepts so that it happens that the objects of the field of knowledge correspond to the concepts, and the assertions regarding the objects to the logical relations between the concepts. Through this mapping, the (logical) investigation becomes entirely detached from concrete reality *(Wirklickkeit)*. The theory has nothing more to do with real objects *(realen Objekten)* or with the intuitive content of knowledge. It becomes a pure construction of thought *(reine Gedankengebilde)*, of which one can no longer say that it is true or false. Nevertheless, this framework of concepts has significance for knowledge of reality in that it presents a possible form of actual connections. The task of mathematics is then to develop this framework of concepts in a logical way, regardless of whether one was led to it by experience or by systematic speculation.[10]

Yet the Mie theory was attractive for a number of other mathematical and philosophical reasons that merit illumination. In particular, Hilbert saw distinct advantages in the Mie theory over the only other rival electromagnetic theory of matter of consequence in 1915, the electron theory, on which Hilbert had lectured in the summer of 1913 and would again in the summer of 1917 (Corry, 1999, 174, 183). Namely, the Mie theory was *a priori* consistent with *the principle of causality* in two ways that the electron theory was not. First, it employed only differential equations, whereas the electron theory, as Hilbert noted in lectures in the summer of 1916 (Hilbert, 1916a, 101–102), was a mixture *(ein Gemisch)* of functional, differential, and integral equations. From the standpoint of consistency with the field-theoretic prohibition against action-at-a-distance laws, the Mie theory was clearly to be preferred to the electron theory.

Second, the Mie world function yielded four electrodynamical equations for the four unknown electrodynamic potentials. From given boundary and initial conditions, one could show that the state of the world at any future time could be univocally determined via these equations through specification of the values of these potentials at any prior time, as required by the principle of causality (as Hilbert understood that principle). Notoriously, the Mie theory purchases its causal determination at the cost of gauge invariance (the Mie potentials have 'absolute' values). Ironically, what current wisdom deems precisely wrong about the Mie theory was thus a philosophical ground in favor of it cited by Hilbert.[11] In sum, in the summer of 1916, and so

---

[10] Hilbert's WS 1921/1922 Lectures on the '*Grundlagen der Mathematik*,' as cited and translated in Hallett (1994), 167–168.

[11] Within the broad framework of Mie's theory, one might hope to find a matter representation based on generalized Maxwell equations following from a Lagrangian containing only gauge invariant terms.

after Einstein's canonical presentation of general relativity (Einstein, 1916), Hilbert continued to regard the standing of the principle of causality in the new physics of Einstein's principle of general invariance as unclear (Hilbert, 1916a, 110). Mie's theory, however, was deemed suitable to be incorporated into Hilbert's axiomatic construction by its *a priori* consistency with the requirement of causality. Finally, we shall see that there were also *a posteriori* reasons justifying Hilbert's incorporation of Mie's theory. Namely, Hilbert would show that the gauge structure of electromagnetism was recovered by his generally covariant generalization of Mie's theory, and that his energy tensor for non-gravitational energy coincided with Mie's energy tensor in the special relativistic limit. Both of these results are crucial to Hilbert's otherwise problematic claim that electrodynamic phenomena are a consequence of gravitation.

8.3 Hilbert's First Note: What Was Hilbert's Aim?

As legend has it, in November 1915, Hilbert engaged with Einstein in a competition to arrive at the generally covariant field equations of gravitation. Certainly, there was some sort of a 'race': no other term quite so well suits the frenzied activities of Einstein and Hilbert in that month. But this can by no means have been Hilbert's only aim, for he postulated an action integral containing a Lagrangian 'world function' for *both* the gravitational *and* the matter fields, from which the fundamental equations of a pure field physics might be derived. In astonishing testimony to his belief in the axiomatic method's power to 'deepen the foundations' of a theory, this objective is stated as the main aim in both published versions of Hilbert's two notes (1915b, 395; 1917, 63–64).

The first note accordingly begins with recognition that the investigations of Einstein and Mie have 'opened new paths for the investigation of the foundation of physics'. Expressing Einstein's theory of gravitation in terms of the 10 independent gravitational 'potentials' $g_{\mu\nu}$, and providing a generally invariant generalization of Mie's theory expressed in terms of the four electromagnetic vector potentials q_s, Hilbert employed sophisticated mathematical techniques to draw out the consequences of his two principal axioms. It is clear that Hilbert was extremely pleased with the axiomatic conjunction of the two theories. The triumphal language at the end of his first note expresses Hilbert's great satisfaction with the illumination gained in revealing unsuspected mathematical relations between the field equations for gravitation and for electrodynamics. In what follows we sketch how this illumination was achieved.

8.3.1 Schematic Outline

The core of Hilbert's approach lies in two axioms, which he states immediately after some preliminary definitions.

- AXIOM I ('Mie's Axiom of the World Function'). Hilbert proposed a variational argument formulated for a 'world function' (Lagrangian density) H, depending upon the 10 gravitational potentials $g_{\mu\nu}$, their first and second derivatives, as well as the four electromagnetic potentials q_s, and their first derivatives:

$$\delta \int H \sqrt{g} d\omega = 0 \ (g = \det|g_{\mu\nu}|, d\omega = dw^1 dw^2 dw^3 dw^4).$$

- AXIOM II ('Axiom of General Invariance'). H is an invariant with respect to arbitrary transformations of the 'world parameters' $w_s (s = 1, 2, 3, 4)$.

The function H is not further specified. But Hilbert's use of the term 'world parameters' in place of the standard locution 'space-time coordinates' is instructive. As expressly stated in his second note, and as Mie noted that same year,[12] it is intended to highlight the analogy Hilbert sought to draw between the arbitrariness of parameter representations of curves in the calculus of variations, and the arbitrariness of coordinates on a space-time manifold. Hilbert was, of course, a grand master of the calculus of variations, as his first note demonstrated. In both cases, objective significance will accrue only to objects invariant under arbitrary transformation of the parameters, respectively, coordinates. Precisely the same language of 'world parameters' is also used in the Proofs, *prima facie* evidence that his views regarding the lack of physical meaningfulness accruing to space-time coordinates were already in place. Similarly, in both versions Hilbert affirms that his second axiom is

> the simplest mathematical expression for the demand that the interconnection of the potentials $g_{\mu\nu}$ and q_s is, in and for itself, completely independent of the way in which one designates the world points through world parameters (1915a, 2; 1915b, 396).

We note that in the 1924 republication of Hilbert's two notes in *Mathematische Annalen*, the term 'world parameters' has been dropped, while the sentence has been reformulated explicitly in terms of the physical meaninglessness of space-time coordinates:

> Axiom II is the simplest mathematical expression for the demand that the coordinates in themselves have no manner of physical meaning, but rather represent only an enumeration of the world points in such a way as is completely independent of the interconnection of the potentials $g_{\mu\nu}$ and q_s (Hilbert, 1924, 4).

But given what is surely a semantic equivalence between the two sentences, we cannot agree with Corry's assessment (2004, 401) that this change ('Hilbert now added a paragraph') represents an alteration 'distancing (Hilbert) from the position that was variously insinuated in his earlier versions'.

---

[12] Hilbert (1917), 61: 'Just as in the theory of curves and surfaces an assertion for which the parameter representation of the curve or surface has been chosen has no geometric meaning for the curve or surface itself, so we must also in physics designate an assertion as *physically meaningless (physikalisch sinnlos)* that does not remain invariant with respect to arbitrary transformation of the coordinate system.' Mie (1917), 599, also stressed this analogy.

Before proceeding further, Hilbert then stated, without proof, a theorem described as the '*Leitmotiv* of my theory', whose content may be more briefly stated as follows:

- THEOREM I ('*Leitmotiv*'). In the system of n Euler–Lagrange differential equations in n variables obtained from a generally covariant variational integral such as in Axiom I, 4 of the n equations are always a consequence of the other $n - 4$ in the sense that 4 linearly independent combinations of the n equations and their total derivatives are always identically satisfied.[13]

One of Hilbert's principal claims, to be discussed below, is that, as a consequence of Theorem I, electromagnetic phenomena may be regarded as consequences of the gravitational. The theorem also gives rise to Hilbert's 'problem of causality' (see Section 8.4.1).

Hilbert next turns to the derivation of the Euler–Lagrange differential equations from his invariant integral, by differentiation of H with respect to the $g_{\mu\nu}$ and their first and second derivatives. This yields ten equations for the gravitational potentials,

$$\frac{\partial \sqrt{g}H}{\partial g^{\mu\nu}} - \sum_k \frac{\partial}{\partial w_k}\frac{\partial \sqrt{g}H}{\partial g_k^{\mu\nu}} + \sum_{k,l} \frac{\partial^2}{\partial w_k \partial w_l}\frac{\partial \sqrt{g}H}{g_{kl}^{\mu\nu}} = 0, \quad \text{or} \quad [\sqrt{g}H]_{\mu\nu} = 0$$

$$\text{(8.1)}$$

$$\left[g_l^{\mu\nu} = \frac{\partial g^{\mu\nu}}{\partial w_l}; g_{lk}^{\mu\nu} = \frac{\partial^2 g^{\mu\nu}}{\partial w_l \partial w_k} \right],$$

while differentiation of H with respect to the electromagnetic potentials q_s and their first derivatives yields four equations,

$$\frac{\partial \sqrt{g}H}{\partial q_h} - \sum_\sigma \frac{\partial}{\partial w_k}\frac{\partial \sqrt{g}\,H}{\partial q_{h\,k}} = 0, \quad \text{or,}$$

$$[\sqrt{g}H]_h = 0 \left[q_{hk} = \frac{\partial q_h}{\partial w_k} \ (h, \, k = 1, 2, 3, 4) \right].^{14} \quad \text{(8.2)}$$

The fourteen equations [(8.1)] and [(8.2)] in Hilbert (1915b) are termed 'the basic equations of gravitation and electrodynamics or generalized Maxwell equations'. On the assumption that the Mie theory renders a viable theory of matter, these equations encompass the entirety of fundamental physics. The remainder of the paper concerns Hilbert's treatment of energy, which includes his demonstration of a connection between the phenomena of gravitation and of electromagnetism. We turn to this issue now.

---

[13] Hilbert (1915a), 2–3; (1915b), 397. However, as Klein (1917), 481, first pointed out, since Hilbert regards the invariant H as the additive sum of *two* general invariants $K + L$, there are then 8 identities between the 14 field equations. According to Klein, 4 of these are a purely mathematical consequence of the 10 gravitational equations. The other 4 permit Hilbert's interpretation of the electromagnetic equations as a consequence of the gravitational equations.

[14] The form of equations [(8.1)] and [(8.2)] is trivially different algebraically between the Proofs and the published version. Here we follow the published version.

8.3.2 The Connection Between Gravitation and Electromagnetism

On the basis of Theorem I, Hilbert concluded that the four equations [(8.2)] are a consequence of the ten equations [(8.1)], such that, '*in the sense indicated, electro-dynamic phenomena are effects of gravitation*' (1915a, 3; 1915b, 397). This claim is certainly not part of the standard lore of general relativity, and it has repeatedly come under severe criticism, most recently by Renn and Stachel (1999, 36–41; 2007, 893–899) and by Corry (2004, 336–337). Since Hilbert relied on a specialized treatment of matter and non-gravitational energy stemming from Mie, we consider only Hilbert's *internal* (to his own theory) justification for this claim.[15] For present purposes, we wish to highlight three results that Hilbert proudly attributed to the use of the axiomatic method:

- general covariance, as we shall prefer to say, is connected with the gauge structure of electromagnetism;
- the electromagnetic energy tensor of Hilbert's generally covariant theory yields that of Mie in the special relativistic limit;
- the gravitational equations entail four mutually independent linear combinations of the electromagnetic equations and their first derivatives.

In our opinion, the first and third of these results express one of the two central outcomes reached by Hilbert, by means of the axiomatic method: for *any* theory which seeks to combine generally covariant theories of gravitation and electromagnetism, there follow strong restrictions on the form of the electromagnetic part of the theory as a consequence of the structure of the gravitational part of the theory.[16] However, we must point out that Hilbert also regarded the second result, concerning the Mie tensor, as a central achievement of his theory, and indeed a bellwether of its general correctness.

The first of the above results is obtained as follows. Hilbert's gravitational equations are expressed as variational derivatives with respect to the metric (1915a, 11; 1915b, 404) $[\sqrt{g}K]_{\mu\nu} + \frac{\partial\sqrt{g}L}{\partial g^{\mu\nu}} = 0$, where the first term is evaluated, in the published version but not in the Proofs, so that the crucial trace term appears, $[\sqrt{g}K]_{\mu\nu} = \sqrt{g}(K_{\mu\nu} - \frac{1}{2}Kg_{\mu\nu})$. Now L is a general invariant that, by Axiom I, is assumed to depend *only* on the $g_{\mu\nu}$, the q_s, and their first derivatives $\frac{\partial q_s}{\partial w^l}$. Hilbert had previously shown that, from Axiom II (the axiom of general invariance) and a supporting theorem (Theorem II, the Lie derivative of the metric), it follows that L must satisfy the relations (1915a, 11; 1915b, 403)

$$\frac{\partial L}{\partial q_{sk}} + \frac{\partial L}{\partial q_{ks}} = 0.$$

Thus, even though the Mie theory assigns 'absolute' values to the electrodynamic potentials q_s, the matter Lagrangian L in Hilbert's theory depends only on the anti-symmetrized derivatives of the q_s

---

[15] Hilbert's treatment of energy is discussed in detail in Sauer (1999), 554–557.

[16] Hilbert's considerations on the tension between general covariance and causality are the other central outcome.

$$M_{ks} = Rot(q_s) \equiv q_{sk} - q_{ks},$$

that is, on the electromagnetic field tensor. As Hilbert did not fail to observe, this is a necessary condition for recovering Maxwell's theory. Only by additional assumption is this also the case with Mie's original theory, but that theory is not generally invariant (Born, 1914, 28). Hilbert has thus shown that the gauge structure of electromagnetism follows from general covariance and the other assumptions for L, emphasizing in italic type that

> *This result [on which the character of Maxwell's equations depends] follows here essentially as a consequence of general invariance, hence on the basis of axiom II.*[17]

The assumption that nothing else beyond the $g_{\mu\nu}$ (but no derivatives of the metric), the q_s, and the so-constrained first derivatives $\frac{\partial q_s}{\partial w^l}$ enter into L also has consequences for the interpretation of the energy-momentum tensor $T_{\mu\nu}$ in Hilbert's theory. Since all non-gravitational energy/matter is contained in L, it is entirely sufficient for forming $T_{\mu\nu}$, i.e.,

$$\frac{\partial \sqrt{g}L}{\partial g^{\mu\nu}} = \sqrt{g}T_{\mu\nu}.$$

In this respect, Hilbert's gravitational field equations differ from Einstein's in their interpretation because Hilbert assumed a particular hypothesis about the electromagnetic constitution of all matter.[18] With this interpretation of $T_{\mu\nu}$, Hilbert is then able to show that the matter tensor of his theory yields the electromagnetic energy tensor of Mie's theory in the special relativistic limit.[19] This is also a fundamental result, and Hilbert emphasized its significance in *Sperrdruck* type:

> *Mie's electromagnetic energy tensor is nothing other than the generally invariant tensor obtained by derivation of the invariant L with respect to the gravitational potentials $g^{\mu\nu}$ in the* [special relativistic] *limit*—a circumstance that first indicated to me the necessary close connection between Einstein's general theory of relativity and Mie's electrodynamics, and which convinced me of the correctness of the theory developed here.[20]

Hilbert regarded this result as a central achievement of his theory, and indeed a bellwether of its general correctness. Moreover, it must be emphasized that the 'necessary close connection' between the two theories has been established through the axiomatic method, and so will count towards the triumph of that method as proclaimed by Hilbert at the end of his paper.

---

[17] Hilbert (1915b), 403; at the corresponding place in Hilbert (1915a), 10, the bracketed expression does not appear.

[18] Earman and Glymour (1978), 303; Sauer (1999), 564.

[19] For details, see Sauer (1999), 555.

[20] Hilbert (1915a), 10; (1915b), 404. Notice that *already in the Proofs*, Hilbert's reference is to 'Einstein's general theory of relativity' *(der Einsteinschen allgemeinen Relativitätstheorie)*, explicitly according due credit to Einstein.

Finally, Hilbert demonstrated the connection between the field equations of gravitation and electromagnetism. Using the Lagrangian form of his gravitational equations in conjunction with a version of the contracted Bianchi identities derived in his Theorem III (and which follow from Theorem I), Hilbert has shown that the gravitational field equations in conjunction with the postulate of general invariance yield four mutually independent combinations of the electromagnetic field equations and their first derivatives. This is the sense in which the electromagnetic phenomena are consequences of the gravitational. Referring back to the assertion that he made following his statement of Theorem I, Hilbert claimed, again in italic type for emphasis:

> *This is the exact mathematical expression of the above generally stated assertion concerning the character of electrodynamics as an accompanying phenomenon (Folgeerscheinuung) of gravitation.*

We wish to stress that Hilbert clearly viewed this result, as well as the just-mentioned recovery of Mie's tensor in the special relativistic limit, as *central achievements of his theory. Neither of these has to do with the explicit formulation of the generally covariant field equations of gravitation.*[21]

8.4 Differences Between the Proofs and the Published Version

There are three main differences between the Proofs and the published version of the first note. First, the explicit form of the field equations of general relativity does not appear in the Proofs. This matter has received considerable attention in the recent literature, and will not be treated here. Our view, following the careful analyses of Sauer (1999, 2005), is that the Proofs already contain the correct gravitational field equations of general relativity in the inexplicit form of a variational principle and the Hilbert action. The two other differences are related, and will be treated here: in the Proofs, but entirely missing from the published version, there is a clear statement of the problem of causality facing any generally covariant theory, as well as a proposed solution that restricts the applicability of space-time coordinates. Our interest lies in identifying a thematic link between this text cut from the Proofs and issues treated in the published second note. This enables us to see that it treats in detail the problem of causality that is addressed in the Proofs, but dealt with unsatisfactorily there.

8.4.1 Hilbert's Target: The 'Problem of Causality'

In the Proofs, but not in the published version of the first note, Hilbert explicitly spells out the implications of Theorem I for his system of fundamental equations of physics.

---

[21] Of course, Hilbert's interpretation of the significance of Theorem I rests on the special choice of H (and L), and the related assumption of the electromagnetic constitution of matter that furnishes the definition of Hilbert's energy-momentum tensor above. Rowe (2001), 404, observes that it was 'microphysics not gravitation that Hilbert saw as the central problem area'. We broadly agree that gravitation was not Hilbert's primary focus.

Our mathematical theorem teaches that for the 14 potentials, the above axioms I and II can yield only 10 equations essentially independent of one another. On the other hand, by upholding general invariance, no more than 10 essentially independent equations for the 14 potentials $g_{\mu\nu}$, q_s, are possible at all. Therefore, if we want to preserve the determinate character of the fundamental equations of physics according to Cauchy's theory of differential equations, the requirement of four additional non-invariant equations supplementing [(8.1)] and [(8.2)] is essential. (1915a 3–4)

Thus, independent of the physical validity of his system of fundamental equations, for which he adduced no evidence whatsoever, Hilbert clearly underscored that the mathematical underdetermination in question (10 independent equations for 14 potentials) is solely a consequence of his axiom of general invariance.

Befitting its preeminent concern with the *consistency* of all axioms and assumptions undergirding a theory, the axiomatic method has revealed a seeming conflict between general covariance and causality in the sense of a failure of univocal determination, a conflict characterized in terms of whether *any* theory satisfying Axioms I and II admits a well-posed Cauchy problem. Theorem I suggests that it is a property of any such theory that it does not. Prior to general relativity, as Hilbert repeatedly emphasized, all physical theories permitting a variational formulation satisfied Cauchy determination in the sense that they yielded precisely as many independent Euler–Lagrange equations as there were independent functions to be determined. However, the situation is complicated in a generally covariant space-time theory by the freedom to make arbitrary coordinate transformations (equivalently, diffeomorphic point transformations) of solutions to the field equations. Formulated for a generally invariant Lagrangian, by Hilbert's Theorem I this is the fact that not all of the Euler–Lagrange equations obtained by variation of the integral invariant with respect to the field quantities and their derivatives are independent. More precisely, four of these are always the result of the remaining $n - 4$ space-time equations. Thus, Theorem I is *a precise mathematical statement of the tension between the postulate of general covariance and the requirement of causality in the mathematical sense of univocal determination.*

Notice that univocal causal determination—in the sense required by a well-posed Cauchy problem—is not an axiom in Hilbert's construction. Nevertheless, it is a requirement satisfied by previous physical theories in variational formulation, and so its seeming failure in the context of general invariance surely sparked Hilbert's interest. But as we have repeatedly stated, in our opinion this is one of the two central outcomes that Hilbert reached by means of the axiomatic method: *any generally covariant theory raises deep questions about causality, in both the mathematical and (as we shall see) the physical sense.*

Hilbert's diagnosis in turn marked out a strategy for resolving the apparent tension between general covariance and failure of univocal determination: to find, if possible, four conditions additional to the ten independent equations that will render the Cauchy problem well posed. Finding the 'missing four space-time' equations is the motivation behind the intricate mathematical construction in the Proofs of an 'energy form'

$$E = \sum_s e_s p^s + \sum_{s,l} e_s^l p_l^s$$

(where e_s is termed the 'energy vector', and p^s is an arbitrary contravariant vector) constructed from the tensor density $\sqrt{g} P_g H$, where P_g is a differential operator on the world function H. A prime consideration both here, and in the different treatment of energy in the published version, will be to recover Mie's energy tensor as a special case. While by construction this 'energy form' is a general invariant, Hilbert finds four supplementary conditions by imposing special coordinate restrictions on his 'energy vector'; namely, that it must satisfy the divergence equation (numbered (15) in Hilbert, 1915b),

$$\sum_l \frac{\partial e_s^l}{\partial w_l} = 0, \quad \text{iff the four quantities } e_s = 0, \qquad (8.3)$$

where the w_l are now special space-time coordinates adapted to this 'energy theorem', as stated in a third, and final axiom appearing only in the Proofs:

- AXIOM III: ('The Axiom of Space and Time'): 'The space-time coordinates are such particular world parameters for which the energy theorem [(8.3)] is valid.'

Elucidating this result, Hilbert clarified the main point, that these four non-covariant equations complete the system of fundamental equations of physics:

> On account of the same number of equations and of definite potentials, the causality principle for physical happenings *(Geschehen)* is also ensured, and with it is unveiled to us the narrowest connection between the energy theorem and the principle of causality, in that each conditions the other. (1915a, 7)

The idea that satisfaction of energy conservation requires four non-covariant conditions is almost certainly taken from Einstein's *Entwurf* theory of 1913–1914 (Renn and Stachel, 1999, 32; 2007, 888). Drawing analogies to Einstein's difficulties in the 'hole argument', Renn and Stachel (1999, 73; 2007, 934) regard Hilbert's energy construction as his 'Proofs argument, based on causality, against general covariance'. But Hilbert's rather more complicated construction has, philosophically and motivationally, a different *raison d'être*.[22] We see that in Hilbert's case, *the aim was to extract a Cauchy-determinate structure within an otherwise generally covariant theory, and not to abandon general covariance.*[23] However, the complex mathematical derivation in the Proofs leading to Hilbert's four energy conditions was

---

[22] Brading and Ryckman (2008), section 7, emphasize the *disanalogies* of Hilbert's and Einstein's respective treatments of the tension between causality and general covariance.

[23] This is also noted by Sauer (2005), n. 5, who writes, 'Hilbert kept the generally covariant field equations as fundamental field equations and only postulated a limitation of the physically admissible coordinate systems.' Yet Sauer does not make enough of this, we think. Earlier in his text he writes that Hilbert's Axiom III is a *restriction* of the general covariance of Hilbert's theory, there seeming to subscribe to the view that Hilbert followed Einstein in seeking to limit the covariance of his theory.

cut, together with *all* of its motivation, from the published version. In the light of the completed form of Einstein's theory, placing coordinate restrictions on the energy term turned out to be the wrong approach for solving the tension between general covariance and Cauchy determination. Hilbert accordingly dropped it altogether, modifying and truncating his treatment of energy. There—almost certainly following the implicitly generally covariant energy in Einstein's final November presentation to the Berlin Academy (Einstein, 1915)—Hilbert derived a generally covariant 'energy equation,' which anyway is consonant with the 'trace' term in the gravitational field equations popping out through explicit calculation from their Lagrangian derivatives.

Nevertheless, the issue of causality in a generally covariant theory does not go away for Hilbert. We claim that the second note contains his much revised, detailed reconsideration of this issue, and is rightly understood only in this light.[24]

8.5 Hilbert's Second Note

This paper is, we argue, principally concerned with providing a satisfactory reconciliation between the principles of general invariance and causality. The impression given by Renn and Stachel is that the second note is a list of special topics within general relativity. Moreover, they allege that the second note shows Hilbert in 'agony' over the 'collapse of his own research program' (Renn and Stachel, 1999, 90; 2007, 953). On the contrary, in our opinion, Hilbert deemed his second note to be its completion.

The treatment of the problem of causality in generally covariant theories here has four principal facets. First, Hilbert observed that arbitrary point transformations (diffeomorphisms) do not respect the relation of cause and effect among world points lying on the same timelike curve. To rectify this, he introduced the notion of *proper coordinate systems*, transformations among which always respect the distinction between spacelike and timelike coordinate axes and can never reverse the temporal order of cause and effect. Next, he pointed to the consequent need to reformulate the causality principle within the 'new physics' of general invariance, showing that here the univocal determination of future states from present states requires coordinate restrictions on the initial data in order to locally describe dynamical evolution off that surface. This is attained by employing a 'Gaussian' coordinate system, a particular type of *proper coordinate system*. The purchase of univocal determination in the 'new physics' at the cost of adopting special coordinate systems prompted Hilbert, thirdly, to state a 'sharper conception' of the principle of general relativity (general invariance) underlying this physics. By means of this sharper conception, he is able to give a clear account of under what conditions a statement of physics is physically meaningful. Finally, though we shall not discuss it here,[25]

---

[24] The topic of energy-momentum in general relativity did not go away either. It was the subject of ongoing discussions between Hilbert, Einstein, and Klein (see Brading, 2005), and remains a delicate issue; for discussion, see Hoefer (2000).

[25] See Brading and Ryckman (2008), section 6.

Hilbert also took up the related issue of the inconsistency of Euclidean geometry (permitting, on account of its globally fixed metrical structure, the concept of action at a distance) with the new physics of fields, which he calls a four-dimensional pseudo-geometry. To this end, he discussed the conditions under which a pseudo-Euclidean (Minkowski) metric arises in the new physics, and he rederived the external Schwarzschild solution corresponding to the solar gravitational field without the assumption that the $g_{\mu v}$ had pseudo-Euclidean values at infinity, that is, that the solar system is embedded in a pseudo-Euclidean world.

8.5.1 The Problem of Causal Order

On the basis of Axiom II of his first note, and with implicit reference to Einstein's requirement of general covariance for the gravitational field equations, all coordinate systems arising from x_s by arbitrary smooth transformations have up to now been regarded as on an equal footing with one another. However, Hilbert observed that a conflict with the causal order will arise if two world *points* lying along the same timelike curve, and standing in the relation of *cause and effect*, can be transformed so that they become *simultaneous* (i.e., lie on the same data hypersurface). The causal order concerns our *experience* of the world in space and time, and thus we have an apparent conflict between the overriding demand of objectivity expressed by general covariance and the *experienced causal ordering* of events.

Although Hilbert speaks (1917, 57) of the need to restrict the arbitrariness *of coordinate systems*, his example concerns *point transformations* (in fact, along one and the same timelike curve) and the fact that diffeomorphism invariance need not preserve the relation of causal order among events. If the new physics is to be compatible with the experienced causal ordering of events, we need to restrict the allowed coordinate systems such that under transformation this causal ordering is preserved. To achieve this end, Hilbert introduced what he called 'proper' coordinate systems.

If x_4 is designated as the 'proper' time coordinate, a 'proper *(eigentlich)* coordinate system' may be defined as one in which the following four inequalities are satisfied by the components of the metric tensor (numbered (31) in Hilbert, 1915b):

$$g_{11} > 0, \quad \begin{vmatrix} g_{11} & g_{12} \\ g_{21} & g_{22} \end{vmatrix} > 0, \quad \begin{vmatrix} g_{11} & g_{12} & g_{13} \\ g_{21} & g_{22} & g_{23} \\ g_{31} & g_{32} & g_{33} \end{vmatrix} > 0, \quad g_{44} < 0. \tag{8.4}$$

These so-called 'reality relations'[26] implement, in the case of general Riemannian geometry, the physical requirement of metrical indefiniteness: that three of the coordinate axes are spacelike, and one timelike. Together, the restrictions imply that $g(= \det |g_{\mu v}|) < 0$, so $\sqrt{-g}$ must replace $\sqrt{g}$ in all tensor formulae. A coordinate

---

[26] Pauli (1921), S22 *'Wirklichkeitsverhältniße'*.

transformation carrying such a proper space-time description into another proper space-time description is called a *proper* space-time coordinate transformation.[27]

The significance of these coordinate conditions for the principle of causality is then clearly spelled out:

> So we see that the concepts of cause and effect lying at the basis of the principle of causality also in the new physics never leads to inner contradiction, as soon as we always take the inequalities [(8.4)] in addition to our fundamental equations; that is, we restrict ourselves to the use of *proper* space-time coordinates.

We observe here (a point we shall return to) that it is not *nature* but the structure of our cognitive experience (preservation of causal relations) that leads to the restriction to proper coordinate systems. Coordinate conditions govern possible objects of experience (for Hilbert, these are physical facts as determined by 'measure threads' and 'light clocks') represented in causal relation within spatio-temporal empirical intuition, but *not* possible objects of physics, governed by the ideal requirement of general invariance, according to Axiom II.

8.5.2 The Problem of Univocal Determination

We reiterate that the scope of Theorem I extends to any generally invariant four-dimensional theory, and is thus broader than either general relativity or Hilbert's theory itself. While the conflict between general invariance and causal determination is only implied, via Theorem I, in the published version of Hilbert's first note (and accordingly downplayed in the literature), in his second note, Hilbert nonetheless claimed that there he had 'especially stressed' this fact. One might conjecture that Hilbert had merely forgotten that explicit reference to the failure of Cauchy determination for his fundamental equations had been excised from the Proofs, along with his non-covariant treatment of energy. However, it is more plausible to think otherwise. Hilbert's first resolution had been cast in terms of finding 'four additional non-invariant equations', a strategy that had not worked. Then, when revising the Proofs in the light of Einstein's published paper of November 25, 1915, it seems he had not yet seen that a solution lay not in four additional non-invariant *equations*, but rather in the coordinate conditions yielding the four *inequalities* [(8.4)]. Uncertain about how the issue was to be resolved, Hilbert had simply buried the entire issue in the published version.

The main point of his second paper is to provide a quite different manner of resolution. Although continuing his interpretation of Theorem I that the four

---

[27] Hilbert's inequalities [(8.4)] apply only to curves that are spacelike or timelike. But points on a *null* curve may be transformed to simultaneity by a coordinate transformation that is 'proper' in the sense that it preserves causal order on all *timelike* curves. In this case the 'reality relations' will be violated even though the transformation satisfies Hilbert's causality principle, and so the inequalities [(8.4)] are merely sufficient, but not necessary, to preserve causal ordering. See Renn and Stachel (1999), 80; (2007), 941–942.

generalized Maxwell equations [(8.2)] are a consequence of the ten gravitational equations [(8.1)], this claim lies well in the background, while the matter of causality takes pride of place. The basic achievement of the paper will be to give the necessary *reformulation* of the causality principle that is required by the new generally invariant field physics.

The need for such a reformulation is explicitly stated. Hilbert observes that up until the present time all physical theories whose laws are written as differential equations have satisfied the requirement of causality, in the sense of univocal determination of future states from present states and their time derivatives. As precisely formulated by Cauchy, causal determination requires that the theory provide an independent equation for each unknown function appearing in the theory, a result secured by 'the well-known Cauchy theorem on the existence of integrals of partial differential equations'. However, the situation is different once the requirement of general invariance is imposed.

> Now the fundamental eqs. [(8.1)] and [(8.2)] set up in my first contribution are, as I there especially stressed, in no way of the above-characterized kind. Rather, according to Theorem I four are a consequence of the remaining ones: We viewed the four Maxwell equations [(8.2)] as a consequence of the ten gravitational equations [(8.1)] and therefore have only the 10 essentially independent equations [(8.1)] for the 14 potentials $g_{\mu v}$ and q_s.
> As soon as we raise the requirement of general invariance for the fundamental equations of physics, the just mentioned circumstance is essential and even necessary.[28]

On the other hand, Hilbert claimed that the situation in the newly emerging generally invariant physics is such that

> from knowledge of physical magnitudes in the present and past, it is no longer possible to univocally deduce their values in the future.

As a result, Hilbert argued, we are driven to *reformulate* the causality principle through 'a sharper grasp' of how the general invariance of the new physics should be understood. The general invariance of the laws, is set as a regulative ideal of physical objectivity that applies to the conceptual structure of fundamental (field) physics. There remains the question of how the principle of general invariance should be understood, not only in the context of laws, but also of individual statements concerning the spatio-temporal evolution of particular systems or objects. Hilbert therefore revisits the question of what is meant by the meaningfulness of physical statements once the principle of causality is taken into account. His solution can be elucidated as follows. A *necessary* condition for such a statement to be physically meaningful is that it has a generally covariant formulation. But of course, this is not *sufficient*. For

---

[28] Hilbert (1917), 60. 'Essential and necessary' because the introduction of a Gaussian (and so, *proper*) space-time coordinate system for the 10 potentials, $g_{\mu v}(\mu, \nu = 1, 2, 3)$; $q_s(s = 1, 2, 3, 4)$, would result in overdetermination of the system (more than ten independent equations), and thus inconsistency.

when such statements are predictions, i.e., concern the future, Hilbert stipulated that their meanings are to be understood in such a way that the requirement of physical causality (viz., that causes precede their effects) is satisfied.

> As now regards the principle of causality, the physical quantities and their time-rates of change may be known at the present time in any given coordinate system; then a statement will have physical meaning only when it is invariant with respect to all those transformations for which precisely those coordinates used for the present time remain unchanged. I declare that statements of this kind for the future are all univocally determined, that is, *the causality principle holds in this formulation: From the knowledge of the 14 physical potentials $g_{\mu\nu}$, q_s in the present, all statements concerning them for the future follow necessarily and univocally, in so far as they have physical meaning.* (1917, 61)

Renn and Stachel (1999, 81; 2007, 942) observe that this is obviously *not* a claim that physically meaningful statements are independent of the choice of a coordinate system. But is this passage evidence for what they go on to suggest, that Hilbert still attaches 'some residual physical meaning to the choice of coordinates'? In our opinion, it is apparent from Hilbert's formulation that the criterion of physical meaningfulness of statements requires satisfaction of the principle of causality in the usual sense that conditions in the present determine those in the future. Furthermore, any such physical statement must be independent of how it is designated by coordinates; i.e., it must be, in Hilbert's terms, an invariant statement.[29]

With this new conception of the causality principle in hand, we can formulate the necessary and sufficient conditions for a proposition to be physically meaningful:

(a) The proposition must have a generally covariant formulation.
(b) When the proposition is expressed with respect to a proper coordinate system, the truth value of that description must be *uniquely* determined by an appropriate spacelike past hypersurface.

In other words, when we express the propositions of physics in terms of possible objects of experience (that is, including the spatio-temporal and causal aspect of how we experience objects), those statements are physically meaningful if and only if they are causally determinate in the sense of condition (b), as well as satisfying condition (a).

From the new point of view, the physical principle of causality, as ensured by the coordinate conditions of a well-posed Cauchy problem, is a lingering but

---

[29] These points are made explicitly in Hilbert's 'Causality Lecture' (1916b), 5–6, given the probable date of November 21, 1916, in Sauer (2001): 'We will prove that the thus formulated causality principle: 'All meaningful assertions are a necessary consequence of what has gone on before *(der vorangegangenen)*' is valid.' Also: 'That is, one must not only say that the world laws are independent of reference system, but rather also that any individual assertion regarding an occurrence or a coincidence *(Zusammentreffen)* of occurrences only has a meaning if it is independent of designation, i.e., if it is invariant.'

not eliminable constraint upon human understanding ('physical meaningfulness'), a necessary condition imposed by the mind in structuring experience. Like the subjectivity of the sense qualities, Hilbert viewed the requirement of physical causality as anthropomorphic, having to do not with the objective world of physics but rather with our experience of that world.[30]

8.6 Hilbert's Revision of Kant in the Light of General Invariance[31]

To many neo-Kantians active in the first quarter of the 20th century, the general theory of relativity showed that Kant's epistemology was still a work in progress, neither a refuted nor a finished edifice, and nearly all were prepared to concede, as did Hilbert, that 'Kant greatly overestimated the role and the extent of the *a priori*' (Hilbert 1930, 961). By the same token, it might be said that Hilbert, through the axiomatic method, was the first neo-Kantian to put his finger on where exactly the general theory of relativity required a modification in the traditional Kantian transcendental framework that expressly bound considerations of objectivity together with conditions of possible experience. In Kant, space and time, as subjective forms of sensibility, are at once also objective conditions for perception of objects— conditions of the possibility of experience. For a cognition to be *objectively valid* (to be a representation pertaining to a possible object *for us*, hence to be *meaningful*) is for it to invoke *our* specifically human type of finite, receptive spatio-temporal sensory intuition of objects.

Hilbert essentially argued that this is no longer the case once the requirement of general invariance is imposed on fundamental physical theory. While retaining part of Kant's linkage of conditions of physical *meaningfulness* to *sensibility*, Hilbert placed general invariance as the superordinate criterion of physical objectivity, explicitly attributing this development to the influence of Einstein's gravitational theory. This is repeatedly affirmed in his lectures, e.g., (1919–1920, 49), and (1921):

> Hitherto, the objectification of our view of the processes of nature took place by emancipation from the subjectivity of human sensations. But a more far reaching objectification is necessary, to be obtained by emancipating ourselves from the *subjective* moments of human *intuition* with respect to space and time. This emancipation, which is at the same time the high-point of scientific objectification, is achieved in Einstein's theory; it means a radical elimination of *anthropomorphic* slag *(Schlacke)*, and leads us to that kind of

---

[30] Hilbert (1919–1920), 85–87, explicitly discusses the problem of causality in the context of general relativity, concluding (87) that "causality *(das Ursächliche)* in the narrower sense [of cause–effect relations] possesses no objective meaning for physics, and that in the search for causes, considerations of the particular conditions of human perception and of human purposes are essentially involved."

[31] See Brading and Ryckman (2008) for a more extensive treatment of this topic.

description of nature which is *independent* of our senses and intuition and is directed purely to the goals of objectivity and systematic unity.[32]

In broad agreement with the cognitive function that the Transcendental Dialectic assigns to reason, of seeking ever more inclusive 'systematic unity', the axiomatic method elevates the *axiom* of general invariance as the guiding principle of a 'description of nature which is *independent* of our senses and intuition'. In this sense, the principle of general invariance *is neither true nor false, but a regulative idea,*

> if, in agreement with Kant's words, we understand by an idea a concept of reason that transcends all experience and through which the concrete is completed so as to form a totality.[33]

Thus, for Hilbert, the axiom of general invariance is the anchor on which the objective scientific description of nature must now rest, even though such a description goes beyond the limits of possible experience, which is always finite and whose conditions are set by sensibility (representation in space and time) and the understanding (causality). By the early 1920s, Hilbert had termed his revised understanding of the *a priori* 'the finite point of view', taking from Kant the methodology or standpoint that objective cognition can only be understood as conditioned by *a priori* structures of the mind, but refashioning the boundaries of the *a priori* somewhat differently:

> We see therefore: in the Kantian theory of the *a priori (Apriori-Theorie)* there is still contained anthropomorphic slag *(Schlacke)*, from which it must be freed, and after such removal only that *a priori* point of view *(apriorische Einstellung)* is left, which also lies at the foundation of pure mathematical knowledge: it is essentially that finite point of view characterized by me in different essays. (Hilbert 1930, 962)

Final Remarks

Before concluding, we offer a brief remark on the alleged 'priority dispute' over the discovery of the generally covariant gravitational field equations. Based on the account of Hilbert's aims and methods given here, it is clear that Einstein and Hilbert were engaged in qualitatively different enterprises that only partially overlapped. In contrast to Einstein, Hilbert's goals were at least as much logical and epistemological as they were physical. We thus concur with the judgment of Felix Klein, who wrote, in 1921, that 'there can be no talk of a question of priority, since both authors pursued entirely different trains of thought (and to be sure, to such an extent that the compatibility of the results did not at once seem assured).'[34]

---

[32] Hilbert, 1921, *Grundgedanken der Relativitätstheorie*, lectures in SS 1921, ed. by Paul Bernays; as cited and translated in Majer (1995, 146).

[33] Hilbert (1925), Engl. trans., 392.

[34] Felix Klein (1917), 566, fn 8. '*Von einer Prioritätsfrage kann dabei keine Rede sein, weil beide Autoren ganz verschiedene Gedankengänge verfolgen (und zwar so, daß die Verträglichkeit der Resultate zunächst nicht einmal sicher schien).*' This remark occurs in a footnote added to the 1921 reprint of Klein (1917).

In conclusion, we have argued that Hilbert's two notes on the Foundations of Physics can be seen as having the common goal of pinpointing, then resolving, the apparent tension between general covariance and causality, and that his approach to this issue should be understood within the logical and epistemological context of his axiomatic method. Adopting general invariance as an axiom while observing, on the basis of his Theorem I, that the initial value problem was not well posed, Hilbert even contemplated surrendering causality (in the sense of Cauchy determination). Yet when he found a way to restore causality in the face of general covariance (essentially by imposing gauge conditions), he subordinated the principle of causality—as a condition of possible experience—to general covariance, an overriding principle of physical objectivity. This achievement, along with the results displaying the connections between general covariance and features of electromagnetic theory, were obtained through application of the axiomatic method. Only in this context can we understand what Hilbert sought to do, and evaluate his success.

References

Born, Max. (1914). "Der Impuls-Energie-Satz in der Elektrodynamik von Gustav Mie", *Nachrichten von der Königliche Gesellschaft der Wissenschaften zu Göttingen, Mathematisch-Physikalisch Klasse* 12, 23–36.

——. (1922). "Hilbert und die Physik", *Die Naturwissenschaften* 10, 88–93.

Brading, Katherine A. (2005). "A Note on General Relativity, Energy Conservation, and Noether's Theorems", in A. J. Kox and J. Eisenstaedt (eds.), *The Universe of General Relativity* (*Einstein Studies* v. 11). Boston: Birkhäuser, 125–35.

Brading, Katherine A. and Thomas A. Ryckman (2008). "Hilbert's 'Foundations of Physics': Gravitation and Electromagnetism Within the Axiomatic Method", *Studies in the History and Philosophy of Modern Physics* 39, 102–53.

Corry, Leo (1999). "Hilbert and Physics (1900–1915)", in Jeremy Gray (ed.), *The Symbolic Universe: Geometry and Physics 1890–1930*. Oxford: Oxford University Press, 145–88.

——. (2004). *David Hilbert and the Axiomatization of Physics (1898–1918): From Grundlagen der Geometrie to Grundlagen der Physik*. Dordrecht: Kluwer Academic Publishers.

Corry, Leo, Jürgen Renn, and John Stachel (1997). "Belated Decision in the Hilbert–Einstein Priority Dispute", *Science* 278, 1270–3.

Earman, John and Clark Glymour (1978). "Einstein and Hilbert: Two Months in the History of General Relativity", *Archive for History of Exact Sciences* 19, 291–308.

Einstein, Albert (1915). "Die Feldgleichungen der Gravitation". *Königlich Preussische Akademie der Wissenschaften* (Berlin). *Sitzungsberichte*, 844–847. Reprinted in A. J. Kox, Martin Klein, and Robert Schulmann (eds.), *The Collected Papers of Albert Einstein, volume 6*. Princeton: Princeton University Press, 1996, 244–9.

——. (1916). *"Die Grundlage der allgemeinen Relativitätstheorie"*, Annalen der Physik 49: 769–822. Reprinted in A.J. Kox, Martin Klein, and Robert Schulmann (eds.), *The Collected Papers of Albert Einstein, volume 6.* Princeton: Princeton University Press, 1996, 283–339.

Ewald, William (1996). *From Kant to Hilbert: A Source Book in the Foundations of Mathematics.* Oxford: Clarendon Press.

Gray, Jeremy (2000). *The Hilbert Challenge.* Oxford: Oxford University Press.

Guth, E. (1970). "Contribution to the History of Einstein's Geometry as a Branch of Physics", in Moshe Carmeli *et al.* (eds.), *Relativity.* New York: Plenum Press, 161–207.

Hallett, Michael (1994). "Hilbert's Axiomatic Method and the Laws of Thought", in Alexander George (ed.), *Mathematics and Mind.* New York: Oxford University Press, 158–200.

Hilbert, David (1899). *Grundlagen der Geometrie.* In *Festschrift zur Feier der Enthullung des Gauss-Weber Denkmals in Göttingen.* Leipzig, Teubner.

——. (1900). *"Mathematische Probleme".* Vortrag, gehalten auf dem internationalen Mathematiker-Kongress zu Paris. Königlichen Gesellschaft der Wissenschaften zu Göttingen, Nachrichten, 253–97. Translation in Jeremy Gray (2000), 240–82.

——. (1915a). *"Die Grundlagen der Physik (Erste Mitteilung)".* Annotated *"Erste Korrektur meiner erste Note"*, printer's stamp date "6 Dez. 1915". 13 pages with omissions. Göttingen, SUB Cod. Ms. 634. Translation in Renn and Schemmel (2007), 989–1001.

——. (1915b). *"Die Grundlagen der Physik: Erste Mitteilung"*, Königliche Gesellschaft der Wissenschaften zu Göttingen. Nachrichten. Mathematische-Physikalische Klasse, 395–407. Translation in Renn and Schemmel (2007), 1003–15.

——. (1916a). *"Die Grundlagen der Physik".* Typescript of summer semester lecture notes. Bibliothek des Mathematisches Institut, Universität Göttingen. 111 pages.

——. (1916b). *"Das Kausalitätsprinzip in der Physik".* Typescript of lectures, dated 21 and 28 November 1916 by Sauer (2001). Bibliothek des Mathematisches Institut, Universität Göttingen. 17 pages.

——. (1917). *"Die Grundlagen der Physik: Zweite Mitteilung"*, Königliche Gesellschaft der Wissenschaften zu Göttingen. Nachrichten. Mathematische-Physikalische Klasse, 53–76. Translation in Renn and Schemmel (2007), 1017–38.

——. (1918). *"Axiomatisches Denken"*, Mathematische Annalen, 78, 405–15. Translated by William Ewald as "Axiomatic Thought", in Ewald (1996), 1105–15.

——. (1919–1920). *Natur und mathematisches Erkennen. Vorlesungen, gehalten 1919–1920 in Göttingen. Nach der Ausarbeitung von P. Bernays.* Published and edited by David Rowe. Basel: Birkhäuser, 1992.

——. (1922). *"Neubegründung der Mathematik. Erste Mitteilung"*, Abhandlungen aus dem mathematischen Seminar der Hamburgischen Universität, 1, 157–77.

Translated by William Ewald as "The New Grounding of Mathematics. First Report", in Ewald (1996), 1115–34.

——. (1924). *"Die Grundlagen der Physik"*, *Mathematische Annalen* 92, 1–32.

——. (1925). *"Über das Unendliche"*, *Mathematische Annalen* 95, 161–90. Translated by Stefan Bauer-Mengelberg as "On the Infinite", in J. van Heijenoort (ed.), *From Frege to Gödel: A Sourcebook in Mathematical Logic, 1879–1931*, Cambridge, MA: Harvard University Press, 1967, 367–92.

——. (1930). *"Naturerkennen und Logik"*, *Die Naturwissenschaften* 18, 959–63. Translation by William Ewald as "Logic and the Knowledge of Nature", in Ewald (1996), 1157–65.

Hoefer, Carl (2000). "Energy Conservation in GTR", *Studies in History and Philosophy of Modern Physics* 31, 187–99.

Klein, Felix (1917). *"Zu Hilberts erster Note über die Grundlagen der Physik"*, *Königliche Gesellschaft der Wissenschaften zu Göttingen. Nachrichten. Mathematisch-Physikalische Klasse.* As reprinted in Felix Klein, *Gesammelte Abhandlungen Bd I.* Berlin: Julius Springer, 1921, 553–67.

Majer, Ulrich (1995). "Hilbert's Finitism and the Concept of Space", in Ulrich Majer and Heinz-Jürgen Schmidt (eds.), *Semantical Aspects of Spacetime Theories.* Mannheim: Wissenschaftsverlag, 145–57.

——. (2001). "The Axiomatic Method and the Foundations of Science: Historical Roots of Mathematical Physics in Göttingen", in M. Redei and M. Stöltzner (eds.), *John von Neumann and the Foundations of Quantum Physics.* Dordrecht: Kluwer Academic Publishers, 11–33.

Mehra, Jagdish (1974). *Einstein, Hilbert and the Theory of Gravitation.* Dordrecht: Reidel.

Mie, Gustav (1917). *"Die Einsteinsche Gravitationstheorie und das Probleme der Materie"*, *Physikalische Zeitschrift* 18, 574–80; 596–602.

Pais, Abraham (1983). *'Subtle is the Lord...' The Science and Life of Albert Einstein.* New York: Oxford University Press.

Pauli, Wolfgang, Jr. (1921). *Relativitätstheorie.* In Arnold Sommerfeld (ed.), *Encyklopädie der mathematischen Wissenschaften, mit Einschluss ihrer Anwedungen.* Leipzig: B.G. Teubner, 539–775. Translated as *The Theory of Relativity.* Oxford: Pergamon Press, 1958.

Peckhaus, Volker (1990). *Hilbertprogramm und Kritische Philosophie.* Göttingen: Vandenhoeck & Ruprecht.

Renn, Jürgen and Matthias Schemmel (eds.). (2007). *The Genesis of General Relativity*, vol. 4. *Gravitation in the Twilight of Classical Physics: The Promise of Mathematics.* Dordrecht: Springer.

Renn, Jürgen and John Stachel (1999). *Hilbert's Foundation of Physics: From a Theory of Everything to a Constituent of General Relativity.* Berlin: Max-Planck-Institut für Wissenschaftsgeschichte, Preprint 118. Reprinted in Renn and Schemmel (2007), 857–973.

Rowe, David (2001). "Einstein Meets Hilbert: At the Crossroads of Mathematics and Physics", *Physics in Perspective* 3, 379–424.

Sauer, Tilman (1999). "The Relativity of Discovery", Hilbert's First Note on the Foundations of Physics", *Archive for History of Exact Sciences* 53, 529–75.

——. (2001). "The Relativity of Elaboration: Hilbert's Second Note on the Foundations of Physics", ms. dated September 10, 2001.

——. (2005). "Einstein Equations and Hilbert Action: What is Missing on p.8 of the Proofs for Hilbert's First Communication on the Foundations of Physics", *Archive for History of Exact Sciences* 59, 577–90.

9

Not Only Because of Theory: Dyson, Eddington, and the Competing Myths of the 1919 Eclipse Expedition

Daniel Kennefick

University of Arkansas, USA

9.1 Introduction

One of the most celebrated physics experiments of the twentieth century, a century of many great breakthroughs in physics, took place on May 29th, 1919, in two remote equatorial locations. One was the town of Sobral in northern Brazil, the other the island of Principe off the west coast of Africa. The experiment in question concerned the problem of whether light rays are deflected by gravitational forces, and took the form of astrometric observations of the positions of stars near the Sun during a total solar eclipse. The expedition to observe the eclipse proved to be one of those infrequent, but recurring, moments when astronomical observations have overthrown the foundations of physics. In this case it helped replace Newton's Law of Gravity with Einstein's theory of General Relativity as the generally accepted fundamental theory of gravity. It also became, almost immediately, one of those uncommon occasions when a scientific endeavor captures and holds the attention of the public throughout the world.

In recent decades, however, questions have been raised about possible bias and poor judgment in the analysis of the data taken on that famous day. It has been alleged that the best-known astronomer involved in the expedition, Arthur Stanley Eddington, was so sure beforehand that the results would vindicate Einstein's theory that, for unjustifiable reasons, he threw out some of the data which did not agree with his preconceptions. This story, that there was something scientifically fishy about one of the most famous examples of an *experimentum crucis* in the history of science, has now become well known, both amongst scientists and laypeople interested in science.

Yet this story has hardly ever been itself subjected to a close examination. It is the contention of this essay that there are no grounds whatsoever for believing that personal bias played any sinister role in the analysis of the eclipse data. Furthermore, there are excellent grounds for believing that the central contention made by the expedition's scientists (including Eddington), namely, that the results were roughly consistent with the prediction of Einstein's theory of General Relativity and firmly ruled out the only other theoretically predicted values (including the so-called "Newtonian" result), was indeed justified by the observations taken.

9.1.1 Overview

The basic outline of the story that I wish to rebut, as with most compelling narratives, is simple. Arthur Stanley Eddington fervently wished for a confirmation of general relativity for two reasons. He was a firm believer in and advocate of the theory, and was utterly convinced that the prediction of light-bending it made was true. He was a pacifist and war resister who earnestly sought postwar reconciliation between scientists in Britain and Germany, and saw the confirmation by a British expedition of the theory of Germany's leading scientist as a heaven-sent opportunity to further this goal. The consequence of his theory-led attitude to the experiment, coupled with his strong political motivation, was that he over-interpreted the data to favor Einstein's theory over Newton's when in fact the data supported no such strong construction. Specifically it is alleged that a sort of data fudging took place when Eddington decided to reject the plates taken by the one instrument (the Greenwich Observatory's Astrographic lens, used at Sobral), whose results tended to support the alternative "Newtonian" prediction of light-bending. Instead the data from the inferior (because of cloud cover) plates taken by Eddington himself at Principe and from the inferior (because of a reduced field of view) 4-inch lens used at Sobral were promoted as confirming the theory. Furthermore, Eddington employed a brilliant, if perhaps somewhat misleading, public relations campaign to stampede scientists and the public into accepting his thesis that the somewhat flimsy and suspect data he had obtained amounted to an epochal contribution to science, encompassing a complete overthrow of the Newtonian world system and its replacement by another (Sponsel 2002).

Those who believe that there is no smoke without fire will not be surprised to hear that nearly every factual statement in the preceding narrative, taken in isolation, is basically true, though there are, of course, caveats. However, once the whole story is fully constructed from such documentation as has survived, I believe it is easily seen that the overall picture presented is completely wrong. Specifically there is no direct link, nor does it seem that one can draw a link, between Eddington's self-admitted predisposition to believe the theory and the story of how the critical data came to be selected as it was.

Let us begin with a few points to restore some balance to the picture of Eddington as a master manipulator. Keep in mind that pacifists and draft resisters like Eddington were a tiny and frequently despised minority during the First World War. Many went to jail, as Eddington was apparently prepared to do himself (Chandrasekhar 1976, 250–251). Within the scientific community in Allied countries there were few voices willing to be heard in favor of postwar reconciliation in 1919. At the time that Eddington was on Principe the final ratification of the Versailles treaty by the German parliament and government was accompanied by unrest in the streets of Einstein's Berlin. The bitterness left by the war was such that even Einstein and Eddington, two of the war's more prominent and steadfast opponents, feared to write to each other lest they give offence to a former "enemy," until after initial attempts to bring them together by scientists in neutral Holland. When, as a result of the eclipse expedition, Einstein was almost awarded the Gold Medal of the Royal Astronomical Society for 1919/20, a heavy backlash amongst members of the Society prevented the award being made at all for that year, in spite of Eddington's best efforts to prevent this.

Not only was Eddington swimming against the tide politically, his view that general relativity was correct placed him in a small minority within the astronomy community (and probably also, but to a less extreme extent, within the physics community). Most astronomers were skeptical of, or frankly hostile to, this theory, insofar as they understood it at all, though Einstein's great reputation and the gradual acceptance which special relativity had achieved demanded that the theory be accorded a very respectful reception. The narrative outlined above demands not only that Eddington achieve a monumental impact on the public and scientific mind, but that he do so from a position of profound weakness. Therefore we should proceed with caution in imagining that this miracle could be accomplished entirely through the oratorical brilliance, public-relations savvy, and scientific prestige of one (or two) individuals, no matter how great.

The usual view is that Eddington hardly thought the experiment worth performing (Chandrasekhar 1976, 250), as he was so sure of the result. There is, indeed, the famous anecdote, told by Eddington against himself, which encapsulates his theory-led outlook:

> As the problem then presented itself to us, there were three possibilities. There might be no deflection at all; that is to say, light might not be subject to gravitation. There might be a 'half-deflection', signifying that light was subject to gravitation, as Newton had suggested, and obeyed the simple Newtonian law. Or there might be a 'full deflection', confirming Einstein's instead of Newton's law. I remember Dyson explaining all of this to my companion Cottingham, who gathered the main idea that the bigger the result, the more exciting it would be. 'What will it mean if we get double the deflection?' 'Then,' said Dyson, 'Eddington will go mad, and you will have to come home alone.' (Chandrasekhar 1976, 250)

One observes, in contradistinction to this, that Eddington was conscious of the possibility of an unexpected result as the date of the eclipse drew near, as witnessed by his closing remarks in an article describing the forthcoming expedition.

> It is superfluous to dwell on the uncertainties which beset eclipse observers; the chance of unfavourable weather is the chief but by no means the only apprehension. Nor can we ignore the possibility that some unknown cause of complication will obscure the plain answer to the question propounded. But, if a plain answer is obtained, it is bound to be of great interest. I have sometimes wondered what must have been the feelings of Prof. Michelson when his wonderfully designed experiment failed to detect the expected signs of our velocity through the aether. It seemed that that elusive quantity was bound to be caught at last; but the result was null. Yet now we can

see that a positive result would have been a very tame conclusion; and the negative result has started a new stream of knowledge revolutionizing the fundamental concepts of physics. A null result is not necessarily a failure. The present eclipse expeditions may for the first time demonstrate the weight of light; or they may confirm Einstein's weird theory of non-Euclidean space or they may lead to a result of yet more far-reaching consequences—no deflection. (Eddington 1919, 122)

It is still noteworthy how theory-led Eddington's viewpoint is. Besides the null result he admits of only two possibilities, the two theoretical predictions made by Einstein at different stages of the development of his theory. I think Eddington was conscious all along of the limitations of the experiment. He and his colleagues would have been utterly over-ambitious to have made very high claims of precision on the order of a repeatable experiment such as Michelson's. Eddington had already shown himself prepared to cast aside Einstein's theory if another one, which might be more in accord with measurement, could be found, as shown by his interest in the unified field theory of the Swiss mathematician Hermann Weyl, at a time when Einstein's theory seemed likely to fail the Solar redshift test (Einstein's theory predicted that light emitted by atoms on the Sun should be shifted or displaced towards the red end of the spectrum relative to light emitted by the same kind of atoms on Earth, because of the stronger gravitational field of the Sun). On December 16, 1918, only a few months before the planned expedition, Eddington had written to Weyl:

One reason for my interest [in your paper] is that it seems to me to reopen the whole question of the displacement of Fraunhofer lines, leaving the theoretical prediction unsettled. (Perhaps you will differ from me as to this). [Charles E.] St. John and [John] Evershed seem to be quite decided that experimental evidence is against the deflection, and this is rather a severe blow to those of us who are attracted by the relativity theory. I venture to think your theory may show a way out of the difficulty—but that is a guess. (ETH-Bibliothek Zurich, Hermann Weyl Nachlass, Hs 91:522)

What follows will largely take the form of a response to one of the contentions in a famous 1981 paper by John Earman and Clark Glymour, one of a landmark trio of papers on the three classical tests of General Relativity (Earman and Glymour 1981, and Earman and Janssen 1993). Earman and Glymour's papers raise a number of thoughtful points about the status of the light-bending test conducted during the 1919 eclipse, most important among them that Eddington's careful framing of the theory test as a showdown between Einstein's theory and an earlier prediction of Einstein's which Eddington labeled the Newtonian one, played a critical and, at one time, little noticed role in the acceptance of the theory. They also observe that it was the phenomenal success of Eddington's campaign to win acceptance for the light-bending falsification of Newton's theory that prompted skeptics of the theory, such as the Solar Astrophysicist Charles St. John, to reverse the negative verdict on the third redshift test of the theory which they had previously pronounced. However, nothing in their paper had the impact of their suggestion that Eddington may have been motivated by bias in favor of Einstein in throwing out some of the data on dubious grounds. I suspect that the outline of this view (not the particulars) did not originate with Earman and Glymour but may have been circulating orally, in one form or another, amongst physicists for some time before the publication of their paper in 1980. Certainly the story has taken on a life of its own since their paper appeared. Sadly as time has gone by there has been little attempt to follow up on the fact that Earman and Glymour themselves offered up some evidence against the thesis that Eddington's personal bias played a significant role in the data analysis.

The trajectory of this story is also an interesting example of how a sufficiently compelling narrative, widely recounted, can quickly evolve from the carefully phrased version found in a scholarly article to a bare bones version found in popular discourse, stripped of all the caveats which originally modulated its dramatic elements. Earman and Glymour's article was used heavily in an account of the eclipse expedition in Harry Collins and Trevor Pinch's *The Golem*, a book that has been much more widely read than the original article. Collins and Pinch are still very careful in not relying on the claim that Eddington fudged any data, their argument being essentially that there is no such thing as a definitive experiment which resolves all doubts and proves one theory over another. This treatment in turn seems to have served as the principal source for a more popular book by John Waller (Waller 2002), whose aim is to debunk some of the well-worn anecdotes of scientific progress which have been scrutinized closely by modern historians of science.

When one finally gets to the reader reviews of Waller's book posted to amazon.com one sees all of the scholarly analysis pared down to the essentials. There it is stated[1] that:

> And we have pioneers like Robert Millikan and Arthur Eddington who made data fit a chosen theory, rather than the other way around. Yet, far from belittling such men, this book shows them in a new and more human light that transforms our understanding of scientific discovery.
> Remember learning in school how Eddington proved Einstein's theory of relativity by comparing the position of stars during and after an eclipse? Actually his images were so poor they proved precisely nothing except that Eddington was a dab hand at faking results. The book catalogues a series of famous scientists whose passion and belief in a theory blinded them to contrary evidence. In fascinating detail the book describes the circumstances surrounding the experiments both in the laboratory and in the wider social context. What links these scientists is that, as it turns out, the theories they were expounding happened to be right—just not for the reasons they gave. This compelling book should be compulsory reading for all students of science and is delightful food for thought for anyone interested in science.

---

[1] (Quoting from several different reviewers, from a version of the website
http://www.amazon.com/gp/product/customer-reviews/0192805673/sr=8-1/qid=
1182979869/ref=cm_cr_dp_all_helpful/104-5654845-3967159?ie=UTF8&n=
283155&qid=1182979869&sr=8-1#customerReviews, accessed on March 22, 2011.)

Thus two main points have come across in the journey from scholarly article, through increasingly popular (though scholarly written) books, to the *vox populi* of the web. They are, firstly, that Eddington fudged, faked, or fit his own results to the theory he believed to be true and, secondly, that one must never take a story at its own evaluation, no matter how plausible it may seem (in this case, the story that Eddington proved that General Relativity was true).

While I will argue that the specific issue of the first point is quite wrong, I certainly cannot fault the attached moral. Indeed it is true to say that the claim that Eddington's data was much dodgier than most people thought, itself probably arose as a reaction to the wrong-headed belief that the 1919 expedition had some-how "proved" General Relativity all by itself. While one can confidently argue that the eclipse measurements were not a very stringent test of relativity, this early and very compelling narrative has now spawned, in reaction, a distorted myth of its own according to which the eclipse results were poor grounds on which to overthrow Newton's theory and replace it with Einstein's. I will argue that, on the contrary, the eclipse results gave rather good grounds for believing that Einstein's theory was better than Newton's when dealing with the strong gravitational fields close to a massive body like the Sun.

9.1.2 Origins of the Expedition

To begin with a brief overview of the eclipse expeditions, Eddington was one of two leaders of the endeavor, the acknowledged senior man being Sir Frank Watson Dyson the Astronomer Royal and, in that capacity, director of Britain's leading observatory at Greenwich. It was Dyson who originally observed that the 1919 expedition would be uniquely suited to make this test, since the Sun would be in the star field of the Hyades, the closest open stellar cluster to the Earth (Dyson 1917). The expeditions were organized by the Joint Permanent Eclipse Committee of the Royal Society and the Royal Astronomical Society, a committee which Dyson chaired. Dyson was, at all times, the principal organizer and director of the two expeditions, each of which was staffed by one of two different observatories. Eddington, as director of the Cambridge Observatory, led the expedition to Principe Island off the west coast of Africa, accompanied by Edwin Turner Cottingham, a Northamptonshire clockmaker who maintained many of the instruments at Cambridge. The other expedition, posted to Sobral, in northern Brazil, was mounted by the Greenwich Observatory and con-sisted of Andrew Claude de la Cherois Crommelin, an assistant at Greenwich who hailed from what was about to become Northern Ireland, and Charles R. Davidson, an experienced computer at Greenwich. The expeditions were devoted exclusively to the test of the light-bending prediction. Although the instrumentation is central to the fudging thesis, I will not discuss it at great length, instead referring readers to Earman and Glymour's excellent article. It does not bear critically on my argument concerning Eddington's culpability and the question of the *motivation* for the alleged fudging. For the moment it suffices to say that Eddington brought one instrument with an "Astrographic lens," chosen for its wide field of view, and borrowed from the Oxford Observatory, and that the Greenwich team at Sobral had two instruments, one employing their own Astrographic lens, the other a 4-inch lens borrowed from the Royal Irish Academy, which may have been brought as a backup instrument because it, alone of the three instruments, had been used on eclipse expeditions previously. Because it is central to modern accusations of bias, I will focus mainly on the data taken with the Sobral Astrographic lens.

For ease of transportation, the astronomers decided not to bring along the telescopes with their complex mountings within which the lenses were normally mounted for observatory work. Instead, the lenses were placed into new tubes fitted with coelostat mirrors at one end. Rather than moving the whole instrument in order to maintain a fixed image of the stars as the Earth turned, only the coelostat mirror would be moved by clockwork during an exposure. After the expeditions Dyson concluded that the coelostat mirrors had performed poorly and should not be used on future expeditions to test the theory. Previous eclipse expeditions had been interested in studying the Sun (a coelostat is usually employed in solar observations), but the 1919 expedition wished to be able to take images of as many stars as possible in the general vicinity of the Sun. Therefore lenses with a wider field of view than those typically employed on eclipse expeditions were vital to the project. Earman and Glymour observed that the only recently concluded First World War greatly restricted the instruments available. The Greenwich observatory had still not recovered the instruments it had sent to a 1914 eclipse in Russia, which had been left behind as a result of the sudden outbreak of the war.

9.1.3 The Data Analysis

Eddington was concerned that a result be announced as soon as possible after the eclipse. Since it would take weeks to travel back to England, and since at least some time would have to be devoted to onsite preparations and the taking of check or comparison plates (about which more later) he wished that data reduction would begin in the field (Chandrasekhar 1976, 253). He urged that both expeditions should bring along a small micrometer to measure star positions. In the event, the Brazilian expedition does not appear to have made any attempts at data reduction until after its return (the probable reasons for this will become apparent later), despite remaining on site for two months in order to take comparison plates of the eclipse star field in the absence of Sunlight. Eddington, on the other hand, began making measurements at once, in spite of bad weather obscuring most of the stars in the field, and in spite of making a comparatively hasty departure for home. He wrote to his mother on board his return ship, giving a glimpse of his uncertainty about his data.

> We developed the photographs 2 ea. night for six nights after the eclipse, and I spent the whole day measuring. The cloudy weather spoilt my plans and I had to treat the measures in a different way from what I intended; consequently I have not been able to make any preliminary announcement of the result. But the one good plate that I measured gave a result agreeing with Einstein and I think I have got a little confirmation from a second plate. (Eddington to his mother, Sarah Ann Eddington, aboard SS. Zaire, June 21, 1919, Trinity College Cambridge Library, Eddington Papers, A4/9)

Eddington arrived home probably sometime in July and there is every reason to believe that he conducted the rest of the data reduction of his plates himself. On the other hand, with only a handful of stars on a few plates to work with, owing to the cloud over Principe, he must have been anxious to learn what was on the Sobral plates, as he was aware that the other expedition had had better luck with the weather. The Sobral expedition arrived back in England on August 25, 1919 (Crommelin 1919b, 281). Reduction of the data on their plates probably began almost immediately. Worksheets documenting the measurements made are preserved in the Royal Greenwich Observatory archives (now housed in the University of Cambridge Library). The first page, dealing with the plates taken by the Astrographic lens, is headed "Total Solar Eclipse – 1919 May 28–29- Sobral – Astro No. 1" and dated September 2, 1919 (Cambridge University Library MS.RGO.8/150). Some of the sheets are initialed CD and HF, for Charles Davidson and Herbert Henry Furner, both longtime computers at Greenwich. The following year, Dyson commented, in a letter to the American Geologist Louis Agricola Bauer (1865–1932) on July 1, 1920 (Cambridge University Library MS.RGO.8/147).

Dear Prof. Bauer,
Your long list of "errata" rather alarmed me, though I could not believe that any serious error had been made in the reduction of the "Einstein' photographs, as both Davidson and I have dealt with some thousands of astronomical photographs in very similar fashion, in fact 'almost' identical fashion except for the inclusion of the term α giving the displacement. [The displacement is that due to gravitational light-bending. Note also that the word almost has been deleted by Dyson in the preceding quote.]

This suggests that Dyson was intimately involved in the reduction process, and presumably took the lead in it. Certainly some of the key pages in it appear to be written in his hand. The data reduction proceeded throughout the month of September, while in Germany Einstein himself waited with bated breath (he wrote to his close friend Paul Ehrenfest in Holland on September 12 to check whether the Dutch scientists, having closer contacts with their English colleagues, had received any news [Einstein 2004, Doc. 103, 154]).

On September 12 Eddington and Cottingham spoke before the British Association for the Advancement of Science at Bournemouth, discussing only briefly the significance of their endeavors for relativity theory. They focussed instead on an interesting sidelight of their observations, the enormous solar prominence shown very clearly on the Principe plates and still today the most recognizable feature of the 1919 eclipse. All Eddington had to say concerning the actual results was described in the report on the conference as follows,

Professor Eddington gave an account of the observations which had been made at Principe during the solar eclipse. The main object in view was to observe the displacement (if any) of stars, the light from which passed through the gravitational field of the sun. To establish the existence of such an effect and the determination of its magnitude gives, as is well known, a crucial test of the theory of gravitation enunciated by Einstein. Professor Eddington explained that the observation had been partially vitiated by the presence of clouds, but the plates already measured indicated the existence of a deflection intermediate between the two theoretically possible values 0.87″ and 1.75″. He hoped that when the measurements were completed the latter figure would prove to be verified. Incidentally Professor Eddington pointed out that the presence of clouds had resulted in a solar prominence being photographed and its history followed in some detail; some very striking photographs were shown. (Eddington and Cottingham 1920, 156)

These photographs showing the prominence most clearly were precisely those, taken through the thicker cloud at the beginning of the eclipse, which were utterly useless for testing Einstein's theory.

Thus Eddington gave nothing away, if indeed he had anything to give away, about the state of the Sobral data reduction, but confined himself to his own plates and cagily committed himself only to a value between the two theoretical predictions which Einstein had made. The first, due to Einstein and dating from 1910, was based merely on the principle of equivalence and the old special relativity theory. Eddington had labeled this the "Newtonian" prediction. The second, dated after Einstein finalized his theory in 1915, derived from the complete theory of general relativity, and differed from the earlier prediction because it took account of the curvature of spacetime in a gravitational field, which is a well-known feature of that theory.

The result from Principe, as it was later published, is much closer to the larger relativistically "correct" 1915 value, but in view of the limitations of his data Eddington was understandably unwilling to claim too much in the way of accuracy in advance of the Sobral results. Nevertheless he was confident at this stage that light-bending, of some magnitude, was an established fact. He had managed to weigh light. It only remained to check whether spacetime was curved.

On or before (almost certainly just before) October 3, Eddington received the long-awaited news from Dyson concerning the reduction of the final plates from Sobral, those from the 4-inch lens, which he refers to as the Cortie plates. These plates are named after Father Aloysius Laurence Cortie, SJ (1859–1926), a Jesuit astronomer who had originally been intended for the party at Sobral, but had to be replaced by Crommelin when his duties prevented him from traveling. The 4-inch lens was loaned to the expedition through the good offices of Father Cortie. Eddington replied to Dyson.

Dear Dyson,
I was very glad to have your letter & measures. I am glad the Cortie plates gave the full deflection not only because of theory, but because I had been worrying over the Principe plates and could not see any possible way of reconciling them with the half deflection.
I thought perhaps I had been rash in adopting my scale from few measures. I have now completed my definite determination of A (5 different Principe v. 5 different Oxford plates), it is not greatly different from the provisional

though it reduces my values of the deflection a little. (Eddington to Dyson, October 3, 1919, Cambridge University Library MS.RGO.8/150)

This suggests that the Sobral Astrographic data was reduced first and that Eddington had been informed that the value derived was rather close to Einstein's prediction of 1910, which Eddington did not expect to be confirmed. This had clearly worried Eddington as he felt his own data tended to support the higher result predicted by general relativity. Indeed he seems to have doubted that his results could be made to agree with the lower figure, despite confessing to some efforts in that direction. Thus he greets the news that the Sobral 4-inch results strongly favor the higher value with relief.

There is much here that can be made to fit the model of Eddington as a partial experimenter, but there is one important detail that throws cold water on the overall data-fudging narrative outlined above. That is that this letter strongly implies that Eddington was not involved in the reduction of the Sobral data. The tenor of this opening paragraph indicates that he is receiving his first news of the 4-inch data reduction by letter, and there is no suggestion that he had any earlier input, or even any prior information beyond the cumulative results from the Astrographic instrument. Indeed since receiving that earlier information he has occupied himself not with any analysis of data from Sobral but with a reanalysis of his own data to see whether it can be made to fit. Coupled with Dyson's comment, quoted above, that he himself was responsible for the Sobral data, along with Davidson, we can conclude that the Sobral data reduction was conducted independently of Eddington. The initials on the data sheets, the fact that the reduction was undoubtedly done at Greenwich and not at Cambridge, where Eddington was, and the fact that the eventual report was written with Dyson solely responsible for discussion of the Sobral data and Eddington for Principe, all reinforce this impression.

It seems likely that Eddington was never present at Greenwich during the Sobral data reduction. In a letter to Dyson on October 21, 1919 (Cambridge University Library MS.RGO.8/150) he refers to having acquired a season railway ticket, rather suggesting he had not been traveling up to see Dyson in the preceding weeks or months. We can be fairly confident that Eddington simply was not privy to the reduction of the Sobral data or to the crucial decision, recorded in the data sheets (in what appears to be Dyson's hand), to reject the Astrographic data and accept the 4-inch data as "the result of the Sobral expedition" (Cambridge University Library MS.RGO.8/150). The eclipse expedition appears to be a reasonably good case of independence being preserved between the two wings of the collaboration.

If Eddington did not make the decision to exclude the Astrographic data, it is clear that it must have been Dyson who did so. It therefore turns out to be largely irrelevant for the central data-fudging claim, what Eddington's motivations were. We must ask instead, what was Dyson's attitude to war, peace, and relativity? Can we claim that he was biased, or overly influenced by Eddington, a younger but more brilliant colleague?

To get a flavor of a viewpoint which was much more common than Eddington's, let me quote from a letter written by Rudolf Moritz (1878–1940), a London barrister and astronomy enthusiast who had made a special study of relativity theory, to Phillip Herbert Cowell (1870–1950), a former assistant at Greenwich who had a reputation as a formidable computer. He was well known for using his calculational skills (in collaboration with Crommelin) to predict the date of reappearance of Halley's comet in 1910. Cowell had apparently asked Moritz for an explanation of relativity theory (special and general), and Moritz' reply (dated March 1, 1918) is preserved in the Greenwich manuscript archive (Cambridge University Library MS.RGO.8/123).

> So much for the first theory of relativity. I can follow it all analytically and physically and I believe it is true. The second theory of Einstein in 1914 is far more speculative and I fear only accord with observations will make me accept it. Besides the analysis is too beastly for words. I can well understand the compatriots of Riemann and Christoffel burning Louvain and sinking the Lusitania.

In other words, the atrocity of inventing the tensor-based mechanisms of differential geometry which underpin general relativity is quite on a par, morally speaking, with the most notorious (to Englishmen) war crimes of World War I.

In contrast with those of Moritz and Eddington, Dyson's views seem to have been moderate, and not unrepresentative, either for an astronomer regarding relativity or for an Englishman regarding reconciliation with Germany (though Moritz' views may have been the most representative of English opinion, at least while the war still progressed). That is to say, I suspect he was skeptical but not intransigent on both counts. We have good evidence for the former, and much more important, issue, concerning his views on relativity. Earman and Glymour have already noted that Dyson was skeptical of the theory, that he "thought it too good to be true" (Earman and Glymour 1981, 85). After the announcement of the eclipse results Dyson sent copies of plates from the 4-inch to a number of leading astronomers, several of whom replied with polite words about the clarity of the images, even though most of them were at least privately not well disposed towards general relativity. In replying to one correspondent, Frank Schlesinger of the Yale observatory, Dyson writes on March 18, 1920:

> We are planning to send an expedition to Christmas Island in 1922; & I hope it may be possible to send one to the Maldives; & that the Australians may do something. Is it likely that there will be an American Expedition? I hope so, in view of the importance of having the point thoroughly settled. The result was contrary to my expectations, but since we obtained it I have tried to understand the Relativity business, & it is certainly very comprehensive, though elusive and difficult. (Cambridge University Library MS.RGO.8/123)

We have confirmation of this from Eddington, who wrote, on August 18, 1920, to the mathematician Hermann Weyl,

> It was Dyson's enthusiasm that got the eclipse expeditions ready to start in spite of very great difficulties. He was at that time very skeptical about the

theory though deeply interested in it; and he realized its very great importance. (ETH-Bibliothek Zurich, Hermann Weyl Nachlass, Hs 91:523)

We do know that Dyson held liberal views on the desirability of postwar reconciliation with Germany. Significantly, his obituary in the Observatory says

after the Great War, when international co-operation in science had lapsed to a considerable extent, Dyson played a prominent part in the reconstitution of international scientific co-operation through the International Research Council (now the International Council of Scientific Unions) and in the formation of the International Astronomical Union... Dyson took an important part in the initial deliberations that resulted in the formation of the Union, which owes much to his wise guidance. (Jones 1939, 186)

Nevertheless, I doubt that he was a crusader for the cause of pacifism, as Eddington was. It is worth observing that the International Astronomical Union, at its birth, did not permit Germany and its allies to enjoy membership in the new body, as is pointed out by Stanley, who places Eddington's internationalist views in context, as representative of a small minority within British society (Stanley 2003, 88). A newspaper clipping preserved in the Greenwich Observatory archives actually discusses a German Zeppelin raid which caused some damage to the Observatory, and denounces an attempt in the German press to justify the targeting of a scientific institution on the grounds that Greenwich, through its meteorological observations, was directly participating in the British war effort.

It seems to me likely that Dyson's views on reconciliation were fairly typical of liberal Englishmen, and very distinct from the radical pacifism of either Eddington or Einstein. It is highly implausible that Dyson would have deliberately fudged data in order to support a theory of which he was skeptical, or to advance an anti-war cause. Nor is it particularly credible that Britain's leading astronomer would have been so eager to please Eddington as to risk his reputation by throwing out data just to please his collaborator. Whatever one thinks of the scientific merit of the work of the 1919 expedition, one must discard the compelling claim that the Greenwich team were motivated by a transparent "political" (in whatever sense of the term) agenda on the part of the experimenters. This conclusion accords well with the obvious fact that in the written report Dyson and Eddington made no attempt to gloss over the discarded data. They gave a full and clear account of it, along with the reasons for throwing it out. Even if one disagrees with their arguments and conclusions, they did not behave as if they had evidence of a conscious fraud to conceal.

Of course in any scientific endeavor there comes the moment at which the scientist becomes convinced of the correct result, after which he or she tends to become a partisan for that result, whatever their original bias. So it is amusing that, in his first draft of the report from Sobral, Dyson endeavored to average the results from the two instruments there (which were $1''.98$ deflection at the limb of the Sun for the 4-inch and $0''.93$ for the Astrographic), having noticed that their average was very close to the Einstein prediction. In his draft of the paper he states: "The mean with these weights is 1."83 and is very close to the value required by Einstein's theory" (Manuscript of report, Cambridge University Library MS.RGO.8/150).

Eddington objected:

> I do not like the combination of the astrographic with the other Sobral results—particularly because it makes the mean come so near the truth. I do not think it can be justified; the probable errors of both are I think below $0''.1$ so they are manifestly discordant. If the results are accepted with the weights assigned, the probable error of the mean (judged from their accordance) is about $\pm 0''.20$, which certainly does not seem to do justice to the results obtained. I would like to omit the last 5 lines of p. 4. It seems arbitrary to combine a result which definitely disagrees with a result which agrees and so obtain still better agreement. (Eddington to Dyson, October 21, 1919, Cambridge University Library MS.RGO.8/150)

There remain many issues raised by Earman and Glymour and others which I will not address, such as the role which Eddington's personal and scientific views did play (especially in his own reduction of the Principe data) and the matter of Eddington's and Dyson's presentation of their results to the public. But one may safely conclude, on the basis of the documentary evidence, that there are no grounds to believe that the most critical decision of the eclipse data reduction was taken from a biased standpoint. Whether that decision was right or wrong, and whether it had an unwarranted influence in the reception of one of the leading physical theories of the twentieth century, are of course quite separate questions, but I will offer some arguments, in the next section suggesting that the decision in favor of Einstein over Newtonian gravity was justified.

9.1.4 A Problem of Scale

The method of comparing positions of stars on the plates was substantially similar for both expeditions. Plates taken during the eclipse were clipped together with plates taken in nighttime conditions, so that the star positions were as close as possible to each other. One of the two plates would be a reversed image (taken with the use of a mirror as opposed to a lens) so that the images could be compared face to face in this way. A micrometer was then used to measure the separation between the positions of identical stars on the two plates. In this way it would be known how far star A on one of the eclipse plates was positioned away from the same star A on a comparison plate.

In practice the Greenwich team faced the difficulty that both their eclipse and comparison plates were reversed (by the use of the coelostat mirrors in their instruments), and so could not be compared face to face. They made use of a third plate, which they called a scale plate, specially taken of the same field, but not reversed (i.e., direct), which was placed against each of the eclipse and comparison plates in turn. The Cambridge team's comparison plates were taken using a telescope at Oxford (the astrographic lens used on Principe was loaned to the expedition by the Oxford

Observatory), and were thus direct. This permitted them to be placed face-to-face with the reversed eclipse plates taken via the coelostat mirror on Principe.

At Greenwich the measurements of the Sobral data were made by Charles Davidson, one of the two Sobral Observers, and Herbert Henry Furner, like Davidson a long-time computer at Greenwich. Both had been "established computers" there since the mid-1890s. Furner was first appointed an established computer in 1897, having been a supernumerary computer since 1889, according to his obituary (Melotte 1953). Davidson had been an established computer for about the same length of time (both men are listed as such in the Annual Report of the Astronomer Royal in 1900, as given in Maunder 1900, Chapter 5). The data reduction at Greenwich was done under Dyson's overall direction, by his own testimony (Dyson *et al.* 1920).

The Greenwich team had been able to take comparison plates of the eclipse star field while still in Brazil. This was possible at Sobral, where the eclipse took place in the morning with the Sun relatively low in the sky. Two months later the Sun had moved far enough away from the Hyades so that the star field had risen almost to the same altitude it had during the eclipse before the Sun itself rose in the morning. This permitted nighttime plates of the field to be taken with the same instrument, at the same location. Although the Greenwich team had originally entertained the possibility of leaving without waiting long enough to take comparison plates in situ, they decided it would be best not to do so after experiencing problems with astigmatism in the coelostat mirror used with their astrographic lens (Dyson *et al.* 1920, 298).

Taking comparison plates on Principe (where the eclipse took place in the afternoon) would have required the Cambridge team to remain on the island for half a year, a considerable inconvenience for any European scientist. This was perhaps especially true of Eddington, who told his mother he was anxious to be back in England before the end of the strawberry season, writing "I suppose I shall be back about July 10 & shall look forward to the strawberries, which are better than anything they have in the tropics" (Arthur Stanley Eddington to his mother, 21 June 1919 from S. S. Zaire, preserved in Trinity College, Cambridge Library, Eddington Papers, A4/9). The Sobral data thus more easily facilitated a direct comparison between the stars on one of the eclipse plates, with the stars on a comparison plate, whereas Eddington's task was somewhat complicated by the fact that the images were taken by a different instrument from a completely different place on the Earth's surface. Therefore Eddington also took images of a different star field at both Oxford and Principe, so he could, by comparison of those plates, make sure there were no systematic differences between the Oxford and Principe eclipse field.

Even so, the task of the Greenwich team did not simply consist in measuring the displacement of the star images between the comparison and eclipse plates and concluding that the resulting raw data was the Einstein displacement. Regardless of the amount of light deflection found, there could be additional differences in star positions between the two plates due to three different kinds of misalignment between them. The first would be whether the centers (more generally the origin) of each plate coincided when clamped together. The second concerned the relative orientation of the two plates, either because of a rotation of the instrument, or simply because of the way the plates were clipped together. Finally there might have been a change of scale on the plates, due to some change in the focus or other property of the instrument between the exposures.

If the star field was photographed at a different altitude in the sky, then there would be differences in stellar positions on the plate due to differential refraction in the Earth's atmosphere. Differential refraction refers to the fact that images at a different altitude experience differing amounts of refraction by the Earth's atmosphere, thus causing relative shifts in position (for this reason the Sun's image is never precisely circular except when it is at the zenith). Even if everything else about the two plates was identical, the lapse of time between taking eclipse and comparison plates (two months in the case of the Sobral expedition) meant that the Earth was moving in a different direction, relative to the direction to the star field, thus creating differences in stellar aberration between the two plates. Stellar aberration refers to a shift in the apparent position of stars due to the relative motion of the Earth compared to the line of sight towards the star. These last two kinds of changes could be calculated theoretically. Both teams carried out these calculations during their data analysis, but other changes in scale and orientation between the plates, if they occurred, could not be predicted in advance, but had to be measured.

Fortunately, changes in the plate position and orientation must, in principle, behave in a way which is characteristically different from the purely radial displacement predicted by Einstein for the light-bending effect. Furthermore, they could be minimized by careful handling of the plates when clipping them together. The most important of the three changes which had to be determined in order to convert raw measurements from the plates into light-bending results, was the difference in *scale* between the two plates, because a change in scale between the plates could mimic the actual light-deflection displacement. The scale could, however, be distinguished from the light-bending deflection, as pointed out by Eddington himself (Eddington 1919, 120), because, if measurements were taken with respect to the position of the Sun at the center of each plate, then the light-bending deflection would be greatest for those stars nearest the Sun, whereas the shift in position due to a scale change would be greatest for those stars farthest away from the origin. Of course, distinguishing scale from light-bending deflection in this manner would require measurement of a variety of stars at different distances from the Sun. Having an insufficient number of stars to measure would make it impossible to distinguish almost any change in the plates from the light deflection. For instance, in the case of the Principe plates, there were so few stars visible that even the orientation could become difficult to distinguish from the deflection, thus causing some plates to be unusable (Dyson *et al.* 1920, 321).

In the jargon of the field, there were six plate constants to be solved for each plate pair (three kinds of possible misalignment times two co-ordinates on each plate). The method used was to set up equations that compared the measured displacements between stars on the eclipse and comparison plates to equations based on the various plate constants and the sought-for displacement. Then over-determination of the plate constants was employed (there are only half-a-dozen plate constants, plus the light-bending factor itself, and seven or more stars exposed on nearly all plates at Sobral) to derive values for all of the plate constants, including the scale factor and the light-bending factor. If no other changes of scale existed, calculation of the theoretical differential refraction and aberration could remove two of the plate constants and thus permit a better determination of the remaining constants.

As is well known, there were problems with the images taken with the Astrographic lens, because, as was noted by the observers on site, it had lost focus during the eclipse (possibly as a result of the change in temperature common during total eclipses). In addition, Eddington claimed that the cloud at Principe might have been actually beneficial in blocking light from the brighter stars, thus making their images on his plates smaller than those on the Sobral plates (Dyson *et al.* 1920, 329). Especially when coupled with smearing or streaking of images, overexposure of a plate may result in star images whose size and shape make measurement of the principal star positions, judged from the center of each star image on those plates, problematic. However, it is not clear that Eddington actually verified whether there really was any overexposure in the Sobral plates.

In the end the Greenwich team decided to measure the star positions on the Astrographic plates only in right ascension, and not in declination (i.e., measuring in celestial longitude, east–west, only and not in celestial latitude, or north–south). The positions of the stars relative to the Sun meant that considerably more of the light-bending effect would be measurable in right ascension than in declination. Given that the data on the Astrographic plates was considered "noisy" it seemed ill advised to try to measure in a coordinate in which the sought-for effect (the "signal") would be smaller compared to the noise. However, as we shall see, this may have resulted in an inaccurate determination of the scale, since so much information about the scale constant was lost by excluding the direction in which the scale would have been more visible, because it would have been larger relative to the light-bending displacement. An inaccurate determination of the scale would obviously result in an inaccurate determination of the Einstein displacement, since the two effects are not very dissimilar.

In the end the result obtained from the Sobral Astrographic plates was discordant with the results obtained from the other two instruments (the 4-inch at Sobral and the Principe Astrographic). It is the particular contention of several modern critics that the decision to discount the results from this instrument must have been largely taken because the result was also discordant with the prediction of General Relativity. I have argued, however, that this seems highly unlikely for the team of astronomers at Greenwich, who were far from being convinced that this theory was correct. The decision to measure only in one coordinate for these plates is clear evidence that the Greenwich team was unhappy with the quality of the data they contained. Indeed, problems had been noted at the time of the eclipse itself. This is noted in the report itself (Dyson *et al.* 1920, 309) and also more or less contemporaneously with the first news from the eclipse, in an account given by Dyson to the Royal Astronomical Society at its June meeting (Fowler 1919).

The Greenwich team went further, however, as we learn from their report. They tried different methods of reducing the data in an effort to better understand it and assess its reliability (for instance a least-squares method for finding the plate orientation, which did not substantially alter the result gained from their preferred method of calculating the plate constants using only the brighter stars on the plates, not surprisingly since it adopted the same average change of scale). The most revealing comment made in the report reads as follows:

> The means of the 16 photographs [taken with the Astrographic lens] treated in this manner [i.e., solving for all plate constants, including scale, and the displacement, from the same data] give
> $\alpha + 243e = +0^r.0435$ [where α is the light-bending displacement and e is the change of scale between the two plates and the angular displacement is given in number of revolutions of the micrometer screw on the plate measuring device] or with the value of the scale $+0^r.082$ from the previous table $\alpha = +0^r.024 = +0''.93$ at the limb.
> It may be noticed that the change of scale arising from difference of refraction and aberration is $+0^r.020$. If this value of e be taken instead of $+0^r.082$ we obtain $\alpha = +0^r.039 = +1''.52$ at the sun's limb. (Dyson *et al.* 1920, 312)

This value is much closer to that recovered from the other two instruments (deflection of $+1.''61$ on Principe, $+1.''98$ from the Sobral 4-inch). It certainly suggests the possibility that the reason for the discrepancy is an inaccurate determination (and exaggeration) of the change of scale undergone by the Astrographic instrument. This alternative result appears nowhere else in the report, but the mere fact that it is mentioned suggests that the author (which means Dyson for this section of the report, written in his hand in the manuscript of the report [Cambridge University Library MS.RGO.8/150]) attached some significance to it. Furthermore, an earlier comment in the paper may be significant in this context.

> These changes [in the focus of the Astrographic lens during the eclipse] must be attributed to the effect of the sun's heat on the mirror, but it is difficult to say whether this caused a real change of scale in the resulting photographs or merely blurred the images. (Dyson *et al.* 1920, 309)

A straightforward interpretation would be that Dyson suspected that the scale value was not accurately determined from the Astrographic data, and he was therefore justified in ignoring any result derived from that data. But what about the possibility that he suspected the $1''.52$ deflection at the limb was the true result from that data?

There is another published mention of this $1''.52$ deflection. It is not in the report itself, but instead in a single-page account of the report as it was given orally to the famous joint meeting of the Royal Society and the Royal Astronomical Society, published by Crommelin in *Nature*. He writes:

> This instrument [the Astrographic] supports the Newtonian shift, the element of which is $0.87''$ at the limb. There is one mode of treatment by which the result comes out in better accord with those of the other instruments. Making the assumption that the bad focus did not alter the scale, and deducing this [scale], from the July plates, the value of the shift becomes $1.52''$. (Crommelin 1919b, 281)

It is noteworthy that the apparently throwaway character of the remark in the published report (two sentences out of over forty pages) is contradicted by Crommelin's decision to devote two sentences to it in the mere page available to him in *Nature*. It is especially interesting that both mentions imply that this method of deriving the deflection is very similar, in important respects, to that employed by Eddington in his reduction of the Principe data, which I will now describe in outline.

According to the published report (Dyson *et al.* 1920, 317) Eddington was solely responsible for the reduction of the Principe data, what there was of it. It will be recalled that his comparison plates were taken in Oxford with a different instrumental setup, making a direct plate-to-plate comparison potentially problematic. Therefore check plates of the starfield around Arcturus were taken with both instruments in both places. Although these were originally intended merely as a safeguard against systematic errors arising out of changes in both instrument and location, in the event his method was to take measurements on the check plates in order to calculate the difference in scale between the two instrumental setups. He could then *assume* that the same change in scale applied to those plates as applied to the eclipse field plates taken in both places. In Eddington's words, from the report:

> As events turned out the check plates were important for another purpose, viz., to determine the difference of scale at Oxford and Principe. As shown in the report of the Sobral expedition, it is not necessary to know the scale of the eclipse photographs, since the reductions can be arranged so as to eliminate the unknown scale. If, however, a trustworthy scale is known and used in the reductions, the equations for the deflection have considerably greater weight, and the result depends on the measurement of a larger displacement. On surveying the meagre material which the clouds permitted us to obtain, it was evident that we must adopt the latter course; and accordingly the first step was to obtain from the check plates a determination of the scale of the Principe photographs. (Dyson *et al.* 1920, 317).

The material available from Principe was meager indeed. Owing to the cloud, which began to clear just as the eclipse was ending, only two plates with five stars on each were usable. This was insufficient to allow all six plate constants to be determined along with the light-bending displacement, and barely sufficient, even if the data for those five stars was perfect (which was far from the case) to calculate four plate constants and the displacement given theoretical calculations of the differential refraction and stellar aberration. Thus, in Eddington's case the need for an independent determination of the change in scale was severe. But of course mere necessity does not provide any answer to the principal charge made by Earman and Glymour, that there was no justification for throwing out the results of the Greenwich Astrographic, while keeping the poor-quality data obtained on Principe by the Oxford Astrographic.

First, let us quote Eddington's own attempt to justify the inclusion of his data:

> ... our result [for the light-bending deflection at the limb of the Sun] may be written $1''.61 \pm 0''.30$.

It will be seen that the error deduced in this way from the residuals is considerably larger than at first seemed likely from the accordance of the four results. Nevertheless the accuracy seems sufficient to give a fairly trustworthy confirmation of **Einstein's** theory, and to render the half-deflection at least very improbable.

It remains to consider the question of systematic error. The results obtained with a similar instrument at Sobral are considered to be largely vitiated by systematic errors. What ground then have we—apart from the agreement with the far superior determination with the 4-inch lens at Sobral—for thinking that the present results are more trustworthy?

At first sight everything is in favour of the Sobral astrographic plates. There are 12 stars shown against 5, and the images though far from perfect are probably superior to the Principe images. The multiplicity of plates is less important, since it is mainly a question of systematic error. Against this must be set the fact that the five stars shown on plates W and X [from Principe] include all the most essential stars; stars 3 and 5 give the extreme range of deflection, and there is no great gain in including extra stars which play a passive part. Further, the gain of nearly two extra magnitudes at Sobral must have meant over-exposure for the brighter stars, which happen to be the really important ones and this would tend to accentuate systematic errors [because it is more difficult to tell where the true center of the star lies when it is over-exposed on the plate and thus has, in effect, become a large blob of emulsion], whilst rendering the defects of the images less easily recognized by the measurer. Perhaps, therefore the cloud was not so unkind to us after all.

Another important difference is made by the use of the extraneous determination of scale for the Principe reductions. Granting its validity, it reduces very considerably both accidental and systematic errors. The weight of the determination from the five stars with known scale is more than 50 percent greater than the weight from the 12 stars with unknown scale. Its effect as regards systematic error may be seen as follow. Knowing the scale, the greatest relative deflection to be measured amounts to $1''.2$ on **Einstein's** theory; but if the scale is unknown and must be eliminated, this is reduced to $0''.67$. As we wish to distinguish between the full deflection and the half-deflection, we must take half these quantities. Evidently with poor images it is much more hopeful to look for a difference of $0''.6$ than for $0''.3$. It is, of course, impossible to assign any precise limit to the possible systematic error in interpretation of the images by the measurer; but we feel fairly confident that the former figure is well outside possibility. (Dyson *et al.* 1920, 328–329)

This last paragraph is a close paraphrase of the closing part of Eddington's letter to Dyson on October 3 (the first paragraph of which was quoted above). Evidently Eddington's first, and most critical audience, for his claims that the Principe result should be accepted as valid was his collaborator Dyson.

The key words in this part of the paper are "granting the validity" of the extraneous determination of scale. Should we, in hindsight, grant Eddington this validity? About this Eddington himself said:

The writer must confess to a change of view with regard to the desirability of using an extraneous determination of scale. In considering the programme it had seemed too risky a proceeding, and it was thought that a self-contained determination would receive more confidence. But this opinion has been modified by the very special circumstances at Principe and it is now difficult to see that any valid objection can be brought against the use of the scale.

The temperature at Principe was remarkably uniform and the extreme range probably did not exceed 4° during our visit—including day and night, warm season and cold season. The temperature ranged generally from $77^{1/2\circ}$ to $79^{1/2\circ}$ in the rainy season, and about 1° colder in the cool gravana. All the check plates and eclipse plates were taken within a degree of the same temperature, and there was, of course, no perceptible fall of temperature preceding totality. To avoid any alteration of scale in the daytime the telescope tube and object-glass were shaded from direct solar radiation by a canvas screen; but even this was scarcely necessary, for the clouds before totality provided a still more efficient screen, and the feeble rays which penetrated could not have done any mischief. A heating of the mirror by the sun's rays could scarcely have produced a true alteration of scale though it might have done harm by altering the definition; the cloud protected us from any trouble of this kind. At the Oxford end of the comparison the scale is evidently the same for both sets of plates, since they were both taken at night and intermingled as regards date.

It thus appears that the check plate is legitimately applicable to the eclipse plates. But the method may not be so satisfactory at future eclipses, since the particular circumstances at Principe are not likely to be reproduced. (Dyson *et al.* 1920, 329–330)

That the Greenwich team also underwent a change of heart in this respect is shown by their preparations for the next eclipse, of 1922. Before the expedition departed, Davidson wrote a paper arguing that the independent determination of scale was not only a superior method to that employed in the reduction of the Sobral data, but would be vital for the 1922 eclipse, which lacked the bright stars close to the Sun that had been a unique feature of the 1919 eclipse. It was, indeed, this happy coincidence which had convinced Dyson that the opportunity of testing Einstein's theory in 1919 could not be passed up (Dyson 1917).

In the Eclipse of 1919, the field of stars was unusually favourable for a determination of the Einstein gravitational displacement of light passing near the Sun—in fact, there is no other field on the Ecliptic with so many bright stars. In the Eclipse of 1922, if exposures are given of sufficient length to photograph faint stars near the Sun, there is grave danger of the images being drowned in the Corona. The brighter stars which are sure to be photographed are at such distances that the *differential* Einstein effect will be small, with consequent uncertainty in the result.

If, however, one had an independent determination of the scale of the photograph, then two stars, each situated at $1°$ distance on opposite sides of the Sun, will show an increase in distance of $0''.88$, a quantity readily measured on good photographs. A scale may be determined by photographs taken the night before or after on a comparison field, as was done by Prof. Eddington in Principe. To this it may be objected that different conditions hold between the day and night observations. (Davidson 1922, 224–225)

Many efforts were made in subsequent eclipses, not always successfully, to get an independent measure of the scale without having to assume that it would be the same during the eclipse as during a later nighttime exposure. Most popular was the taking of check plates actually during the eclipse, as advocated by Davidson in his paper, even though this required changing the pointing of the instrument towards a different star field.

Thus, for most subsequent eclipses, an independent determination of scale was employed, precisely because, as noted by Davidson, the smaller the observed effect when deprived of stars close to the Sun, the more danger there was in effectively halving the size of the effect to be sought by calculating the scale change from the same data used for the deflection.

It is important to keep in mind that the Sobral 4-inch plates remain highly unusual, even after many subsequent eclipse expeditions to test Einstein's theory. They are almost unique in being taken with a working instrument, in clear weather, with several bright stars relatively close to the Sun. Even so the experimenters were conscious of the difficulties of dealing with a possibly unknown change of scale. Much space is devoted to a discussion showing that the values adopted for the scale are consistent from one image to another across a plate (Dyson *et al.* 1920, 306–309).

In the case of the Principe plates only the presence of unusually bright stars permits any kind of measurement at all, given the cloudy conditions. The fact that, luckily enough, check plates were available, made it possible for Eddington to derive a result which he, at least, was reasonably happy with. I accept, however, that his admitted biases might have made him especially anxious to extract a result from data that another experimenter would have been tempted to discard altogether.

In the case of the Sobral Astrographic plates, the excellent conditions were compromised by the poor performance of the instrument. Although the Sobral plates show more stars than Eddington's at Principe, in practice only the five brightest are used for the determination of the deflection (Dyson *et al.* 1920, 310–311). Recall also that these star positions are measured only in one co-ordinate, thus halving the amount of information available to correctly determine the scale (on top of the "halving" of reliability claimed by Eddington as a consequence of not having an independent determination of the scale). In the circumstances something might have been done had check plates been available to permit an independent determination of the scale. But note that Eddington's use of the scale change calculated from his check plates presumes that the scale did not change between the time of the eclipse and the taking of check plates. Eddington claims that the extraordinary evenness of temperature in the unusual equatorial conditions of Principe justifies this. It might also be justifiable at Sobral, even though two months passed between eclipse and the taking of the comparison plates. If one assumes the scale did not change (apart from what is theoretically predictable on the basis of differential refraction and aberration), then one obtains, perhaps significantly, a value for the deflection very similar to what Eddington derived from his data.

Does this mean that the 1919 report should have argued that all three instruments favored Einstein's theory over the "Newtonian" result? Undoubtedly the decision to discount the suspect data altogether was the correct one, but the existence of an alternative result mentioned in the report suggests that the authors attached some significance to it. It certainly constituted evidence that the data was unreliable. The logic of accepting the "Newtonian" value obtained by the main data analysis demanded that a large part of the deflection observed at Sobral by the Astrographic was due to a significant change of scale *within the instrument itself*, presumably due to the change of focus during the eclipse. Accepting the lower deflection result had as a corollary that the instrument had performed unexpectedly and perversely, which would tend to suggest that data taken with it may not have been terribly reliable.

I suspect that the Greenwich team actually came to believe that the $1''.52$ value was more likely the correct one, on the grounds that, in hindsight, they considered Eddington's data reduction method superior. But in the absence of check plates there was no way to prove that the instrument had experienced no significant change of scale, so they decided to publicly discount the Astrographic data. Psychologically, I claim, it must have been significant to them that there was reason to believe that this instrument also might have yielded a result which was inconsistent with the "Newtonian" value, had matters been handled differently.

9.1.5 The 1978 Reanalysis

Prolonged examination of the 1919 report by Dyson, Eddington, and Davidson, the fruits of which are presented in the previous section, led me to consider the possibility that the Sobral Astrographic data might actually have supported the full Einstein deflection after all. Although not myself an astronomer I am married to one, and my wife, Julia Kennefick, pointed out that the original plates might still be preserved at the Royal Greenwich Observatory (RGO). Could more be done, she observed, with modern plate-measuring machines and astrometric data-reduction software, than had been possible in the days of merely human computers like Davidson and Furner?

In June 2003 I made a trip to Cambridge, primarily for the purpose of looking over Eddington's papers at Trinity College and the RGO's manuscript archive, now housed at the Cambridge University Library, but with the idea of also making enquiries about the original plates. To my surprise I learned from Adam Perkins, the curator of the RGO archives at the Cambridge UL, that a modern reanalysis of the data had already been done, in 1978, to commemorate the Einstein centenary of 1979. He gave me the name of the man who had set the project in motion, Andrew Murray, who had been in charge of the RGO's astrometry team at that time. It turned out that Murray's name was not on the published paper, but happily he had occasion to write a letter to *The Observatory* on the topic ten years later which I found (Murray and Wayman 1989), and this led me to the original paper (Harvey 1979). Thanks to Perkins, and to Donald Lynden-Bell, former director of the Cambridge Institute of Astronomy (successor to the Cambridge Observatory), I eventually got in touch with Murray himself, who very kindly corresponded with me in some detail about the process of the 1978 data reanalysis. It seems clear that it is to Murray himself, and to Francis Graham Smith, then director of the RGO, that we owe thanks for the inspired idea to subject the plates to a modern astrometric analysis (Geoffrey M. Harvey, private communication to the author).

In 1978 most of the Sobral plates still survived intact (it appears, by the way, that the Principe plates have not survived). One eclipse plate and one comparison plate taken with the 4-inch lens were missing, and one of its eclipse plates was broken. A few of the Astrographic plates were discolored. In general nothing stood in the way of the project. The plate measurements were made by E. D. Clements (known as "Clem") on the Zeiss Ascorecord instrument at the RGO (by then relocated to Herstmonceux Castle in Sussex). The data reduction software was written by Murray and the process of reduction was carried out by Geoffrey M. Harvey, author of the published paper. Both Clements and Harvey were members of Murray's staff at the RGO. The first item of note is that the modern methods had no particular difficulty with the Astrographic data, although the images were sufficiently non-circular, on many of the plates from both instruments, to prevent the use of a completely auto- mated plate-measuring machine, such as the RGO's GALAXY. The fact that the Astrographic plates were measured in both declination and right-ascension during the reanalysis does not even call for particular comment in Harvey's paper. Murray remarks, in an e-mail to me, written November 22, 2003:

> In 1978 the plates were re-measured individually on the ASCORECORD machine at Herstmonceux; each image was centered in a square graticule and the (X, Y) co-ordinates were recorded by means of moiré fringe grat- ings. The reasonable results obtained, particularly for the "inferior" Sobral astrographic images, would seem to indicate that the problem with the 1919 measurements was not so much in the quality of the images, but rather in the reduction method, which relied very heavily on the experimental determination of the scale constant e. The Herstmonceux plate reductions of course included both co-ordinates on each plate, giving a much better separation of the plate scale from the deflection on the eclipse plates.

In his 1979 paper Harvey compares the 1919 results with those he recovered using modern techniques.

Gravitational displacement at the Sun's limb in seconds of arc

Determination	Displacement
Predicted from Einstein's Theory	1.75
4-inch plates reduced by Dyson *et al.*	1.98 ± 0.18
4-inch plates measured on the Zeiss	1.90 ± 0.11
Astrographic plates reduced by Dyson *et al.*	0.93
Astrographic plates measured on the Zeiss	1.55 ± 0.34

Harvey comments (p. 198):

> For the 4-inch plates there is no great difference between the value obtained
> by Dyson *et al.* and that from the new measurements, but the error has
> been considerably reduced. For the Astrographic plates, however, a signi-
> ficant improvement has been achieved by the new measurements. Where
> the previous reduction yielded a value of $0''.93$ with an unspecified, large
> error, the new determination is $1''.55 \pm 0''.34$. This is still a weak result,
> but does provide support for that from the 4-inch plates. Combining the
> two fresh determinations, weighted according to their standard errors, gives
> $1''.87 \pm 0''.13$, a result which is just within one standard error of the
> predicted value.

Note that the revised 4-inch result still does not place the Einstein prediction
within the limits of the error bars for this instrument. In general eclipse expeditions
often recovered a value in excess of the GR prediction and were only sporadically
successful in wrestling Einstein's prediction within their error bars. This situation
had barely improved even by the time of the last such expeditions in the mid-1970s.
It is only with radio telescopes measuring quasars being occulted by the relatively
radio-quiet Sun, thus with no need for an eclipse, that the Einstein value has been
precisely confirmed (Will 1993). Of course, the so-called Newtonian result fares even
worse compared to the high value obtained by some eclipse expeditions.

It is remarkable that the alternative value, using the constant scale assumption,
for the Sobral Astrographic from 1919 happens to have an almost identical value
($1.''52$ compared to $1.''55$) to that obtained by the modern reanalysis. Is this mere
coincidence, or are there grounds for believing that it is more than that? Certainly
the modern value casts grave doubt on the $0.93''$ value obtained by the original team.
Murray, in the same e-mail to me, comments:

> We have to remember than in those days, the labour of computation was
> a problem, so short-cuts had to be taken. In particular, the Greenwich
> astrographic plates were only measured in one co-ordinate (declination).
> The general philosophy, both at Greenwich and Cambridge, seems to have
> been to determine the relative plate scale and individual orientations by some

means or other, and then apply these to the measured displacements of individual stars to derive the deflection obtained from each of them.

There can always be a problem with trying to determine independently too many plate constants simultaneously in one solution, because of correlations between them due to the actual geometrical distribution of the stars in the field. Presumably there must be some such effect which has affected the scale derived in Table IX [i.e., the derived scale which apparently gives rise to the erroneous result from the Astrographic plates in Dyson *et al.*]. We should note that, although the deflection is greater in declination, there is a lot of information on the plates scale in the right ascension direction which has been completely ignored.

He adds in a later letter (November 27, 2003):

I can only infer that the inclusion [by Harvey in 1978] of the right ascension measures on the astrographic plates (which were ignored by Dyson *et al.*) has considerably improved the solution for the deflection, in spite of its smaller effect.

There is a wealth of extra information in the modern analysis. Rather than being obliged to simply compare pairs of images plate by plate, the astrometric software permitted the construction of a database of positions from the comparison plates, against which the positions of the stars on the eclipse plates could be compared. Thus, the displacement of each star could be compared to the position of every other star on every comparison plate, not just to its own position on one or two comparison plates. In addition, the plate-measuring machine was able to provide reliable measurements of position in both co-ordinates, for the Astrographic plates. This wealth of extra data meant that there was no difficulty in calculating the scale change.

Since it now seems likely that the low "Newtonian" value was due to errors in reducing the data, it seems plausible that the problem lay in an incorrect value for the scale, given that an incorrect value of other plate constants, such as the orientation, would not have affected the final value so much. Therefore it seems likely that the assumption that the scale did not change much between eclipse and comparison plates was the correct one, though obviously it had to remain only an assumption in the context of 1919.

In hindsight, there seem to be excellent grounds for believing that the Greenwich team did make a justifiable assessment of their data in concluding that it definitely supported the prediction of General Relativity over the "Newtonian" prediction. Nonetheless, most astronomers of the day, including Dyson, wished to see the observation repeated at the eclipse of 1922, before treating the case as closed.

It seems, as one might expect, that the teams who took and handled the data knew best after all. But it is hard to stop a good story once it gets going. Because it is fair to say that the eclipse results over the years were never in perfect accord with general relativity's prediction (even given that they definitely were sufficient to falsify the "Newtonian" value), people had begun to question the stock account of the *experimentum crucis* that vindicated Einstein. Some were not very impressed with the theoretician's swagger with which Eddington and Einstein had greeted the verdict of the observations (I suspect both men were playing up to this image of the cocky and self-assured theorist). Gradually a notion seems to have arisen that there was something highly suspect about the 1919 results, and the new mood is seen in two papers published around 1980, by the philosophers Earman and Glymour, and by the physicist C. W. Francis Everitt. Everitt, like Earman and Glymour, and apparently independently of them, concluded that the Astrographic data was excluded largely on grounds of failure to agree with the theoretically predicted value of GR.

> Others again from an astrographic camera at Sobral gave a very-reliable looking measurement of 0.93 ± 0.05 arc-sec—the scaling coefficient must have been wrong, so they were thrown out though the evidence for them was much better than that for the 1.61 ± 0.30 arc-sec measurement at Principe. It is impossible to avoid the impression—indeed Eddington virtually says so—that the experimenters approached their work with a determination to prove Einstein right. Only Eddington's disarming way of spinning a yarn could have convinced anyone that here was a good check of General Relativity. (Everitt, 1980, 534)

Nevertheless, as I have argued above, I believe that close inspection of the totality of information now available to us since 1979, suggests that the 1919 experimenters probably were justified in concluding that they had at least falsified the lower Newtonian prediction.

So successful was the new counter-myth of bias on the part of Eddington and colleagues that in 1988 Stephen Hawking included the following passage in his famous book *A Brief History of Time*.

> It is normally very difficult to see this [light-bending] effect, because the light from the sun makes it impossible to observe stars that appear near to the sun in the sky. However, it is possible to do so during an eclipse of the sun, when the sun's light is blocked out by the moon. Einstein's prediction of light deflection could not be tested immediately in 1915, because the First World War was in progress, and it was not until 1919 that a British expedition, observing an eclipse from West Africa, showed that light was indeed deflected by the sun, just as predicted by the theory. This proof of a German theory by British scientists was hailed as a great act of reconciliation between the two countries after the war. It is ironic, therefore, that later examination of the photographs taken on that expedition showed the errors were as great as the effect they were trying to measure. Their measurement had been sheer luck, or a case of knowing the result they wanted to get, not an uncommon occurrence in science. The light deflection has, however, been accurately confirmed by a number of later observations. (p. 32)

It appears that Hawking was aware that widespread suspicions concerning the original data reduction existed. He may have read Earman and Glymour's paper, or

Everitt's, or encountered similar stories which I suspect had been circulating orally amongst physicists for some time. I myself have heard physicists claim that the size of the error in the early eclipse expeditions was the same size as the effect itself, which is not a claim made by Earman and Glymour, and not one which, as far as I can tell, is supported by the evidence. Hawking remembered that a reanalysis had been done, which in itself makes him nearly unique, since I have not found even one paper that cites Harvey's publication (the Science Citation Index lists only the Murray and Wayman letter discussed below). It seems the GRO reanalysis team's results were reported at a meeting of the Royal Society in 1979, which might be where Hawking became aware of their efforts.

Knowing that some reanalysis had been undertaken, and recalling the many stories of a dubious data analysis by the original team, he may naturally have jumped to the conclusion, when writing his book nearly a decade later, that the reanalysis must have given birth to the story he remembered. Nothing, of course, could be further from the truth, as far as the reanalysis goes, and it was for this reason that Murray and Patrick A. Wayman of Dunsink Observatory in Dublin (possessors of what remains of the 4-inch instrumentation used at Sobral) wrote to the Observatory in 1989 to object.

> The result from the 4-inch plates was thus confirmed, with a smaller standard error, and even the very low-weight 13-inch plates gave a significant result. The last attempt to measure the deflection by optical observations at an eclipse was in 1973, by a team from the University of Texas. The result from that expedition was: Deflection (arcseconds) $= 1.66 \pm 0.19$.
> Conditions were very far from ideal on that occasion, but by comparison, the results from the 1919 eclipse were very respectable. It is completely unjustifiable to dismiss them as Hawking has done. A recent assessment by Will (Was Einstein Right?, p. 78) that '...these expeditions were triumphs for observational astronomy and produced a victory for general relativity...'would seem to be much fairer.

I suspect there are three myths which arose at different epochs concerning this famous experiment. Initially Eddington and Dyson experienced extraordinary success in winning public acceptance for their thesis that the experiment should be read as an *experimentum crucis* falsifying Newton's theory and vindicating Einstein. This gave birth to a widespread belief that the eclipse observations had proved general relativity correct. In fact, eclipse observations never imposed particularly stringent tests on general relativity, and even the fairly wide error bars that existed were not sufficient to always bring GR's prediction within the range of the measured deflection. Once more accurate tests of relativity became possible from the mid-1950s on, a backlash began to take hold, at least amongst physicists. It was realized that one ought to be able to do far better and gradually the notion that eclipses were not much of a test of relativity led to a version of the story, heard by myself in the 1980s, and apparently also by Hawking, that in fact the error bars in such tests were as wide as the effect to be measured. This in turn helped give birth to the third and most recent myth, that the results reported were an outright fraud.

I should emphasize at this point that I do not regard the opinion of the physicists as irrelevant because it was orally transmitted. Obviously, oral transmission of knowledge plays a critical role in physics and other sciences. Nevertheless, just as we have seen how texts lose nuance as we move from those written by the core group of experts (recall Earman and Glymour's (1981) paper and Collins and Pinch's (1993) book), to those written by non-experts (recall comments quoted earlier from amazon.com's website), so the same is true for oral transmission. Orally transmitted knowledge within the core group of experts is perhaps the most nuanced knowledge of all. I would argue that even texts written by the members of the core group themselves cannot completely capture all of the nuance of their expert opinion, if only because the reader may lack the expertise to correctly assess it. But as oral knowledge moves outside the core group, it too loses nuance and accuracy. In this instance I do not feel that the opinion that "the errors were as great as the effect they [the eclipse observers] were trying to measure" can be regarded as having a factual foundation, though it is based on the very reasonable contention that most eclipse teams tended to underestimate their reported errors, probably a reflection of the difficulty of excluding systematic errors when all measurements must be taken during one brief period of a few minutes' duration.

Dennis Sciama's book *The Physical Foundations of General Relativity* (Sciama 1969) is a good example of a text which had, within the physics community, the same kind of influence which Earman and Glymour's papers had on the science studies community (that is to say, the historians, philosophers, and sociologists of science). It is a considered opinion of the true place of the 1919 eclipse expeditions in this history of relativity which may be sometimes taken as a warrant for completely dismissing them from that history. Sciama says:

Eddington himself later referred to it as 'the most exciting event I recall in my own connection with astronomy.' Ironically enough, we shall see that Einstein's prediction has not been verified as decisively as was once believed. Between 1919 and 1966 there have been fewer than thirty [total solar] eclipses, giving altogether a total observing time of not more than about two hours. (The longest possible duration of a total eclipse is about $71/2$ minutes, and such an occasion occurs very seldom). . . . In fact results have been published for only six eclipses. (Sciama 1969, 69–70)

Looking over the results from all these eclipses (including that of 1919), Sciama concludes:

One might suspect that if the observers did not know what value they were 'supposed' to obtain, their published results might vary over a greater range than they actually do; there are several cases in astronomy where knowing the 'right' answer has led to observed results later shown to be beyond the power of the apparatus to detect. (Sciama 1969, 70)

To be sure, Sciama comes quite close here to charging that one might achieve any result from an eclipse measurement of light-bending. Indeed it is interesting how the viewpoints of the physicists and the sociologists, such as Collins and Pinch (1993), dovetail here. Admittedly Sciama regards the eclipse measurements as pathological because of their interpretative flexibility, whereas the sociologists see this as typical of much of science. Nevertheless, they both have a point. There may very well be a sense in which the position of general relativity was much stronger, in the social sphere, than one would expect for a new theory, and that the position of Newtonian theory was far weaker than it had been for two centuries, and far weaker than many non-experts appreciated. Despite appearances, the old theory of gravity may have been sociologically primed to fall at this moment. I suspect this is true myself. Nevertheless, we must be careful to put this social nexus, which permitted the overthrow of Newtonianism at this historical juncture, in context. In this instance I believe that context is the context of reception, more so than the context of discovery. I doubt that the most critical aspects of the eclipse data reduction were taken in an atmosphere of pro-relativity bias, and I believe that the best evidence suggests that the decisions which were made were done so judiciously, for scientifically defensible reasons. Dyson and his colleagues may have been readier to find in favor of Einstein than they would have been for many another alternative theory of gravity, but this is far from saying that they were looking for excuses to skew their data against Newton. I think they strongly believed that their measurements could not be reconciled with the "Newtonian" prediction. The existence of a viable alternative theory undoubtedly played a significant role in their willingness to unambiguously falsify Newton, but this is a far cry from the accusations of bias which have become fashionable in recent years.

In any case Sciama's real point is that the eclipse measurements, all of them, not just the 1919 ones, do not particularly vindicate general relativity. He does not claim that the data does not support Eddington and Dyson's main contention, which is to say that the results falsify the "Newtonian" result. I contend that much of the change in narrative concerning the eclipse is due to this change in focus, from a contemporary one which was excited by a test, the first ever to do so, which falsified Newton while leaving a rival theory standing, to our modern one which sees Einstein's theory as the undefeated theory fending off all rivals. With the benefit of hindsight, looking back from a time when Einstein's theory occupies the position once held by Newton's, we are struck by how threadbare seem the emperor's old suit of clothes. Fortunately for relativity theory it has much less revealing clothes to wear nowadays.

Earman and Glymour's 1981 paper, the first proper historical study of the expedition, raised several interesting points. They observed that there were questions that could be raised about the long-forgotten third instrument which had actually agreed with the so-called Newtonian result. This led to a new version of the story, in which Eddington began to emerge almost as a villain, and this version has clearly gained some traction and become quite widespread. There are elements of truth to all these stories, but in my opinion, the closest we can come to the truth on this matter is to say the following: The 1919 eclipse expeditions established clearly that light-bending in a gravitational field was a real effect. This was, after all, their principal goal. As Eddington put it, their mission was to "weigh light." Secondly they showed that of the two predictions made by Einstein, the evidence strongly supported the higher general relativistic value and appeared to falsify the lower. This conclusion was supported by subsequent eclipse expeditions, by radio observation of occulting quasars from the 1970s on, and by the data reanalysis of the Sobral plates in 1978. Nevertheless, to paraphrase Eddington, myths have weight. They are not easily suppressed or replaced, and mere written accounts cannot necessarily hope to halt the spread of a good story. Excessive nuance is generally alien to myth. For instance, it seems that "fudging" and "faking" are the popular words which spring to mind when the more nuanced sociological term "interpretive flexibility" is invoked. Should my own account have the extraordinary good fortune to spawn a myth of its own, who knows what will be made out of it?

Acknowledgements

I would like to thank Diana Buchwald for suggesting that I look into this subject, and for providing every kind of support throughout the research, logistical, financial, and, most importantly, moral. The first two kinds of support were provided in her capacity as director of the Einstein Papers Project, who also employed me, in one capacity or another, throughout the research. I also thank her gratefully for carefully reading and critiquing the final manuscript. My colleague Harry Collins also read and commented perceptively upon the manuscript. As stated in the paper I received help from a number of others during my work, especially Adam Perkins of the Cambridge University Library, and Andrew Murray, formerly of the Royal Greenwich Observatory. My research was enormously facilitated by the very generous help of Charlie Johnston of Flint, Michigan, who was able to direct me to the various archives with material relevant to the expedition, based on his own interest on the eclipse, and showed me copies of items he had already located. Jean Eisenstaedt of the Observatoire de Paris kindly showed me copies of the minutes of the Joint Permanent Eclipse Committee, the originals of which are in the Royal Society Archives in London. I am grateful to the Science and Technology Facilities Council and the Syndics of the Cambridge University Library, for granting permission to quote from letters and manuscripts in the Royal Greenwich Observatory archive, which are state papers of the United Kingdom. I am also grateful to the Master and Fellows of Trinity College, Cambridge for permission to quote from letters of Arthur Stanley Eddington in their possession. The Library of the Eidgenoessische Technische Hochschule in Zurich, in particular Ms. Yvonne Voegeli, very kindly facilitated my use of correspondence of Eddington and Hermann Weyl in their collection.

References

Chandrasekhar, Subrahmanyan (1976). "Verifying the theory of relativity." *Notes and Records of the Royal Society of London* 30, 249–260.

Collins, Harry M. and Pinch, Trevor (1993). *The Golem: What Everyone Should Know about Science.* Cambridge University Press, Cambridge.

Crommelin, Andrew Claude de la Cherois (1919a). "The Eclipse Expedition to Sobral." *The Observatory* 42, 368–371.

——. (1919b). "Results of the Total Solar Eclipse of May 29 and the Relativity Theory." *Nature* 104, 280–281.

Davidson, Charles R. (1922). "Observation of the Einstein Displacement in Eclipses of the Sun." *The Observatory* 45, 224–225.

Dyson, Frank Watson (1917). "On the Opportunity afforded by the Eclipse of 1919 May 29 of verifying Einstein's Theory of Gravitation." *Monthly Notices of the Royal Astronomical Society* 77, 445–447.

Dyson, Frank Watson, Eddington, Arthur Stanley and Davidson, Charles R. (1920). "A Determination of the Deflection of Light by the Sun' Gravitational Field, from Observations made at the Total Solar Eclipse of May 29, 1919." *Philosophical Transactions of the Royal Society, Series A* 220, 291–330.

Earman, John and Glymour, Clark (1980). "The Gravitational Redshift as a Test of General Relativity: History and Analysis." *Studies in the History and Philosophy of Science* 11, 175–214.

Earman, John and Glymour, Clark (1981). "Relativity and Eclipses: The British Eclipse Expeditions of 1919 and their Predecessors." *Historical Studies in the Physical Sciences* 11, 49–86.

Earman, John and Janssen, Michel (1993). "Einstein's Explanation of the Motion of Mercury's Perihelion." In *The Attraction of Gravitation. New Studies in the History of General Relativity*, 129–172. John Earman, Michel Janssen, and John D. Norton, eds. Birkhäuser, Boston.

Eddington, Arthur Stanley (1919). "The Total Eclipse of 1919 May 29 and the Influence of Gravitation on Light." *The Observatory* 42, 119–122.

Eddington, Arthur Stanley and Cottingham, Edwin Turner (1920). "Photographs taken at Principe during the Total Eclipse of the Sun, May 29th." *Report of the Eighty-Seventh Meeting of the British Association for the Advancement of Science*. John Murray, London.

Einstein, Albert (2004). *The Collected Papers of Albert Einstein, vol. 9*, eds. Diana Kormos Buchwald, Robert Schulmann, Jozsef Illy, Daniel Kennefick and Tilman Sauer. Princeton University Press, Princeton, New Jersey.

Everitt, C. W. Francis (1980). "Experimental Tests of General Relativity: Past, Present and Future," in *Physics and Contemporary Needs*, vol. 4, ed. Riazuddin, 529–555.

Fowler, Alfred (1919). Meeting of the Royal Astronomical Society, Friday, June 13, 1919. *The Observatory* 42, 261–263.

Harvey, G. M. (1979). "Gravitational Deflection of Light: A Re-examination of the observations of the solar eclipse of 1919." *The Observatory* 99, 195–198.

Hawking, Stephen (1988). *A Brief History of Time*. Bantam Press, London.

Jones, H. Spencer (1939). "Sir Frank Watson Dyson." *The Observatory* 62, 179–187.

Maunder, E. Walter (1900). *The Royal Observatory Greenwich: A glance at its history and work*. The Religious Tract Society, London. (Published in electronic form on the internet by Eric Hutton, 2000. Accessed on July 6, 2007 at http://atschool.eduweb.co.uk/bookman/library/ROG/INDEX.HTM)

Melotte, P. J. (1953). "Herbert Henry Furner." *Monthly Notices of the Royal Astronomical Society* 113, 305–306.

Murray, C. Andrew and Wayman, P. A. (1989). "Relativistic Light Deflections." *The Observatory* 109, 189–191.

Sciama, Dennis William (1969). *The Physical Foundations of General Relativity.* Doubleday, New York.

Sponsel, Alistair (2002). "Constructing a 'revolution in science': the campaign to promote a favourable reception for the 1919 solar eclipse experiments." *British Journal for the History of Science* 35, 439–467.

Stanley, Mathew (2003). "An Expedition to Heal the Wounds of War: The 1919 Eclipse and Eddington and Quaker Adventurer." *ISIS* 94, 57–89.

Waller, John (2002). *Fabulous Science: Fact and Fiction in the History of Scientific Discovery.* Oxford University Press, Oxford. Published in the United States as *Einstein's Luck: The Truth Behind Some of the Greatest Scientific Discoveries.* Oxford University Press, New York.

Will, Clifford (1993). *Was Einstein Right?* Basic Books, New York.

10

Peter Havas
(1916–2004)

Hubert F. Goenner

University of Göttingen, Germany

With the death in June 2004 of our colleague *Peter Havas* near Philadelphia, Pennsylvania, we have lost one of the renowned peers in the field of classical relativistic field and particle theory, with a particular focus on Einstein's special and general relativity theories.

As a child of Hungarian parents who had fled Hungary, at the time ruled by fascists (Horthy), Peter grew up in Vienna, Austria from 1921 and received his schooling and part of his university training there. His professional life was spent predominantly in the United States of America. Because of the deplorable state of political affairs in Europe during his university education, he had to endure two abortive attempts at a Ph.D. until he was finally able to succeed. He started out at the Technische Hochschule in Vienna as an experimental physicist by working in nuclear physics (mass spectroscopy) with Josef Mattauch in 1937/38. In 1938, after the Nazi takeover of Austria, he had to leave the country because of his pro-socialist political activities battling the clerical Austro-fascism (Dollfuss) and also, possibly, because of his classification as being of Jewish descent by the decrees of the Hitler administration (he did not consider himself to be Jewish). Starting afresh in Lyon, France, under the guidance of Jean Thibault, director of the Institut de Physique Atomique, again in experimental nuclear physics, he met theoretical physicist Guido Beck, also exiled, who convinced him to switch to theory. Peter's first scientific publications resulted from this interaction.[1] After the outbreak of World War II, his stay in Lyon from 1939 to 1941 was interrupted by his internment by the French authoritie, as a "hostile" (German) alien without a work permit—since Austria had then become part of the German "Reich"—and by the subsequent occupation of parts of France by German troops (Lyon remained a "free" city). He was fortunate enough to be freed from captivity due to the efforts of his wife, and to receive an entry visa for the United States; together with their baby daughter

---

[1] Guido Beck and Peter Havas: "La dissymétrie de la rupture de l'uranium." *Comptes Rendus de l'Academie des Sciences Paris* 208, 1084–86 (1939). Guido Beck and Peter Havas: "Sur le ralentissement dans l'air des fragments atomiques resultant de l'explosion de l'uranium." *Comptes Rendus de l'Academie des Sciences Paris* 208, 1643–45 (1939).

they arrived in New York in June 1941—on the day the German troops invaded the USSR. At Columbia University, under the formal supervision of *Willis Lamb* but more or less on his own, he wrote his thesis in the field of quantum electrodynamics, at that time still in its infancy.[2]

Peter Havas first taught at Columbia University as an assistant to Enrico Fermi and Edward Teller. After he had obtained his doctoral degree in 1944, (Havas 1944) he became an instructor of physics at Cornell University in 1945 as a substitute for Hans Bethe—who at the time was doing war research in Los Alamos. In 1946, he became an assistant professor and, later, full professor of physics at Lehigh University in Bethlehem, Pennsylvania (1946–1964). Starting with the term 1964/65 he changed to Temple University in Philadelphia. Peter spent sabbatical leaves at the Institute for Advanced Studies in Princeton, the Niels Bohr Institute in Copenhagen, and Argonne National Laboratory; he was a visiting professor at Birkbeck College of the University of London and at the University of Goettingen, Germany. After his retirement from Temple University in 1981 he continued his scientific work as an adjunct professor of physics at the University of Pennsylvania, Philadelphia (1982–1988), and at the University of Utah (1987–1992). He also served on the editorial board of *The Collected Papers of Albert Einstein*. His wife of sixty-five years, Helga Francis, née Hoellering, a professor of microbiology and cancer research at Temple University from 1963, not only shared his rewarding life—though at times uneasy and endangered—in Austria, France, and the United States, but was also his equal in the political fight: she was imprisoned in Vienna for her anti-fascist activities.

Peter's refusal to undertake weapons research during wartime barred him from being accepted into the inner circle of theoreticians involved in the eventual development of quantum field theory—many of whom later became famous. Essentially, this circle was formed by theoreticians who had returned from working on the atomic bomb in Los Alamos. Being excluded from the network's prepublication exchange of relevant scientific information concerning quantum field theory, Peter decided to leave this field.

Nevertheless, his original interest in quantum fields and elementary particles became a partial motivation for his subsequent research in classical theory: he aimed at a possible uncovering of the difficulties of quantum field theory through the study of its underlying classical basis.[3] This connection with quantum physics carried on when, in the 1950s and 1960s, Peter Havas and his students dealt with the classical scattering of mesons, and, in addition, with the equations of motion of point

---

[2] The thesis is entitled: "On the interaction of radiation and two electrons."

[3] Peter Havas: "Generalized Lagrange Formalism and Quantization Rules." *Bulletin of the America Physical Society* 1, 337–338 (1956); Jack E. Chatelain and Peter Havas. "The Classical Limit of the Quantum Theory of Radiation Damping." *Physical Review* 129, 1459–1463 (1963).

particles[4] including classical Yang–Mills–Higgs fields[5] in the 1970s and 1980s. In a well-known paper with Joshua Goldberg, the (relativistic) "fast-motion approximation" for the alternate solution of field equations and equations of motion in General Relativity was introduced, a method well adapted to a relativistic theory of gravitation and complementing the post-Newtonian expansion used in the well-known EIH (Einstein–Infeld–Hoffmann) paper.[6] The problem of determining the motion of particles is relevant for the (approximate) calculation of gravitational radiation emitted by a (point) source and its back-reaction upon the source; the debate concerning a convincing theoretical fundament for Einstein's quadrupole formula lasted well into the 1970s.[7] In both fields, equations of motion and radiation reaction, Peter Havas wrote numerous papers and gave many lectures at various venues including summer schools and international conferences.[8]

His broad interests also brought him into applied or even mathematical physics; for example, he investigated which systems of PDEs could be derived from a Lagrangian, or in which coordinate systems important classical equations like the Hamilton–Jacobi or the Schrödinger equation could be separated.[9] The connection between equations of motion, conservation laws, and symmetries led to interest-

---

[4] We note a few sample papers; Clarence R. Mehl and Peter Havas: "The Classical Scattering of Neutral Mesons. I." *Physical Review* 91, 393–397 (1953); Alba D. Craft and Peter Havas: "The Classical Scattering of Neutral Mesons. II." *Physical Review* 154, 1460–1468 (1967); Peter Havas: "A Note on the Absorber Theory of Radiation." *Physical Review* 86, 974–975 (1952); "Multipole Singularities of Classical Scalar and Pseudoscalar Meson Fields." *Physical Review* 93, 1400–1411 (1954).

[5] William C. Schieve, Arnold Rosenblum, and Peter Havas: "Classical Theory of Particles Interacting with Electromagnetic and Mesonic Fields. II." *Physical Review* D 6, 1501–1522 (1972); Wolfgang Drechsler, Peter Havas, and Arnold Rosenblum: "Theory of Motion for Monopole-dipole Singularities of Classical Yang–Mills–Higgs Fields. I. Laws of motion"; II. "Approximation Scheme and Equations of Motion." *Physical Review* D 29, 658–667, 668–686 (1984).

[6] Peter Havas and Joshua N. Goldberg: "Lorentz-invariant Equations of Motion of Point Masses in the General Theory of Relativity." *Physical Review* 128, 398–414 (1962).

[7] Jürgen Ehlers, Arnold Rosenblum, Joshua N. Goldberg, and Peter Havas: "Comments on Gravitational Radiation Damping and Energy Loss in Binary Systems." *Astrophysical Journal* 208, L77–L81 (1976).

[8] For example, Stanley F. Smith and Peter Havas: "Effects of Gravitational Radiation Reaction in the General Relativistic Two-body Problem by a Lorentz-invariant Approximation Method." *Physical Review* 138, B495–B508 (1965); *Ondes et radiations gravitationelles*, Colloques internationaux du CNRS, vol. 220, "Equations of Motion, Radiation Reaction, and Gravitational Radiation," pp. 383–392 (1974); "Equations of Motion and Radiation Reaction," in: *Isolated Gravitating Systems in General Relativity*, Enrico Fermi Summer School, 67th course (J. Ehlers ed.), 74–155 (1974).

[9] Peter Havas: "Separation of Variables in the Hamilton-Jacobi, Schrödinger, and Related Equations. I. Complete Separation," II. Partial Separation. *Journal of Mathematical Physics* 16, 1461–1167; 2476–2489 (1975).

ing publications by Peter quoted abundantly in the literature.[10] As to conservation laws, he considered not only the Lorentz group but also the Galilean group,
its possible invariants, and the Lagrangians to be built from them. This research
included post-Galilean and post-Lorentzian approximation, in connection with the
center-of-mass law, and also including particle theories.[11] In his endeavor to precisely formulate Newton's gravitational theory as a limit to Einstein's, he developed
a four-dimensional representation of what now is called the Newton–Cartan theory.[12]
Peter also did research on special relativistic particle theories (direct interaction), and
on a relativistic formulation of statistical mechanics.[13]

Although no specialist in the generation of exact solutions, Peter found these both
in the realm of Newtonian gravitational theory and in general relativity.[14] He also
became interested in "Birkhoff's theorem" and tried to extend its range to alternative
theories of gravitation, for example, those with Lagrangians of higher-order in the
curvature tensor. In this context, his interest in a (local) classification of spaces with
constant Ricci scalar and, in connection with it a very special PDE like Emden's
equation emerged.[15] In fact, revisiting the results of research obtained years earlier
led to Peter's next-to-last research papers—just before his eyesight became severely
afflicted by macular degeneration.[16] His very last publication, in collaboration with

[10] Peter Havas: "The Range of Application of the Lagrange Formalism—I." *Nuovo Cimento
(Suppl.)* 5, 363–388 (1957); "The Connection between Conservation Laws and Invariance
Groups: Folklore, Fiction, and Fact." *Acta Physica Austriaca* 38, 145–167 (1973).

[11] Richard B. Hoffman and Peter Havas: "Qualitative Aspects of the N-Body Problem
for Approximately Relativistic Equations of Motion." *Physical Review* 140, Issue 4B,
1162–1173 (1965); Peter Havas and John Stachel: "Invariances of Approximately Relativistic Lagrangians and the Center-of-mass Theorem. I." *Physical Review* 185, 1636–1647
(1969); Harry W. Woodcock and Peter Havas: "Approximately Relativistic Lagrangians for
Classical Interacting Particles." *Physical Review* D 6, 3422–3444 (1972); John Stachel and
Peter Havas: "Invariances of Approximately Relativistic Hamiltonians and the Center-of-
mass Theorem." *Physical Review* D 13, 1598–1613 (1976); Thomas Pascoe, John Stachel,
and Peter Havas: "Center-of Mass-Theorem in Post-Newtonian Hydrodynamics." *Physical
Review* D 14, 917–921 (1976).

[12] Peter Havas: "Four-dimensional Formulation of Newtonian Mechanics and Gravitation and
their Relation to the Special and the General Theory of Relativity." *Reviews of Modern
Physics* 36, 938–965 (1964).

[13] John E. Krizan and Peter Havas: "Relativistic Corrections in the Statistical Mechanics of
Interacting Particles," *Physical Review* 128, 2916–2924 (1962).

[14] Peter Havas: "Homographic Motions of a Newtonian System of Point Masses." *Celes-
tial Mechanics and Dynamical Astronomy* 7, 321–346 (1973); De-Ching Chern and Peter
Havas: "Exact Nonplanar Solutions of the Classical Relativistic Three-body Problem."
Journal of Mathematical Physics 14, 470–478 (1973); Angelo Armenti, Jr. and Peter Havas:
"A Class of Exact Solutions for the Motion of a Particle in a Monopole-Prolate Quadrupole
Field." In: *Relativity and Gravitation.* Eds. C. G. Kuper and A. Peres, pp. 1–15. London:
Gordon & Breach 1971.

[15] Hubert Goenner and Peter Havas: "Spherically-Symmetric Space-Times with Vanishing
Curvature Scalar." *Journal of Mathematical Physics* 21, 1159–1167 (1980).

[16] Hubert Goenner and Peter Havas: "Exact Solutions of the Generalized Lane-Emden Equation." *Journal of Mathematical Physics* 41, 7029–7042 (2000); "Spherically Symmetric

his former student Harry Woodcock, was printed only after his death; it is concerned with a Galilei-invariant particle mechanics incorporating non-instantaneous interaction.[17]

Conceptual clarity, curiosity toward the evolution of concepts, and simplicity in the representation of results were dear to Peter Havas. He also showed a keen interest in the philosophy and history of his fields of research. He wrote about causality, determinism, and other concepts important to relativity.[18] His later papers in the history of physics are concerned with conceptual developments such as the early history of equations of motion, and with prosopographical and biographical presentations.[19] We could have expected many more interesting contributions had Peter Havas lived longer.[20]

Peter was a prolific writer, both as a single author and as a coauthor with well-known fellow physicists or with his many doctoral students. Some of his students have held, or are still holding, research and teaching positions at respected universities; some have become college or university (vice-)presidents. He mastered several languages, and he strove for completeness, both in solving the problems posed and in consulting as much of the existing literature on a subject as possible. From some of the papers referred to, we note that he could revisit a problem after many years. At times, working with him could be rather demanding because he insisted on completeness and perfection. More than once Peter Havas was frustrated by the level of linguistic, historical, and cultural ignorance of some of his students and fellow physicists in the United States.

Politically, Peter was a liberal in the best sense of the word; in numerous comments, appeals, and letters to governors and congressmen he spoke up against violations of human and civil rights. He and his wife worked for the implementation

---

Spacetimes with Constant Curvature Scalar." *Journal of Mathematical Physics* 42, 1837–1860 (2001).

[17] Harry W. Woodcock and P. Havas: "Generalized Galilei-Invariant Classical Mechanics." *International Journal of Modern Physics A* 20, 4259–4289 (2005).

[18] Peter Havas: "Causality Requirements and the Theory of Relativity." *Synthese* 18, 75–102 (1968); "Foundation Problems in General Relativity." In: *Delaware Seminar in the Foundations of Physics.* Ed. M. Bunge, pp. 124–148 (1967); "Simultaneity, Conventionalism, General Covariance, and the Special Theory of Relativity." *General Relativity and Gravitation* 19, 435–453 (1987).

[19] Peter Havas: "The Early History of the 'Problem of Motion' in General Relativity." In: *Einstein and the History of General Relativity.* Eds. Don Howard and John Stachel, *Einstein Studies* 1, 234–276 (1989); "The General-Relativistic Two-Body Problem and the Einstein-Silberstein Controversy." In: *The Attraction of Gravitation.* Eds. J. Earman, M. Janssen, and J. D. Norton. *Einstein Studies* 5, 88–125 (1993); "Einstein, Relativity and Gravitation Research in Vienna before 1938." In: *The Expanding Worlds of General Relativity.* Eds. H. Goenner, J. Renn, J. Ritter, and T. Sauer. *Einstein Studies* 7, 161–206 (1999); "The Life and Work of Guido Beck: The European Years: 1903–1943." *Anais Academia Brasileira de Ciências.* 67 (Supl. 1), 11–36 (1995).

[20] Over a period of many years, with much archival work already done, he prepared a book on the Viennese physicist and socialist politician *Friedrich Adler* with whom his father had been well acquainted; unfortunately, its realization was thwarted by Peter's eye problems.

of social justice and also for the preservation of nature. My political education was essentially due to him and took place during the Vietnam war. It thus included, whenever necessary, political action embracing involvement in public protest through political demonstrations. Francis Havas passed away only two months after her husband. I will always remember Peter Havas as a superb scientist and an upright man. I cherish both Peter and Francis as mentors and very good old friends.

Appendix 1: A list of Ph.D. students of Peter Havas

James Alan McLennan (1924–2002) (Ph.D. 1952) → professor, Lehigh University, Bethlehem, PA
Frederic R. Crownfield, Jr. (Ph.D. 1952)
Clarence R. Mehl (Ph.D. 1954)
Jack E. Chatelain (1922–1999) (Ph.D. 1957) → professor, Utah State University, Logan, UT
Alba D. Craft (Ph.D. 1959)
William C. Schieve (Ph.D. 1960) → professor, University of Texas, Center for Complex Quantum Systems, Austin, TX
Stanley F. Smith (Ph.D. 1960) → professor, Onondaga Community College, Syracuse, NY
John E. Krizan (Ph.D. 1962) → professor, University of Vermont, Burlington, VT
George Fulton Roemhild (Ph.D. 1965)
Richard B. Hoffman (Ph.D. 1967) → vice president, Franklin & Marshall College, Lancaster, PA
Angelo Armenti, Jr. (Ph.D. 1970) → president, California University, California, PA
Arnold Rosenblum (1943–1991) (Ph.D. 1970) → professor, Utah State University, Logan, UT
De-Ching Chern (Ph.D. 1971) → professor, National Central University, Chung-li, Taiwan
Harry W. Woodcock (Ph.D. 1971) → professor, Philadelphia University, PA
Warren N. Hermann (Ph.D. 1976)
Angel M. Salas (Collante) (Ph.D. 1978) → professor, Universidad de Zaragoza, Spain
Laurence I. Gould (Ph.D. 1981) → professor, University of Hartford, West Hartford, CT
Walter Surgis Koroljow (Ph.D. 1981)
David Kerrick (Ph.D. 1990) → professor, University of the Sciences, Philadelphia, PA

Appendix 2: List of publications of Peter Havas

1939

(With Guido Beck) "La dissymétrie de la rupture de l'uranium." *Comptes Rendus de l'Academie des Sciences Paris* 208, 1084–1086.

(With Guido Beck) "Sur le ralentissement dans l'air des fragments atomiques resultant de l'explosion de l'uranium." *Comptes Rendus de l'Academie des Sciences Paris* 208, 1643–1645.

1940

"Sur le ralentissement des ions lourds dans la matière: Application à la rupture de l'Uranium." *Journal de Physique et le Radium* 1, 146.

1944

"On the interaction of radiation and two electrons." *Physical Review* 66, 69–67.

1945

"On the interaction of radiation and matter." *Physical Review* 68, 214–226.

1948

"On the Classical Motion of Point Charges." *Physical Review* 74, 456–463.

1949

"Bemerkungen zum Zweikörperproblem der Elektrodynamik." *Acta Physica Austriaca* 3, 342–351.

1951

"On the Range of Application of the Lagrange and Hamilton Formalism." *Physical Review* 83, 224–225.

1952

"A Note on the Absorber Theory of Radiation." *Physical Review* 86, 974–975.

"The Classical Equations of Motion, I." *Physical Review* 87, 309–318.

(With James A. McLennan) "Conservation Laws for Fields of Zero Rest Mass. I." *Physical Review* 87, 898.

"Conservation Laws for Fields of Zero Rest Mass. II." *Physical Review* 87, 899.

1953

(With Clarence R. Mehl) "The classical scattering of neutral mesons. I." *Physical Review* 91, 393–397.

"The Classical Equations of Motion," II. *Physical Review* 91, 997–1007.

1954

"On the Classical Theory of Particles Interacting with Electromagnetic and Mesonic Fields. I." *Physical Review* 93, 882–888.

"Multipole Singularities of Classical Scalar and Pseudoscalar Meson Fields." *Physical Review* 93, 1400–1411.

(With Frederic R. Crownfield, Jr.) "The Classical Equations of Motion of Point Particles in Neutral Meson Fields." *Physical Review* 94, 471–477.

1956

"Generalized Lagrange Formalism and Quantization Rules." *Bulletin of the American Physical Society* 1, 337–338.

1957
"Theory of Fields of Integral Spin Greater than One." *Bulletin of the American Physical Society* 2, 189–190.
"Radiation Damping in General Relativity." *Physical Review* 108, 1351–1352.
"The range of application of the Lagrange formalism—I." *Nuovo Cimento (Suppl.)* 5, 363–388.

1959
"Equations of Motion of Point Particles in Fields of Nonzero Restmass and Spin." *Physical Review* 113, 732–740.
"Multipole Singularities of Classical Vector and Pseudovector Fields." *Physical Review* 116, 202–217.
"Classical Relativistic Theory of Elementary Particles." In: *Argonne National Laboratory Summer Lectures in Theoretical Physics 1958*. ANL-5982: pp. 124–220.

1960
(With Stanley F. Smith) "Influence of Gravitational Radiation Reaction on the Motion of Two Bodies." *Bulletin of the American Physical Society* 5, 53.
(With J. Plebanski) "Relativistic Dynamics and Newtonian Causality." *Bulletin of the American Physical Society* 5, 433.

1961
"Radiation Reaction in Linear Theories of Gravitation." *Bulletin of the American Physical Society* 6, 346.

1962
"General Relativity and the Special Relativistic Equations of Motion of Point Particles." In: *Recent Developments in General Relativity*, pp. 259–277. New York: Pergamon Press.
(With Joshua N. Goldberg) "Lorentz-invariant equations of motion of point masses in the general theory of relativity." *Physical Review* 128, 398–414.
(With John E. Krizan) "Relativistic Corrections in the Statistical Mechanics of Interacting Particles." *Physical Review* 128, 2916–2924.
"Equations of Motion in Special Relativistic Nonlinear Field Theories." *Bulletin of the American Physical Society* 7, 299.

1963
(With Jack E. Chatelain) "The Classical Limit of the Quantum Theory of Radiation Damping." *Physical Review* 129, 1459–1463.

1964
"Four-dimensional formulation of Newtonian mechanics and gravitation and their relation to the special and the general theory of relativity." *Reviews of Modern Physics* 36, 938–965.
"The Connection between Conservation Laws and Laws of Motion in Affine Spaces." *Journal of Mathematical Physics* 5, 373–378.

1965
(With Stanley F. Smith) "Effects of Gravitational Radiation Reaction in the General Relativistic Two-Body Problem by a Lorentz-Invariant Approximation Method." *Physical Review* 138, B495–B508.

"Some Basic Problems in the Formulation of a Relativistic Statistical Mechanics of Interacting Particles." In: *Statistical Mechanics of Equilibrium and Non-Equilibrium.* Ed. J. Meixner, pp. 1–19. Amsterdam: North-Holland.

"Relativity and Causality." In: *Proceedings of the 1964 International Congress of Logic, Methodology, and Philosophy of Science*, Jerusalem. Ed. Y. Bar-Hillel, pp. 347–362. Amsterdam: North-Holland.

(With Richard B. Hoffman) "Qualitative Aspects of the N-Body Problem for Approximately Relativistic Equations of Motion." *Physical Review* 140, B1162–B1173.

1966

"A Note on Hertz's 'Derivation' of Maxwell's Equations." *American Journal of Physics* 34, 667–669.

"Are Zero Rest Mass Particles Necessarily Stable?" *American Journal of Physics* 34, 753–757.

1967

"Foundation Problems in General Relativity." In: *Delaware Seminar in the Foundations of Physics.* Ed. M. Bunge, pp. 124–148.

(With Alba D. Craft) "The classical scattering of neutral mesons. II." *Physical Review* 154, 1460–1468.

1968

"Causality Requirements and the Theory of Relativity." *Synthese* 18, 75–102.

1969

(With John Stachel) "Invariances of approximately relativistic Lagrangians and the center-of-mass theorem. I." *Physical Review* 185, 1636–1647.

"Causality Requirements and the Theory of Relativity." In: *Boston Studies in the Philosophy of Science* 5. Eds. Robert S. Cohen and M. Wartofsky, pp. 151–178. Berlin: Springer.

1971

(With Angelo Armenti, Jr.) "A class of exact solutions for the motion of a particle in a monopole-prolate quadrupole field." In: *Relativity and Gravitation.* Eds. C.G. Kuper and A. Peres, pp. 1–15. London: Gordon & Breach.

"Galilei- and Lorentz-Invariant Particle Systems and their Conservation Laws." In: *Problems in the Foundations of Physics.* Ed. M. Bunge, pp. 31–48. Berlin: Springer.

1972

"Multipole Singularities of Classical Scalar and Pseudoscalar Meson Fields." *Physical Review* D 5, 3048–3065.

(With William C. Schieve and Arnold Rosenblum) "Classical Theory of Particles Interacting with Electromagnetic and Mesonic Fields. II." *Physical Review* D 6, 1501–1522.

(With Harry W. Woodcock) "Approximately Relativistic Lagrangians for Classical Interacting Particles." *Physical Review* D 6, 3422–3444.

(With Arnold Rosenblum) "Classical Theory of Particles Interacting with Electromagnetic and Mesonic Fields. III." *Physical Review* D 6, 1501–1522.

1973

"The connection between conservation laws and invariance groups: folklore, fiction, and fact." *Acta Physica Austriaca* 38, 145–167.

"Homographic motions of a Newtonian system of point masses." *Celestial Mechanics and Dynamical Astronomy* 7, 321–346.

(With De-Ching Chern) "Exact nonplanar solutions of the classical relativistic three-body problem." *Journal of Mathematical Physics* 14, 470–478.

"Causality and Physical Theories." In: *AIP Conference Proceedings* 16. Ed. W. B. Rollnick, pp. 23–47. New York: American Physical Society.

1974

"Equations of motion, radiation reaction, and gravitational radiation." In: *Ondes et radiations gravitationelles*, Colloques internationaux du CNRS, vol. 220, pp. 383–392.

"Equations of motion and radiation reaction." In: *Isolated gravitating systems in general relativity*, Enrico Fermi Summer School, 67th course. Ed. J. Ehlers, pp. 74–155.

1975

"Separation of variables in the Hamilton-Jacobi, Schrödinger, and related equations. I. Complete separation. II. Partial separation." *Journal of Mathematical Physics* 16, 1461–1167; 2476–2489.

1976

(With Jürgen Ehlers, Arnold Rosenblum, and Joshua N. Goldberg) "Comments on gravitational radiation damping and energy loss in binary systems." *Astrophysical Journal* 208, L77–L81.

(With John Stachel) "Invariances of approximately relativistic Hamiltonians and the center-of-mass theorem." *Physical Review* D 13, 1598–1613.

(With Thomas Pascoe and John Stachel) "Center-of mass-theorem in post-Newtonian hydrodynamics." *Physical Review* D 14, 917–921.

(with F. Coester) "Approximately relativistic Hamiltonians for interacting particles." *Physical Review* D 14, 2556–2569.

1977

"On Theories of Gravitation with Higher-Order Field Equations."*General Relativity and Gravitation* 8, 631–645.

1978

(With Jerzy Plebanski) "Conformal extensions of the Galilei group and their relation to the Schrödinger group." *Journal of Mathematical Physics* 19, 482–488.

"The conservation laws of nonrelativistic classical and quantum mechanics for a system of interacting particles." *Helvetica Physica Acta* 51, 394–411.

1980

(With Hubert Goenner) "Spherically-Symmetric Space-Times with Vanishing Curvature Scalar." *Journal of Mathematical Physics* 21, 1159–1167.

"Einstein és Kora." *Klönlenyomat a fizikai szemle* 30 Éveolyam 2. Számából [in Hungarian]

1982

(With A. Rosenblum and R. E. Kates) "Use of Hyperfunctions for Classical Radiation-Reaction Calculations." *Physical Review* D 26, 2707–2712.

(With Hubert Goenner) "Theories of Gravitation with Higher-Order Field Equations and Birkhoff's Theorem." In: *Proceedings of the Third Latin-American Symposium for General Relativity and Gravitation, Mexico City, Aug. 1978*. Eds. S. Hojman, M. Rosenbaum, and M.P. Ryan, Jr., pp. 145–166. Mexico: Universidad Nacional Autonoma.

1984

(With Wolfgang Drechsler and Arnold Rosenblum) "Theory of motion for monopole-dipole singularities of classical Yang-Mills-Higgs fields. I. Laws of motion." II. "Approximation scheme and equations of motion." *Physical Review* D 29, 658–667, 668–686.

1987

"Simultaneity, Conventionalism, General Covariance, and the Special Theory of Relativity." *General Relativity and Gravitation* 19, 435–453.

1988

"Translation of Einstein's Papers." *Science* 240, 24 June, p. 1716.

1989

"The Early History of the 'Problem of Motion' in General Relativity." In: *Einstein and the History of General Relativity*. Eds. Don Howard and John Stachel. *Einstein Studies* 1, 234–276.

1990

"Energy-Momentum Tensors in Special and General Relativity." In: *Developments in General Relativity, Astrophysics, and Quantum Theory*. A Jubilee Volume in Honour of Nathan Rosen. *Annals of the Israel Physical Society* 9, 131–153.

1992

"Shearfree spherically symmetric perfect fluid solutions with conformal symmetry." *General Relativity and Gravitation* 24, 599–615.

1993

"The General-Relativistic Two-Body Problem and the Einstein-Silberstein Controversy." In: *The Attraction of Gravitation*. Eds. J. Earman, M. Janssen, and J. D. Norton. *Einstein Studies* 5, 88–125.

1995

"The Life and Work of Guido Beck: The European Years: 1903–1943." *Anais Academia Brasileira de Ciências* 67 (Suppl. 1), 11–36.

1999

"Einstein, Relativity and Gravitation Research in Vienna before 1938." In: *The Expanding Worlds of General Relativity*. Eds. H. Goenner, J. Renn, J. Ritter, and T. Sauer. *Einstein Studies* 7, 161–206.

2000

(With Hubert Goenner) "Exact solutions of the generalized Lane-Emden equation." *Journal of Mathematical Physics* 41, 7029–7042.

2001

(With Hubert Goenner) "Spherically symmetric spacetimes with constant curvature scalar." *Journal of Mathematical Physics* 42, 1837–1860.

2005

(With Harry W. Woodcock) "Generalized Galilei-Invariant Classical Mechanics." *International Journal of Modern Physics A* 20, 4259–4289.

11

Peter Bergmann and the Invention of Constrained Hamiltonian Dynamics

D. C. Salisbury

Austin College, TX, USA

11.1 Introduction

It has always been the practice of those of us associated with the Syracuse "school" to identify the algorithm for constructing a canonical phase space description of singular Lagrangian systems as the Dirac–Bergmann procedure. I learned the procedure as a student of Peter Bergmann, and I should point out that he never employed that terminology. Yet it was clear from the published record at the time (the 1970s) that his contribution was essential. Constrained Hamiltonian dynamics constitutes the route to canonical quantization of all local gauge theories, including not only conventional general relativity, but also grand unified theories of elementary particle interactions, superstrings, and branes. Given its importance and my suspicion that Bergmann has never received adequate recognition from the wider community for his role in the development of the technique, I have long intended to explore this history in depth. This paper is merely a tentative first step, in which I will focus principally on the work of Peter Bergmann and his collaborators in the late 1940s and early 1950s, indicating where appropriate the relation of this work to later developments. I begin with a brief survey of the prehistory of work on singular Lagrangians, followed by some comments on the life of Peter Bergmann. These are included in part to commemorate Peter in this first meeting on the History of General Relativity since his death in October 2002. Then I will address what I perceive to be the principal innovations of his early Syracuse career. Josh Goldberg has already covered some of this ground in his 2005 report (Goldberg 2005), but I hope to contribute some new perspectives. I shall conclude with a partial list of historical issues that remain to be explored.

11.2 Singular Lagrangian Prehistory

All attempts to invent a Hamiltonian version of singular Lagrangian models are based either explicitly or implicitly on Emmy Noether's remarkable second theorem (Noether 1918). I state the theorem using the notation for variables employed by

Bergmann in his first treatment of singular systems (Bergmann 1949b). Denote field variables by $y_A(A = 1, \ldots, N)$, where N is the number of algebraically independent components, and let x represent the independent variables or coordinates. Noether assumes that n is the highest order of derivatives of y_A appearing in the Lagrangian, $L(y_A, y_{A,\mu}, \ldots, y_{A,\mu_1 \ldots \mu_n})$, but I will assume that $n = 1$. The extension of the theorem to higher derivatives is straightforward. Then for an arbitrary variation $\delta y_A(x)$, after an integration by parts we have the usual identity

$$L^A \delta y_A \equiv \delta L - \frac{\partial}{\partial x^\mu}\left(\frac{\partial L}{\partial y_{A,\mu}}\delta y_A\right), \tag{11.1}$$

where the Euler–Lagrange equations are

$$L^A := \frac{\partial L}{\partial y_A} - \frac{\partial}{\partial x^\mu}\left(\frac{\partial L}{\partial y_{A,\mu}}\right) = 0. \tag{11.2}$$

Now suppose that the action is invariant under the infinitesimal coordinate transformation $x'^\mu = x^\mu + \xi^\mu(x)$. Invariance is defined by Noether as follows:

$$\int_{\mathcal{R}'} L(y'_A, y'_{A,\mu})d^4x' = \int_{\mathcal{R}} L(y_A, y_{A,\mu})d^4x. \tag{11.3}$$

(The notion of invariance was extended later, as we shall see below, to include a possible surface integral.) Crucial to this definition is the fact that the Lagrangian is assumed not to have changed its functional form, guaranteeing that this transformation does not change the form of the equations of motion, i.e., it is a symmetry transformation. Noether writes $\delta y_A(x) := y'_A(x') - y_A(x)$, and therefore $y'_A(x) = y_A(x - \xi) + \delta y_A(x)$. She then defines

$$\bar{\delta} y_A(x) := y'_A(x) - y_A(x) = \delta y_A(x) - y_{A,\mu}(x)\xi^\mu(x). \tag{11.4}$$

This $\bar{\delta}$ notation was appropriated by Bergmann in his 1949 paper and retained throughout his life. It is, of course, the Lie derivative with respect to the vector field $-\xi^\mu$, a terminology introduced in (Sledbodzinski 1931). Returning to the elaboration of Noether's theorem and using this notation, we may rewrite the invariance assumption (11.3) as

$$\bar{\delta} L \equiv -\frac{\partial}{\partial x^\mu}\left(L\xi^\mu\right), \tag{11.5}$$

so that under a symmetry transformation the identity (11.1) becomes

$$L^A \bar{\delta} y_A \equiv \frac{\partial}{\partial x^\mu}\left(-\frac{\partial L}{\partial y_{A,\mu}}\bar{\delta} y_A - L\xi^\mu\right). \tag{11.6}$$

Next, let us assume that $\bar{\delta}$ variations of y_A are of the form

$$\bar{\delta} y_A = {}^0\!f_{Ai}(x, y, \ldots)\xi^i(x) + {}^1\!f_{Ai}^\nu(x, y, \ldots)\xi^i_{,\nu}(x), \tag{11.7}$$

where we have admitted the possibility of additional non-coordinate gauge symmetries by letting the index i range beyond 3. We are finally in position to state (and prove) Noether's second theorem: Perform an integration by parts on the left-hand side of (11.6) using (11.7); then for functions ξ^i that vanish on the integration boundary it follows that

$$L^{A\,0}f_{Ai} - \frac{\partial}{\partial x^\nu}\left(L^{A\,1}f_{Ai}^\nu\right) \equiv 0. \tag{11.8}$$

In vacuum general relativity these are the contracted Bianchi identities. The derivation from general coordinate symmetry had already been anticipated by Hilbert in 1915 in a unified field-theoretic context (Hilbert 1915).[1] Weyl applied a similar symmetry argument in 1918 (Weyl 1918). He adapted Noether's theorem to a gravitational Lagrangian $\mathcal{L}_W$ from which a total divergence has been subtracted so that the highest order of derivatives appearing in it are $g_{\mu\nu,\alpha}$. $\mathcal{L}_W$ is no longer a scalar density, but the extra divergence term can easily be incorporated in its variation,

$$\mathcal{L}_W = \sqrt{-g}R - \left(\sqrt{-g}g^{\mu\nu}\Gamma_{\mu\rho}^\rho - \sqrt{-g}g^{\mu\rho}\Gamma_{\mu\rho}^\nu\right)_{,\nu} = \sqrt{-g}g^{\mu\nu}\left(\Gamma_{\rho\sigma}^\sigma\Gamma_{\mu\nu}^\rho - \Gamma_{\mu\rho}^\sigma\Gamma_{\nu\sigma}^\rho\right). \tag{11.9}$$

Bergmann and his collaborators later worked with this Lagrangian. It appears in his 1942 textbook (Bergmann 1942). In 1921 Pauli applied similar symmetry arguments, citing Hilbert and Weyl, but curiously never mentioning Noether (Pauli 1921). Pauli is an important link in this story. In his groundbreaking paper on constrained Hamiltonian dynamics, Leon Rosenfeld writes that it was Pauli who suggested to him a method for constructing a Hamiltonian procedure in the presence of identities (Rosenfeld 1930).

> Bei der näheren Untersuchung dieser Verhältnisse an Hand des besonders lehrreichen Beispieles der Gravitationstheorie, wurde ich nun von Prof. Pauli auf das Prinzip einer neuen Methode freundlichst hingewiesen, die in durchaus einfacher und natürlicher Weise gestattet, das Hamiltonsche Verfahren beim Vorhandensein von Identitäten auszubilden [. . .]

Rosenfeld did indeed make astounding progress in constructing a gravitational Hamiltonian. Full details are reported elsewhere (Salisbury 2009), but its relevance specifically to the work of the Syracuse "school" will be addressed below.

11.3 A Brief Bergmann Biography

Peter Bergmann was born in 1915 in Berlin Charlottenburg. His mother, Emmy (Grunwald) Bergmann, was one of the first female pediatricians in Germany. In 1925 she was also the founder of the second Montessori school in Germany, in Freiburg, where she moved with her son and daughter in 1922. She had taken a course in Amsterdam in the winter of 1923/1924 with Maria Montessori. The chemist Max Bergmann, Peter's father, was a student of and collaborator with the 1902 Nobel prize winner in chemistry, Emil Fischer. In 1923 he was appointed the first Director of the Kaiser Wilhelm Institut für Lederforschung in Dresden. Despite the personal

---

[1] For a thorough discussion see (Renn and Stachel 2007).

intervention of then-president of the Kaiser Wilhelm Gesellschaft, Max Planck, he was removed from this position by the new Hitler regime in 1933. He then assumed a position in New York City at what was to become Rockefeller University in 1936. Max Bergmann is recognized as one of the founders of modern biochemistry. Peter's aunt, Clara Grunwald, was the founder of the Montessori movement in Germany. He had fond memories of visits with his mother's eldest sister in Berlin.[2] He clearly had benefited from Montessori methods, as attested by his aunt in references to him in her letters written from what had become a forced labor camp near Fürstenwald just outside of Berlin (Grunwald 1985). In 1943 Clara Grunwald perished with her students in Auschwitz.

After completing his doctorate in physics at Charles University in Prague in 1936, Peter Bergmann was appointed an assistant to Albert Einstein at the Institute for Advanced Study in Princeton. He worked with Einstein on unified field theory until 1941. There followed brief appointments at Black Mountain College, Lehigh University, Columbia University, and the Woods Hole Oceanographic Institute. In 1947 he joined Syracuse University, where he remained until his retirement in 1982. From 1963 to 1964 he also held an appointment as Chair of the Department of Physics of the Belfer Graduate School of Science of Yeshiva University. He remained active for many years after his retirement with a research appointment at New York University. Syracuse became the center for relativity research in the United States during the 1950s and early 1960s, bringing virtually all the leading relativists in the world for either brief or prolonged interaction with Bergmann and his collaborators. Bergmann concentrated from the beginning on the conceptual and technical challenges of attempts to quantize general relativity. Not unlike Einstein himself, his deep physical intuition was founded on hands-on laboratory experience, in his case extending back to "enjoyable" laboratory courses in physics and chemistry in 1932 as an undergraduate at the Technical University in Dresden (Bergmann Archive). Later on he expressed appreciation for the opportunity that teaching at the graduate level had given him to explore domains outside of relativity. His two-volume set of lectures on theoretical physics provides magisterial, lucid surveys of the field (Bergmann 1949a, 1951), and it is lamentable that it is now out of print. In fact, the purely mathematical aspect of relativity was not especially appealing to him, and he tended not to work closely with visitors in the 1960s who approached the subject from this perspective.[3] For additional biographical material see my short sketch (Salisbury 2005) and a longer discussion by Halpern (2005).

11.4 1949–1951

Bergmann's aim in his 1949 paper is to establish a general, classical, canonical framework for dealing with a fairly narrow set of generally covariant dynamical systems, but a set that includes as a special case general relativity described by the

---

[2] Personal communication.

[3] Engelbert Schucking, personal communication.

Lagrangian $\mathcal{L}_W$ above. He assumes that, under the infinitesmal general coordinate transformation $x'^\mu = x^\mu + \xi^\mu(x)$, the $\bar{\delta}$ transformations are given by

$$\bar{\delta}y_A = F_{A\mu}{}^{B\nu}y_B\xi^\mu_{,\nu} - y_{A,\mu}\xi^\mu, \tag{11.10}$$

where the $F_{A\mu}{}^{B\nu}$ are constants. Noether is not cited in this paper, surely because at this time her theorem was common knowledge.[4] A principal concern from the start is with the group structure of these symmetry transformations, and with the requirement that canonically-realized variations faithfully reproduce the $\bar{\delta}$ variations.

Due to the intended use of the Lagrangian $\mathcal{L}_W$, an additional term will appear on the right-hand side of the invariance assumption (11.5). This eventuality is accommodated by Bergmann with the assumption that $\bar{\delta}L \equiv Q^\mu_{,\mu}$. Rather than consider ξ^μ that vanish on the integration boundaries, he equivalently requires the identical vanishing of that contribution to the duly-rewritten (11.6), which now cannot be written as a total divergence. Thus he obtains the generalized contracted Bianchi identity (11.8), which for the variations (11.10) takes the form

$$\left(F_{A\mu}{}^{B\nu}y_B L^A\right)_{,\nu} + y_{A,\mu}L^A \equiv 0. \tag{11.11}$$

It is at this stage that new information is mined from the invariance of the Lagrangian. In particular, Bergmann noted that the third time derivative of y_A appears linearly in (11.11), and its coefficient must vanish identically,

$$F_{A\mu}{}^{B0}\Lambda^{AC}y_B \equiv 0, \tag{11.12}$$

where

$$\Lambda^{AB} := -\frac{\partial^2 L}{\partial\dot{y}_A\partial\dot{y}_B} \tag{11.13}$$

is minus the Legendre matrix.[5] Thus the Legendre matrix possesses null vectors. This is the signature of singular Lagrangians.

Bergmann deduces several interrelated consequences. Firstly, since by assumption the Euler–Lagrange equations are linear in $\ddot{y}_A$, with the linear term of the form $\Lambda^{AC}\ddot{y}_C$, the following four linear combinations of the equations of motion do not contain accelerations:

$$y_B F_{A\mu}{}^{B0}L^A = 0. \tag{11.14}$$

Therefore, the evolution from an initial time will not be uniquely fixed through an initial choice of y_A and $\dot{y}_A$. Secondly, it will not be possible to solve for velocities in terms of canonical momenta $\pi^A := \partial L/\partial\dot{y}_A$ since the matrix Λ^{AB} cannot be inverted. Thirdly, since

$$y_B F_{A\mu}{}^{B0}\frac{\partial\pi^A}{\partial\dot{y}_C} \equiv \frac{\partial}{\partial\dot{y}_C}\left(y_B F_{A\mu}{}^{B0}\pi^A\right) \equiv 0, \tag{11.15}$$

---

[4] As far as I can tell, Bergmann's first explicit published reference to Noether's theorem occurs in his *Handbuch der Physik* article on general relativity (Bergmann 1962).

[5] A "dot" signifies a derivative with respect to time.

straightforward integration yields a constraining relation among the momentum π^A and configuration variables y_B.

Bergmann was not aware that Léon Rosenfeld had obtained the same results in 1930, but working directly with the identity (5). (Rosenfeld 1930) Rosenfeld used the fact that in this identity the coefficients of each order of time derivative of the arbitrary functions ξ^i must vanish. He did not make explicit use of the Bianchi identities.

Although the stated central objective of Bergmann's paper was to prepare the ground for a full-scale quantization of the gravitational field, he did note that the canonical phase space approach offered a potentially new method for solving the classical particle equation of motion problem. Indeed, he expressed a hope shared by Einstein that, by avoiding singular field sources at the locations of point particles, it might be possible to eliminate singularities in an eventual quantum-gravitational field theory. This hope led in the second paper, henceforth denoted BB49, co-authored with Brunings, to the introduction of a parameterized formalism, in which spacetime coordinates x^μ themselves became dynamical variables (Bergmann and Brunings 1949). For the further development of the constrained dynamical formalism this was an unnecessary computational complication, yet several important results were obtained. In the parameter formalism, the Lagrangian is homogeneous of degree one in the velocities. Consequently, the Hamiltonian density $\mathcal{H}$ vanishes identically. It was possible to find immediately seven functions of the y_a and conjugate momenta π^b whose vanishing follows from the Legendre map $\pi^a(y, \dot{y}) := \partial L / \partial \dot{y}_a$. (The range of the index a has been expanded by four to accommodate the spacetime coordinates.) BB49 recognized that the pullback of the Hamiltonian under the Legendre map yielded a null vector of the Legendre matrix,

$$0 \equiv \frac{\partial}{\partial \dot{y}_a} \mathcal{H}\left(y, \pi(y, \dot{y})\right) = \frac{\partial \mathcal{H}}{\partial y_b} \Lambda^{ba}. \tag{11.16}$$

But the homogeneity of the Lagrangian implies that the velocities are also components of a null vector. It follows that one may set $\dot{y}_a = \partial \mathcal{H} / \partial \pi^a$. Dirac would soon reach the same conclusion in his parameterized flat spacetime models (Dirac 1950). Apparently unbeknownst to either of the parties, Rosenfeld had already shown in 1930 that a relation of this form more generally reflects the freedom to alter the velocities without affecting the momenta, albeit in models with Lagrangians quadratic in the velocities (Rosenfeld 1930). Next, considering variations of $\mathcal{H}$ at a fixed time, and using the Euler–Lagrange equations, Bergmann and Brunings obtained the "usual" additional Hamiltonian dynamical equations $\dot{\pi}^a = -\frac{\partial \mathcal{H}}{\partial y_a}$. BB49 do note that there is considerable freedom in the choice of the vanishing Hamiltonian: Given any $\mathcal{H}$ resulting from the homogeneity of the Lagrangian, one may multiply by an arbitrary function of the spacetime coordinates and add arbitrary linear combinations of the remaining seven constraints without altering the canonical form of the Hamiltonian equations. However, they do appear to claim erroneously that the vanishing of all of these possibilities is preserved under the evolution of a fixed Hamiltonian. Unfortunately, this renders untenable the proposed Heisenberg picture quantization, in which the quantum states are annihilated by all of the constraints $\mathcal{L}_\mathcal{W}$.

In this paper, we find the first statement of the requirement of projectability under the Legendre transformation from configuration-velocity space to phase space. Only those functions Ψ that are constant along the null directions u_a of Λ^{ab} have a unique counterpart in phase space since

$$u_a \frac{\partial}{\partial \dot{y}_a} \Psi(y, \pi(y, \dot{y})) = u_a \frac{\partial \Psi}{\partial p^b} \Lambda^{ba} = 0. \tag{11.17}$$

This requirement remained a concern until it appeared to have been resolved in 1958 through the elimination of lapse and shift variables, as described below. Only much later was its relevance to the canonical symmetry group understood (Lee and Wald 1990; Pons et al. 1997).

The explicit expression for the Hamiltonian was obtained by Bergmann, Penfield, Schiller, and Zatzkis in the next paper, henceforth denoted BPSZ50 (Bergmann et al. 1950). Because of the ensuing complications in the parameterized formalism, the solution was a daunting task. The work focuses on an algorithm for transforming the Legendre matrix into a "bordered" form in which the final eight rows and columns are zero. We will not address the details here since much of the technology was rendered superfluous by the discovery by Penfield, one of Bergmann's students, that the parameterization could be profitably dispensed with. Josh Goldberg vividly recalls the ensuing excitement; it was he who communicated the news to their approving mentor.[6] Penfield worked with a quadratic Lagrangian of the form

$$L = \Lambda^{A\rho B\sigma}(y) y_{A,\rho} y_{B,\sigma}, \tag{11.18}$$

so

$$\pi^A = 2\Lambda^{A0Ba} y_{B,a} + \Lambda^{AB} \dot{y}_B, \tag{11.19}$$

where $\Lambda^{AB} := 2\Lambda^{A0B0}$ is the Legendre matrix (Penfield 1951). His task was to find the appropriate linear combination of the $\dot{y}_A$ such that Λ^{AB} is transformed into a bordered matrix. In somewhat more technical terms, he sought a linear transformation in the tangent space of the configuration-velocity space such that each null vector acquires a single non-vanishing component. This procedure had already been undertaken in BPSZ50, but its implementation in this context was much simpler.

Indeed, it is immediately clear from (11.19) that, once a particular solution for $\mathcal{H}$ is found, resulting in a fixed $\dot{y}_A$, any linear combination of the remaining constraints may be added to H since, as also noted in BPSZ50, the additional terms do not change π^B. (Recall that the gradients of constraints with respect to momenta are null vectors.)

As pointed out already in BB49, additional gauge symmetry can easily be incorporated into the formalism, resulting in as many new constraints as there are new gauge functions. Thus both BB49 and BPSZ50 produced Hamiltonians for gravity coupled to electromagetism.

---

[6] J. Goldberg, personal communication.

At some time in 1950, the Syracuse group became aware of the pioneering work of Leon Rosenfeld. Reference to Rosenfeld appears in a 1950 *Proceedings* abstract (Bergmann 1950). James Anderson thinks it is possible that he brought the work to Bergmann's attention,[7] and Bergmann showed the paper to Ralph Schiller. In fact, according to Schiller the paper inspired his doctoral thesis.[8] In any case, the culminating paper of this period by Bergmann and Anderson, henceforth denoted BA51 (Bergmann and Anderson 1951), was written after this discovery, and it does appear that the authors were motivated by it to broaden the scope of their published investigations of constrained Hamiltonian dynamics. In particular, in addition to abandoning the parameterized theory, BA51 contemplated more general symmetry transformations, similar to those of Rosenfeld:

$$\bar{\delta} y_A = {}^0f_{Ai}(x, y, \ldots)\xi^i(x) + \cdots + {}^Pf_{Ai}^{\nu_1 \cdots \nu_P}(x, y, \ldots)\xi^i_{,\nu_1 \cdots \nu_P}(x). \qquad (11.20)$$

The BA51 collaboration was a watershed, in which most of the basic elements of the formalism were presented. For the first time in this paper the question was asked whether coordinate-transformation-induced variations of the momenta are realizable as canonical transformations. BA51 assumed that the canonical generator density $\mathcal{C}$ of these symmetry transformations could be written as

$$\mathcal{C} = {}^0A_i\xi^i + {}^1A_i\frac{\partial \xi^i}{\partial t} + \cdots + {}^PA_i\frac{\partial^P \xi^i}{\partial t^P}, \qquad (11.21)$$

where the ${}^MA_i$ are phase space functions and t is the coordinate time. Thus it was necessary to show, as they did, that the momenta variations do not depend on time derivatives of ξ of order higher than P; the potential offending term in $\bar{\delta}\pi^A$ is $2\Lambda^{AB}(-1)^P \, {}^Pf_{Bi}^{\nu_1 \cdots \nu_P}\frac{\partial^P \xi^i}{\partial t^P}$, but $\Lambda^{AB} \, {}^Pf_{Bi}^{\nu_1 \cdots \nu_P}$ vanishes identically, extending the null vector relation (11.12) to this model. Most importantly, BA51 argued that, since the commutator of transformations generated by $\mathcal{C}$'s must be of the same form, the ${}^MA_i$'s must form a closed Poisson bracket algebra. Furthermore, they were able to show that the ${}^PA_i$ are the constraints that follow from the momenta definitions. For these they introduced the term "primary constraints." They showed that, in order for these constraints to be preserved under time evolution, all of the ${}^MA_i$'s must be required to vanish; again, according to their terminology, ${}^{P-1}A_i$ is a secondary constraint, ${}^{P-2}A_i$ tertiary, etc. The argument employed here is similar to one used by Rosenfeld. Up to this point Rosenfeld's results are similar. He does not, however, take the next step, in which BA51 derives a partial set of Poisson relations among the ${}^MA_i$'s. All of these results are displayed explicitly for gravity and a generic generally-covariant model that includes Einstein's gravity as a special case.

11.5 Preview of Some Later Developments

It is not possible to do justice to Bergmann's complete oeuvre in constrained Hamiltonian dynamics in this paper. I will just briefly mention two important developments

---

[7] Personal communication.
[8] Personal communication.

that are treated in detail elsewhere, and then conclude with a teaser of contemporary importance. Much effort was expended in Syracuse in the 1950s in constructing gravitational observables: functions of the canonical variables that are invariant under the full group of general coordinate transformations. In 1958 Paul Dirac published his simplified gravitational Hamiltonian, achieved through a subtraction from the Lagrangian that resulted in the vanishing of four momenta (Dirac 1958).[9] He argued that the corresponding configuration variables, the lapse and shift functions, could then simply be eliminated as canonical variables. There remained a puzzle over the precise nature of the canonical general coordinate symmetry group. Bergmann and Komar made considerable headway in describing this group in 1972 (Bergmann and Komar 1972). They showed in particular that the group must be understood as a transformation group on the metric field. This view was forced on them by the observation that the group involved a compulsory dependence on the metric, and it was manifested in part by the appearance of metric components in the group Poisson bracket algebra. A close relation exists between these developments and the "problem of time" in general relativity. Are invariants under the action of the group necessarily independent of time? The issue is addressed in an early exchange between Bergmann and Dirac, with which I will close (Bergmann Archive). In a letter to Dirac dated October 9, 1959, Bergmann wrote: "When I discussed your paper at a Stevens conference yesterday, two more questions arose, which I should like to submit to you: To me it appeared that because you use the Hamiltonian constraint H_L to eliminate one of the non-substantive field variables K, in the final formulation of the theory your Hamiltonian vanishes strongly, and hence all the final variables, i.e. $\tilde{e}^{rs}, \tilde{p}^{rs}$, are 'frozen,' (constants of the motion). I should not consider that as a source of embarrassment, but Jim Anderson says that in talking to you he found that you now look at the situation a bit differently. Could you enlighten me?" Here is Dirac's response, dated November 11, 1959: "If the conditions that you introduce to fix the surface are such that only one surface satisfies the condition, then the surface cannot move at all, the Hamiltonian will vanish strongly and the dynamical variables will be frozen. However, one may introduce conditions which allow an infinity of roughly parallel surfaces. The surface can then move with one degree of freedom and there must be one non-vanishing Hamiltonian that generates this motion. I believe my condition $g_{rs}p^{rs}$ is of this second type, or maybe it allows also a more general motion of the surface corresponding roughly to Lorentz transformations. The non-vanishing Hamiltonian one would get by subtracting a divergence from the density of the Hamiltonian."

---

[9] At about the same time, the same Hamiltonian was obtained independently by B. DeWitt and also by J. Anderson. Beginning with their first paper in 1959, and culminating in 1962, Arnowitt, Deser, and Misner produced an equivalent geometrically based gravitational Hamiltonian formalism (Arnowitt et al. 1962).

Acknowledgements

I would like to thank the Instituto Orotava for its hospitality and the Max Planck Institute für Wissenschaftsgeschichte for inviting me to contribute to this meeting. Thanks also to Josh Goldberg for his critical reading of this paper and helpful comments.

References

Anderson, J. L. and Bergmann, P. G. 1951. "Constraints in covariant field theories", *Phys. Rev.* 83: 1018–1025.

Arnowitt, R., Deser, S., and Misner, C. 1962. "The dynamics of general relativity", in *Gravitation: an Introduction to Current Research*, ed. L. Witten. New York: Wiley.

Bergmann Archive, Syracuse University.
http://archives.syr.edu/collections/faculty/bergmann.html

Bergmann, P. G. 1942. *Introduction to the Theory of Relativity*. New Jersey: Prentice-Hall.

——. 1949a. *Basic Theories of Physics: Mechanics and Electrodynamics*. New Jersey: Prentice-Hall (Dover, 1962).

——. 1949b. "Non-linear field theories", *Phys. Rev.* 75: 680–685.

——. 1950. "Covariant quantization of nonlinear field theories", *Proceedings of the Int. Congress of Mathematicians*, vol. 1. Providence, RI: American Mathematical Society.

——. 1951. *Basic Theories of Physics: Heat and Quanta*. New Jersey: Prentice-Hall (Dover, 1962).

——. 1962. "The general theory of relativity", in *Handbuch der Physik*, vol. 4, ed. S. Flügge. Berlin: Springer-Verlag.

Bergmann, P. G. and Brunings, J. H. M. 1949. "Non-linear field theories II. Canonical equations and quantization", *Rev. Mod. Phys.* 21: 480–487.

Bergmann, P. G. and Komar, A. 1972. "The coordinate group symmetry of general relativity", *Int. J. Theor. Phys.* 5: 15.

Bergmann, P. G., Penfield, R., Schiller, R. and Zatzkis, H. 1950. "The Hamiltonian of the general theory of relativity with electromagnetic field", *Phys. Rev.* 30: 81–88.

Dirac, P. A. M. 1950. "Generalized Hamiltonian dynamics", *Can. J. Math.* 2: 129–148.

——. 1958. "The theory of gravitation in Hamiltonian form", *Proc. R. Soc. London* A246, 333–343.

Goldberg, J. 2005. "Syracuse: 1949–1952", in *The Universe of General Relativity*, eds. A. J. Kox and J. Eisenstaedt. Boston: Birkhäuser.

Grunwald, C. 1985. *"Und doch gefällt mir das Leben": Die Briefe der Clara Grunwald 1941–1943*, ed. E. Larsen. Mannheim: Persona Verlag.

Halpern, P. 2005. "Peter Bergmann: the education of a physicist", *Phys. Perspect.* 7: 390–403.

Hilbert, D. 1915. "Grundlagen der Physik", *Nachr. Ges. Wiss. Göttingen*, 395.

Lee, J. and Wald, R. M. 1990. *J. Math. Phys.* 31: 725.

Noether, E. 1918. "Invariante Variationsprobleme", *Nachr. Ges. Wiss. Göttingen*, 235–257. (The original article is available for download from the Göttingen Digitalisierungs-Zentrum at http://gdz.sub.uni-goettingen.de/de/index.html. The original German and an English translation by M. A. Tavel are also available online at http: //www.physics.ucla.edu/~cwp/articles/noether.trans/german/emmy235.html.)

Pauli, W. 1921. "Relativitätstheorie", I *Enzyklopädie der Mathematischen Wissenschaften*, vol. 219. Leipzig: Teubner.

Penfield, R. 1951. "Hamiltonians without parametrization", *Phys. Rev.* 34: 737–743.

Pons, J. M., Salisbury, D. C. and Shepley, L. C. 1997. "Gauge transformations in the Lagrangian and Hamiltonian formalisms of generally covariant theories". *Phys. Rev.* D55, 658–668 [gr-qc/9612037].

Renn, J. and Stachel, J. 2007. "Hilbert's foundation of physics: from a theory of everything to a constituent of general relativity", in *Gravitation in the Twilight of Classical Physics: The Promise of Mathematics*, eds. J. Renn and M. Schemmel. (Genesis of General Relativity vol. 4.) Dordrecht: Springer.

Rosenfeld, L. 1930. "Zur Quantelung der Wellenfelder", *Ann. Phys.* 5: 113–152.

Salisbury, D. C. 2005. "Albert Einstein and Peter Bergmann", in *Albert Einstein: Engineer of the Universe: One Hundred Authors for Einstein*, ed. J. Renn. Weinheim: Wiley-VCH.

Salisbury, D. C. 2009. "Translation and commentary of Léon Rosenfeld's 'Zur Quantelung der Wellenfelder', Annalen der Physik 397, 113 (1930), "Max Planck Institute for the History of Science Preprint 381".

Sledbodzinski, W. 1931. "Sur des equations de Hamilton", *Bulletin de l'Academie Royale de Belgique* 5: 864–870.

Weyl, H. 1918. *Raum, Zeit, Materie*. Berlin: Springer.

12

Thoughts About a Conceptual Framework for Relativistic Gravity

Bernard F. Schutz

Max Planck Institute for Gravitational Physics (Albert Einstein Institute), Germany

12.1 Introduction

Mine is one of several talks at this meeting that consider the revival of relativity and its integration into the mainstream of physics beginning in the 1950s. Ted Newman has described the physics problems that created confusion during the slow period 1930–1950, and how eventually a new generation of young physicists pulled the theory out of its mire. Silvio Bergia has emphasized the changes of thinking that were required, and the importance of the physical insight and especially the geometrical perspective that John Wheeler, among others, brought to the subject. I want to focus on the gulf that opened up during the slow period between relativists and the rest of what I will call mainstream theoretical physics. This gulf is important not just because of the negative influence it exerted on the development of relativity. It also has much to teach us about what physicists expect from a theory of physics, and especially about the role of heuristic concepts in physicists' communication with one another.

My thesis is that general relativity, despite its essential *mathematical* completeness in 1916, did not become a complete theory of *physics* until the 1970s. In order to understand this period, scholars of relativity need to look, not just at progress in understanding the mathematical theory, but also at the slow development of heuristic concepts that were needed to enable relativists to talk to other physicists in a common language.

Today we have a fairly secure set of heuristic concepts: for example, we know what a black hole is, we know what gravitational waves do, we know how gravitational lenses work. These concepts—black holes, gravitational waves, gravitational lenses—have gained a kind of concrete physical reality although if you take them apart, they are just ideas that rest ultimately on rather complex (and usually approximate) solutions of Einstein's field equations. Very importantly, they are concepts that relativists can communicate to nonrelativists who may need them (astronomers, experimental physicists, historians, the general public) without needing to pass on all their mathematical underpinnings.

The key accomplishment of the generation of physicists who revived relativity is that they created a wide range of useful concepts like these out of the confusions that plagued the previous generation. This took a huge amount of work, but the work was not done randomly. Rather, a handful of creative and senior physicists, many of whom came to relativity from other branches of physics, very deliberately shaped the directions of research toward developing these paradigmatic concepts, thereby adding the physics to the mathematical skeleton of the theory. In my view, in the dark period, the absence of such a vision of how to make relativity into a working theory of physics was what led to the increasing isolation of relativity from the mainstream.

12.2 The Gulf of Relativity

The gulf between mainstream physics and relativity between 1930 and 1960 is remarkable for its hugeness (Eisenstaedt 1989, 2006). Rarely has an important sub-field of physics enjoyed such a poor reputation. Very few physicists moved back and forth across the gulf or even made an effort to communicate across the divide.

Equally remarkable has been the subsequent huge turnaround. Gravitational physics is mainstream physics today: massive amounts of money fund gravitational wave experiments; the holy grail of theoretical particle physics is to unify the nuclear and electromagnetic forces with gravity; a course in general relativity is standard for physics graduate students. A few short anecdotes serve to illustrate the depths to which relativity sank and the heights to which it has subsequently risen.

Anecdote 1. The Nobel prize-winning astrophysicist Subrahmanyan Chandrasekhar kept a remarkable scientific diary, in which at the end of each year he summarized his scientific work and decisions of that year. Shortly after Chandra's death in 1995, Norman Lebovitz (private communication) showed me some of the entries. Very interestingly, Chandra writes that, during the 1930s, he considered starting to do research in relativity, in order to explore what would happen to a compact star that exceeded the maximum white-dwarf mass that Chandra himself had recently established. He consulted other physicists, who strongly advised him against doing this. General relativity, one told him, had proved to be a "graveyard of many theoretical astronomers." Chandra particularly mentions that Niels Bohr discouraged him from making a move into relativity. (Considering Bohr's exchanges with Einstein on the interpretation of quantum mechanics, this is a tantalizing remark!) Chandra's reputation and career were by no means secure in the 1930s, and so he looked (very productively) elsewhere for research problems. It was not until after 1960 that he felt confident enough of his reputation to finally indulge his long-postponed wish to work on general relativity. [Kip Thorne, one of the dominant figures in modern relativity research, reports (Thorne 1994) that he received similarly negative advice when he was contemplating doing graduate work in relativity in the early 1960s.]

Anecdote 2. It is worth looking at Richard Feynman's famous reaction (in a letter to his wife) to the 1962 Warsaw relativity meeting (Feynman 1988):

I am not getting anything out of the meeting. I am learning nothing. Because there are no experiments this field is not an active one, so few of the best men are doing work in it. The result is that there are hosts of dopes here and it is not good for my blood pressure: such inane things are said and seriously discussed that I get into arguments outside the formal sessions (say, at lunch) whenever anyone asks me a question or starts to tell me about his "work." The "work" is always: (1) completely un-understandable, (2) vague and indefinite, (3) something correct that is obvious and self-evident, but worked out by a long and difficult analysis, and presented as an important discovery, or (4) a claim based on the stupidity of the author that some obvious and correct fact, accepted and checked for years, is, in fact, false (these are the worst: no argument will convince the idiot), (5) an attempt to do something probably impossible, but certainly of no utility, which, it is finally revealed at the end, fails, or (6) just plain wrong. There is a great deal of "activity in the field" these days, but this "activity" is mainly in showing that the previous "activity" of somebody else resulted in an error or in nothing useful or in something promising. It is like a lot of worms trying to get out of a bottle by crawling all over each other. It is not that the subject is hard; it is that the good men are occupied elsewhere. Remind me not to come to any more gravity conferences!

This is, of course, typical Feynman hyperbole. We know that at that meeting (Infeld 1964) a core group of relativists was already coming to grips with issues like energy, black holes, and the reality of gravitational waves. And, ironically, it took place just a year before the first Texas Symposium in Relativistic Astrophysics (Robinson *et al.* 1965), which is often regarded as the moment at which relativity began to have real interest for astrophysicists. Nevertheless, Feynman's remarks show why it would still be another couple of decades before mainstream theoretical physics would completely drop its prejudices against the relativity community. The two sides were not communicating.

Anecdote 3. I vividly remember my own personal experiences as a young relativist. In the 1970s; if I mentioned black holes to an astronomer, the best I could usually hope for was a patronizing smile. And this was after the discovery of what we now know was the first black hole in a binary system, Cyg X-1, by the Uhuru satellite (which led to the award of the 2002 Nobel prize to Riccardo Giacconi). Later, during the 1980s, when I moved into gravitational wave detection, many astronomers told me I was throwing away my career. And they were the sympathetic ones; others just saw me as a misguided threat to their own research funding!

Anecdote 4. If the low point of relativity was very low, the current high point is indeed very high. Nothing illustrates the dramatic nature of this turnaround better than money. By 2020 at least 3–4 billion dollars will have been invested by a dozen national and international scientific organizations in building gravitational wave detectors on the ground and in space. Most of this money has already been committed, at least in a planning sense, and this has all happened even before the first direct detection of a gravitational wave!

Where did today's immense faith in general relativity originate? How did relativity establish such strong credentials after being in such disrepute? It seems to me that to answer this question we need to do more than simply catalog the details of what happened in relativity and astrophysics to get us where we are today. We have to understand how physicists judge the credibility of other physicists. The tortured development of gravitational physics is a good case study of how physicists decide that other people are really doing physics, even though they may not understand the mathematical and technical details.

12.3 Heuristics in General Relativity

I won't attempt to give a complete set of answers to the questions I have just posed, but I think a key to answering them lies in the fact that physicists have a characteristic way of thinking, which they call physical intuition. Physicists think in terms of models, of heuristic concepts that they connect by using this physical intuition. Physicists' models must of course be founded on the mathematical expression of a theory, but physicists are typically not happy if all they have are mathematical links between their models. They want concepts that enable them to understand essential parts of theories, even if they have not developed a facility with the mathematics of those theories. They need to have models they can exchange with physicists in other specialties, which allow those physicists to work with the concepts without being expert in their underlying theory.

I will illustrate the conceptual changes in relativity between the "dark ages" and the modern era by considering two key issues that were also listed by Ted Newman in his talk as key problems that were not solved during the dark years of relativity. The first is the meaning of the Schwarzschild solution. In the 1930s people talked about the "Schwarzschild singularity" (by which they meant the horizon, not the crunch at the center). Today we use the term "black hole." There is a world of difference between the ideas behind these different terminologies. If you think you have a singularity, then you can't use it in a physical model. You don't know how to include such an object in a physical system, either as the outcome of gravitational collapse or as an object that might affect other objects with its gravitational field. On the other hand, the term "black hole" is a shorthand description of a real object, one which you can confidently include in models for some physical systems: as the constituent of a model for an X-ray binary system, for example, or as a gravitating center in the middle of a galaxy. In the first case you are paralyzed by incomprehension. In the second you can hide away all the nonlinear general relativity, if you wish, and treat the object as just another member of the vast zoo of objects that make up our fascinating universe.

My second example is gravitational radiation. In the low period, people worried about the reality of the radiation itself. Doubting that waves could remove energy from sources and/or deposit it in detectors, relativists were unable to draw the clear parallels with electromagnetic radiation that would have emphasized the natural place that general relativity has in theoretical physics. By resolving these issues, relativists were finally in a position by 1980 to take advantage of the discovery of the Hulse–Taylor binary pulsar to show that observations supported the dynamical sector of Einstein's theory. Not all the mathematical problems associated with gravitational waves have been solved even today; but the field has enough confidence in its approximation methods and its control over the remaining outstanding issues that it has been able to develop a thoroughly convincing physical picture of gravitational waves.

12.4 Why Did the Gulf Drift Open?

Why did relativists find themselves excluded from the rest of theoretical physics in the 1930s to 1950s? Apart from a few notable exceptions, such as J. Robert Oppenheimer and Lev Landau, hardly anyone worked in relativity and other areas of theoretical physics between 1930 and 1960. And Oppenheimer and Landau were largely ignored by relativists (Thorne 1994). Let me list some explanations that are often offered and indicate why I don't find them adequate.

1. General relativity is mathematically very difficult. The combination of nonlinearity and coordinate freedom made it difficult to make definite statements. This certainly underlay the problem that relativists had, and it explains why progress in understanding the theory was slow. But it does not explain the low regard that "real" physicists had for relativists. Indeed, one might have expected the relativists to have gained respect from the rest of physics for making even slight progress with such a difficult theory.

2. As Feynman remarked, there was little experimental data. This meant that progress relied especially strongly on the ability to ask and resolve the right kinds of theoretical questions. But one might have expected the field to have exploited the few observational hints that did exist. Chandrasekhar's upper limit on the mass of white dwarfs, coupled with Fritz Zwicky's suggestion that supernova explosions led to neutron stars, was a clear invitation to explore gravitational collapse and the Schwarzschild solution. But only Oppenheimer and Landau seem to have found this interesting. Importantly, they were physicists who approached relativity from outside, from the point of view of the mainstream theoretical community. Moreover, it is significant that the revival of relativity started during the 1950s without the stimulus of any new experimental or observational data. So, while an abundance of data would certainly have driven the field in the right direction had it been available, I don't think that its absence explains why the field slipped into such a low state.

3. Relativity had to compete with quantum theory for good people. As Feynman says, "few of the best men are doing work in it." The competition was certainly there, but I don't believe that physics was that compartmentalized in the 1930s to 1950s. Leading quantum theorists had a deep interest in general relativity; Wolfgang Pauli wrote a beautiful textbook on it. The theory was widely regarded as the supreme achievement of 20th-century theoretical physics. One would

think that if relativists had made their own work interesting to mainstream physicists, then they would not have worked in such isolation. There might have been many more Oppenheimers and Landaus crossing the gulf if the relativity community had welcomed them and worked with them, or at least been able to communicate with them.

4. The Second World War got in the way. There is no doubt that the war seriously retarded research, removing young people from research and inhibiting international scientific communication. The cold war afterwards did not help. Nuclear physics had proved so useful to the military that it (including particle physics) was well funded after the war, whereas relativity fell into a theoretical backwater. But I am not convinced that this should have caused relativists to lose their way. Attacking the key problems of this period did not require a lot of money. A small field can still earn the respect of the majority. And the relativity community seems to have suffered less than other fields from the divisions of the cold war. It seems clear to me that, once the revival started, it went significantly more rapidly because of the relatively free intellectual interchange between Western and Soviet-bloc scientists working in relativity.[1]

I believe that the gulf opened between relativity and mainstream physics, not directly because of the problems listed above, but because the relativity community's response to at least the first two problems was to ask the wrong questions. For example, one of the serious mathematical challenges that they faced was coordinate freedom. Ted Newman in his talk at this meeting cataloged the way that the community clearly missed opportunities to understand that the so-called Schwarzschild singularity is just a coordinate effect. Today this episode seems baffling. Relativists do not seem to have understood the importance of controlling the effects of coordinates on their results despite Einstein's emphasis that the physics should be coordinate invariant. They even had coordinate systems available at that time (from the work of Sir Arthur Eddington and Georges Lemaître) that went across the horizon in a nonsingular way.

During the same years, quantum physicists were (at times painfully) revolutionizing their physical thinking, agreeing that they should only concern themselves with the results of measurements, which they called observables, and that they should not try to create physical models for what can't be measured, such as the "paths" of quantum particles. Special relativity already had a similar tradition, going back to Einstein's *Gedankenexperimente*, which were designed to focus attention on the outcome of experimental measurements rather than phrase the predictions of special relativity in terms of observer-dependent notions of time and space. Yet this trend

---

[1] The International Society for General Relativity and Gravitation (known as the GRG Society), which today is the main professional society for relativists worldwide, is one of the few societies adhering directly to IUPAP, which has individual scientists as members, not national organizations. During the cold war this structure enabled the society and its predecessor (the International Committee for General Relativity and Gravitation) to organize relatively apolitical meetings that scientists from both sides of the Iron Curtain attended. An example was the famous meeting in Warsaw that Feynman criticized.

did not seem to influence research in general relativity in the period 1930–1950 as much as it should have.

Ted Newman also mentioned another example: the deep confusion over the concept of energy in space-times containing gravitational waves. The resolution of this issue only began when Hermann Bondi and his successors, who included Roger Penrose and Ted himself, discovered how to treat radiated energy far from its source. It is easy to understand why relativists felt that they needed to clarify the concept of energy: energy is one of the key heuristics of mainstream physics. However, I confess that I don't understand why relativists allowed the genuine difficulties of defining gravitational wave energy to stop their developing a physical understanding of gravitational waves themselves. It appears that, because it was difficult to define the energy of a radiating system or to localize the energy carried by waves, relativists during this period were unable to develop any kind of useful physical model for gravitational waves.

We know today that it is perfectly possible to describe the generation of gravitational waves and their action on a simple detector without once referring to energy. The quadruple formula for the generation of the waves and the geodesic equation for their action on a simple detector are all one needs, and these tools were available in 1918. It is also possible to show that gravitational waves certainly deposit energy in some kinds of detectors, without having a full global energy conservation law. Indeed, at the earlier relativity meeting in Chapel Hill in 1957 (DeWitt & Rickles 2011) Feynman presented a simple argument to show how a gravitational wave would heat a detector that has internal friction. The argument is so direct that I used a version of it myself in my undergraduate-level relativity textbook (Schutz 2009), and I extended it there to derive the standard expression (first put on a firm foundation by Isaacson 1968) for the local average energy flux in gravitational waves. I think that Feynman was right to be disappointed that his argument at Chapel Hill seemed to impress no one and was not taken up and developed by relativists at the time.

I think this example goes to the heart of the question. Feynman was asking a physicist's question, about how gravitational waves act. All he wanted was a convincing intuitive argument that the waves were real and that he could treat them as part of the rest of physics, for example, by extracting thermal energy from them. The relativists of his day, on the other hand, were not interested in this kind of physicist's answer, apparently not even as a first step toward a more complete understanding of gravitational waves. Instead, they seemed to want to transplant as much of the apparatus of energy conservation as they could from the rest of classical physics. Energy conservation is of course a key concept in theoretical physics. But the work of Emmy Noether had shown long before that one should not expect exact energy conservation in the absence of time invariance, e.g., in a space-time containing gravitational waves. In relativity energy will always be a subsidiary concept, valid in some circumstances and useless in others. I believe Feynman found it intensely frustrating that relativists seemed more interested in the pure-radiation energy concept—in other words, relativity for its own sake—than in exploring the interaction of gravitational waves with material systems—gravitational waves as part of physics.

Feynman's example was more than just symptomatic of the way mainstream physics reacted to relativity. Feynman was one of the few mainstream physicists who attempted to cross over the gulf in the 1950s, and he was a prominent opinion-former. Relativity might have been accepted back into the mainstream physics community much earlier if relativists had succeeded in establishing a fruitful dialog with Feynman. Instead, his well-publicized scorn surely damaged the standing of the relativity community materially.

12.5 Einstein and the Gulf

It is hard to escape the conclusion that Einstein himself was one of the main reasons that the relativity community found itself excluded from mainstream physics. His influence on relativity research was naturally enormous. He appears to have rejected the idea of gravitational collapse, for reasons that today are hard to understand. He also appears not to have been comfortable with gravitational waves, troubled by the coordinate problems. Coordinates were a particular issue, as Silvio Bergia emphasized at this meeting in connection with the issue of general covariance, a principle that seems to have inhibited the development of heuristic concepts until Wheeler began emphasizing a more explicitly geometrical perspective on gravity. Perhaps most importantly, Einstein was focused mainly on finding a unified field theory. He does not seem to have been interested in the importance that general relativity had in classical theoretical physics, still less in its potential in astronomy. Einstein's key bridge to mainstream physics was the search for unified field theory. Its failure seems to have left relativity without any other bridges.

One further aspect of Einstein's position that I believe may have been important was his rejection of the standard interpretation of quantum mechanics. Naturally, this isolated him from mainstream physics thinking. Perhaps Bohr's advice to Chandrasekhar not to go into relativity had at least something to do with this. But I think there was a more profound way in which Einstein's rejection of quantum heuristics hurt relativity. As I mentioned earlier, quantum physics changed the philosophy of theoretical physics. The key objective of quantum theory became the observable: don't try to describe or understand something that you cannot measure. Relativity could have benefited in the period 1930–1950 from this imperative to focus only on what is—at least in principle—measurable.

It is paradoxical that quantum physicists focused on the importance of observables long before relativists did. After all, coordinate invariance was a key tenet of general relativity. The difference between quantum theorists and relativists is that, in the quantum field the principle was practiced, while in relativity there seems to have been no systematic effort to focus on measurables as a way to solve coordinate difficulties until the "revival" began.

I remember, as a graduate student, my supervisor Kip Thorne emphasizing that if coordinate confusion threatened, then one should construct a thought experiment and worry only about what the experimenter could in principle measure; and he made it clear that his own supervisor, John Archibald Wheeler, had emphasized this to him. Wheeler, of course, had worked extensively on quantum physics before taking up relativity in the mid-1950s. The idea of focusing on observables was natural to him and to other physicists of his generation. It had unfortunately not developed sufficiently in relativity, and it seems clear to me that introducing the strict discipline of observability was essential to ending relativity's isolation.

12.6 The Gulf Closes

It is arguable that a key reason that relativity pulled out of its doldrums was that new blood entered the field with this maxim from quantum theory deeply ingrained in their physical thinking. For people like Bondi, Pascual Jordan, Wheeler, and Yakov Zel'dovich, among others, it was natural to test any question about general relativity with the demand that it be phrased in terms of observables. Is there something you can measure, at least an experiment in principle?

This way of thinking forces you, for example, to look for physical features of the black hole horizon other than just the bad behavior of some metric components. Do the local tidal stretching forces near the Schwarzschild "singularity" remain finite? Can a real body reach and cross this surface in a finite amount of its own time?

Regarding gravitational waves, this perspective leads you to ask whether a radiating body experiences a back reaction that changes its observable behavior, and whether the radiated gravitational waves in turn produce a measurable effect in the detector. It is then natural to ask under what circumstances it is reasonable to expect that a definition of energy exists that plays a role in self-gravitating systems analogous to what physicists are used to in nonrelativistic physics; but the energy question does not stop you from answering the questions about observable physical effects of gravitational waves.

Unfortunately, it may not be a coincidence that relativity began climbing out of the doldrums at about the same time that Einstein died. His disappearance left the subject open for people to come in who had a background in mainstream physics and who were asking different kinds of questions. A large number of people working actively in classical theoretical relativity today (leaving aside the quantum gravity and string theory communities) can trace their lineage back to a handful of key physicists who entered relativity between about 1950 and 1960. These physicists reinvigorated the subject by asking the right kinds of questions, and they answered these questions with new heuristic notions that enabled relativity to communicate with and fit into the rest of physics.

Nothing illustrates this change better than the evolution of the black hole concept, to which I referred earlier. The term "black hole" was coined as late as 1967 by Wheeler to describe something whose reality he initially also doubted, but which he finally came to understand was the likely endpoint for the evolution of a large range of massive systems. Today we talk about black holes, and not just the Kerr metric or the Schwarzschild solution, because the concept of a black hole is wider than just these time-independent exact solutions of Einstein's vacuum field equations. Wheeler himself took a major step toward our present picture by showing, with Tullio Regge, that the Schwarzschild horizon and exterior are stable against small perturbations. Immediately this meant that the idealized Schwarzschild solution was robust enough to be included in models of more complicated physical systems: it would retain its essential properties even when disturbed. This robust object is what we call the black hole.

Once the new generation of mathematical physicists recognized that their job was to develop a heuristic understanding of this object, they set to work. An immense number of research papers between 1960 and 1990, including some remarkably elegant mathematics, led to the modern concept of a black hole.

This concept is far wider than the exact Kerr solution of Einstein's equations. Black holes can have accretion disks around them, in which case they are not Kerr. They can have matter falling into them, so they need not even be time independent. They can convert matter into energy, as Penrose showed. They radiate thermal radiation, as Stephen Hawking showed. They even obey the laws of thermodynamics.

When an astronomer and a relativist talk about black holes, they need this common heuristic concept. In order to use black holes in models of astronomical systems, the astronomer needs to regard the black hole as a kind of black box, an object whose inputs and outputs are known but whose inner workings can be ignored. The astronomer wants to feel that it is safe to put a black hole into a binary system without worrying about the details of the horizon or the curvature singularity inside. Relativists today are able to provide astronomers with this black-hole black box.

12.7 General Relativity Is Part of Physics

This is absolutely typical of physical thinking in other fields. Astronomers talk about stars, by which they mean a synthesis of a huge amount of physics. Nobody can even write down the complete mathematics needed to give an adequate description of a star. Nevertheless, an astrophysicist knows pretty well what a star is. The same could be said about a laser, a superconductor, the plasma in a tokamak, or even about relatively simple composite systems like atoms, protons, and neutrons. Even in front-line research, where such concepts are not settled, physicists work hard to develop them. The string theory community uses very visual and geometrical heuristics to describe their work. The extension of strings to multi-dimensional branes has opened up a rich source of possible phenomenology, and it is striking to me that, when I listen to talks given by theorists about the applications of brane theory to cosmology and to gravitation theory, the speakers often skip completely over the mathematics in favor of drawings that condense the mathematics into visual relations.

I believe that this is a basic aspect of the way physicists think about physics. The mathematical representation of the laws of physics is their foundation, but physicists would generally be paralyzed if they could not package physical systems into heuristic black boxes, confident that they know (or at least someone knows!) enough about their internal complexity to understand how they will interact with each other.

General relativity has a reasonably well-developed set of physical constructs today. This was the reason, for example, that Ted Newman could give his talk without showing any equations: when he talked about black holes and gravitational waves, we all knew what he meant. Or at least those of you who are not specialists in general relativity knew something about what he meant, and you had faith that all of us who are specialists knew sufficiently more about what he meant for it to be safe for us to talk about these concepts as physical reality! Without that faith, physics would simply not be possible. In the 1930s relativity had few such heuristic concepts to offer, and it did not look like it was moving toward constructing many more of them. I suggest that this is what led to the big gulf between relativists and mainstream theoretical physicists between 1930 and 1950. If this picture is right, then general relativity emerged mathematically complete in 1916, but as a theory of physics it was not completed until the 1980s. This must be one of the most gradual of Kuhnian revolutions ever!

References

DeWitt, Cécile M., and Rickles, Dean (eds.), *The Role of Gravitation in Physics: Report from the 1957 Chapel Hill Conference*, Max Planck Research Library for the History and Development of Knowledge (Sources 5) (Max Planck Institute for the History of Science, Berlin, Germany, 2011).

Eisenstaedt, Jean (1989), "The low water mark of general relativity, 1925–1955", in *Einstein and the History of General Relativity*, eds. Howard, Don; and Stachel, John; Birkhäuser, Boston, 277–292.

——. (2006), *The Curious History of Relativity: How Einstein's Theory of Gravity Was Lost and Found Again*, Princeton University Press, Princeton.

Feynman, Richard P. (1988), *What Do You Care What Other People Think?* W.W. Norton, New York.

Infeld, Leopold (1964), *Relativistic Theories of Gravitation*, Pergamon Press, Oxford.

Isaacson, Richard (1968), Gravitational Radiation in the Limit of High Frequency. 11. Nonlinear Terms and the Effective Stress Tensor. *Physical Review* **166**, 1272–1280.

Robinson, Ivor; Schild, Alfred; and Schücking, Engelbert L. (1965), *Quasi-stellar Sources and Gravitational Collapse*, University of Chicago Press, Chicago.

Schutz, Bernard F. (2009), *A First Course in General Relativity* (2nd ed.) Cambridge University Press, Cambridge, UK.

Thorne, Kip S. (1994), *Black Holes and Time Warps*, W.W. Norton, New York.

Part IV

A New Worldview in the Making

13

Observational Tests of General Relativity: An Historical Look at Measurements Prior to the Advent of Modern Space-Borne Instruments

J. E. Beckman

Instituto de Astrofísica de Canarias & Consejo Superior de Investigaciones Científicas, Spain

13.1 The Importance of the Equivalence Principle

This chapter gives a brief summary of the main current lines of observational (and in a few cases experimental) research designed to test the predictions of general relativity (GR). The basic pieces of work described include laboratory tests of the principle of equivalence and also astronomical tests within the solar system, light deflection and light delays both within the solar system and on larger scales within the Galaxy, and binary pulsars as the most powerful current probes of GR. Contrasts between the numerical limits on the accuracy of the methods, the predictions of GR, and predictions of alternative theories are brought out, with the basic conclusion that so far GR gives an entirely adequate framework for the results of each different test applied. Finally, a very short survey of sensitive space-based tests is added, to give a perspective on current experimental trends.

The principle of equivalence lies at the heart of GR. Both Newton and Einstein used the principle of equivalence as a cornerstone of their theories of gravitation. Now it can be seen as the foundation of the geometrical approach to spacetime, as we will see in the development of this paper. The weak equivalence principle (WEP) of Newton can be summarized as saying that weight is proportional to mass, or more formally that the trajectory of a freely falling body (not acted on significantly by other forces such as electromagnetism, and too small to be affected tidally by gravitation) is independent of its internal structure or composition. This leads directly to the consequence that all bodies in free fall suffer identical acceleration. Einstein generalized this and his formulation is usually referred to as the Einstein equivalence principle (EEP). This can be summarized in three statements:

- The WEP is valid.
- The outcome of any local non-gravitational experiment is independent of the velocity of the freely falling frame in which it is performed.
- The outcome of any local non-gravitational experiment is independent of where and when in the universe it is performed. A measurement of the electric force between two charged bodies is a local non-gravitational experiment, whereas a

measurement of the gravitational force between two bodies (e.g., the measurement of G by Cavendish) is not.

The EEP leads directly to the conclusion that gravitation is a curved spacetime phenomenon, i.e., the effects of gravity are the same as those of living in curved spacetime. So if the EEP is valid, the only theories of gravity which can work are those satisfying the postulates of "metric theories of gravity":

- Spacetime has a symmetric metric.
- The trajectories of freely falling bodies are geodesics of that metric.
- In local freely falling reference frames the non-gravitational laws of physics (e.g., Maxwell's equations for electromagnetism) are those described by the formalism of special relativity.

GR is a metric theory of gravity, but there are others, e.g., the Brans–Dicke theory (Brans and Dicke 1961). In its current form superstring theory is not a metric theory, and some of its postulates violate the WEP at very low levels.

It is thus highly important to keep testing the EEP with increasing precision, both in the laboratory and from astrophysical observations.

13.2 Historical Tests of the Equivalence Principle (Mainly Laboratory but with Some Astronomical Input)

Galileo first qualitatively tested the WEP in his falling body experiments (beautifully confirmed as a demonstration by the Apollo 15 astronaut David Scott, who dropped a feather and a hammer onto the moon's surface to show them landing simultaneously) and Newton followed him fifty years later with pendulum timing using bobs of different metals, confirming the WEP with a precision of one part in a thousand. Potter (1923) reached a limiting accuracy with the pendulum method of one part in 10^6. Eötvös (1890) used two objects of the same weight but different materials suspended from a torsion fibre at either end of a rigid bar to test the WEP. The acceleration due to the Earth's gravitation plus its centrifugal rotational force were used to act on the two masses, and also the gravitational force of the Sun. If the gravitational forces act differently on two types of material, when the mount is rotated exactly 180 degrees, the angle made by the bar with the fiducial line on the mount should change slighly but measurably. Eötvös tested the WEP this way to five parts in 10^8. Newton had already realised that the Earth–Moon system and also Jupiter and its satellites are systems which can be used to test the WEP. These systems are subject to orbital accelerations and to solar attraction. Any violation of the WEP would result in orbital perturbations which can be calculated but which are not observed. Newton could put a precision limit to the violation of the WEP at one part in a thousand this way, and Laplace (1774) using better observational and new mathematical techniques for orbital analysis could set a limit of one part in 10^7 to WEP violation, i.e., the Earth and the Moon are accelerated towards the Sun equally within this error limit. The Eötvös method has given rise to a number of modern ground-based laboratory tests of the WEP, with increasing accuracy as techniques have improved. Roll, Krotkov, and Dicke (1964) used not only technical improvements but also allowed the Earth to turn the fiducial mount through 180 degrees with respect to the Sun via its diurnal rotation, thus minimizing hysteresis problems with the torsion fibre, and looked for a diurnal signal, which they failed to find (for aluminium and gold test masses) at a level of two parts in 10^{11} at the 95% confidence limit. Subsequent improvements by Braginsky and Panov (1972) with aluminium and platinum test masses reached a precision of one part in 10^{12} while Su et al. (1994) used a continually rotating torsion balance and obtained an accuracy of five parts in 10^{13}, the most accurate ground-based test of the WEP to date.

13.3 Lunar Laser Ranging Tests of General Relativity: WEP, EEP, and Beyond

Ever since Armstrong and Aldrin deployed prisms on the surface of the Moon in 1969, the technique of lunar laser ranging (LLR) has been available for (inter alia) tests of gravitational theory. Averaging tens of minutes of photon returns allows a distance measurement accurate to 2–3 cm. The LLR technique in fact allows two distinct tests of the EP. The first tests the effect of gravity on two bodies of different element composition. The Earth is principally iron/nickel, and the Moon principally silicate. The tidal oscillations induced as the Earth and Moon orbit each other in the field of the Sun (and the planets) are large but calculable with great accuracy. Subtracting the theoretically computed travel time for each epoch of observation from the observed time allows a test of the WEP with an accuracy of one part in 10^{13} (Williams et al. 2004) already considerably better than the precision now available in the laboratory, and with perspectives of a little improvement as time progresses and the number of observations increases steadily.

However, LLR gives an additional advantage for EP testing. It also can test the fact that in GR four parts in 10^{-10} of the mass of the Earth is attributable to gravitational self-energy. Although the Earth is not highly compact by astrophysical standards (see later how neutron stars give a much greater self-energy contribution), the effect, first theoretically quantified by Nordtvedt (1995) for solar system light travel times, allows a test of the EEP (also sometimes called the SEP or strong equivalence principle) which is an effective limit to the gravitational effect of self-energy mass (the self-energy mass of the Moon is negligible in comparison to that of the Earth) at a level of two parts in 10^{13}. The conventional way of expressing a possible departure from the SEP due to self-energy is to use the expression

$$[M_G/M_I]SEP = 1 + \eta(U/Mc^2) \tag{1}$$

where M_G is the gravitational mass and M_I the inertial mass due to the gravitationally induced mass (self-energy) U and Mc^2 is the total mass energy of the body. The value of U depends on the mass and geometrical compactness of the body in question. The constant η is a dimensionless parameter which quantifies any departure from the EP in this case. Application to the Moon–Earth system gives

$$[M_G/M_I E - M_G/M_I M]SEP = [U(E)/M(E)c^2 - U(M)/M(M)c^2]\eta$$

$$= -4.45 \times 10^- 10\eta \qquad (2)$$

The best accumulated measurement of the total differential acceleration of the Earth and Moon in the Sun's gravitational field is: $[M_G/M_I E - M_G/M_I M]EP$ is $-1(\pm1.4) \times 10^{-13}$ and from laboratory experiments the fraction of this effect due to composition differences is $1.0(\pm1.4) \times 10^{-13}$. Combining these results allows us to get a best measurement of the differential effect due to self-gravitation:

$$[M_G/M_I E - M_G/M_I M]SEP = -2.0(\pm2.0) \times 10^{-13} \qquad (3)$$

Combining eqns. (2) and (3) we obtain the best current estimate of η from LLR to date as

$$\eta = 4.4(\pm4.5) \times 10^{-4} \qquad (4)$$

The parameter η can be shown to be a linear function of parameters termed "post-Newtonian" in the relativistic formulation of the equations of motion under gravity. The two largest of these parameters, β and γ, are related to η via the expression

$$\eta = 4\beta - \gamma - 3 \qquad (5)$$

The strong equivalence principle relates to the fact that a fraction of the mass of an object is a result of its gravitational interaction with itself, i.e., non-linear behaviour, and the β parameter is an indicator of the degree of non-linearity. As we will see below, an upper limit to the parameter γ has been determined using time delay and gravitational deflection techniques, and combining the upper limits for η and γ we find the result:

$$\beta - 1 = 1.2(\pm1.1) \times 10^{-4} \qquad (6)$$

This result is not a significant departure of β from unity.

13.4 Other Results from LLR with Implications for Theories of Gravity

One of the more interesting results of LLR has been to set new limits on the rate of change of Newton's constant of universal gravitation G. This has been taken seriously by Dirac, with his large number hypothesis, and also by Brans and Dicke in their scalar-tensor theory of gravitation. Interest in the possibility of such theories is linked to modern superstring theories, where G is a dynamical quantity. If G varies with time, this will give rise to a steady variation in the mean Moon–Earth distance, which can be measured with LLR. Variations in G, written to first order as $\dot{G}/G$, could be related to the expansion of the universe using the expression

$$\dot{G}/G = \sigma H_0 \qquad (7)$$

where H_0 is the value of the Hubble parameter and σ a dimensionless parameter, whose value depends on both G and the cosmological model. The best practical LLR results on this at the end of 2004 (Williams et al. 2004) give

$$\dot{G}/G = 4(\pm 9) \times 10^{-13} \, \mathrm{yr}^{-1} \tag{8}$$

The uncertainty in $\dot{G}/G$ is 83 times smaller than the inverse age of the universe to $1/H_0 = 13.4 \, \mathrm{Gyr}$ (flat universe with H_0 determined at $72 \, \mathrm{km \, s^{-1} \, Mpc^{-1}}$) from the Hubble Key Project (Freedman et al. 2001) or the WMAP Cosmic Microwave Background Anisotropy satellite (Spergel et al. 2003). This uncertainty is decreasing with time as the baseline for the LLR measurements grows.

A further interesting limit set by LLR is that for the geodetic precession of the lunar orbit. Without going into detail, if K_{gp} is a parameter expressing the relative deviation of the geodetic precession rate from its predicted GR value, whose value should be zero in GR the best results to date give

$$K_{gr} = -0.0019(\pm 0.0064) \tag{9}$$

which is indistinguishable from zero. Here more work will have to be done to evaluate the effect of lunar core oblateness if any improvement in sensitivity is to be achieved.

13.5 Light Deflections and Light Delays in the Solar System

13.5.1 Light Deflections

The Eddington (1919) observation of light deflection of starlight by the Sun which "put Einstein on the map" was an excellent result, but accurate to only 30%, and subsequent similar experiments did not do much better. The development of very long baseline interferometry (VLBI) by radio astronomers enabled radio sources to have angular separations and deflections measured to within 100 microarcseconds. The earlier measurements using this method used the passage of QSOs (strongly radio-emitting point sources) near the Sun as test sources. The measured angle $\delta\theta(t)$ as a function of time between pairs of QSOs as one of them passes close to the Sun can be fitted to a theoretical prediction with one of the fitted parameters being $0.5(1+\gamma)$.

The best value from single measurements was obtained by Lebach et al. (1995) using 3C273 and 3C279, giving

$$(1+\gamma)/2 = 0.9996(\pm 0.0017)$$

which is compatible with a GR predicted value of 1. A more recent all-sky study of light deflection via VLBI by Shapiro et al. (2004) using deflections over the full celestial sphere (at 90 degrees from the Sun the deflection is still 4 milliarcseconds!) with over 2 million VLBI observations, yielded

$$(1+\gamma)/2 = 0.99992(\pm 0.00014) \tag{10}$$

which is the best solar system value to date.

13.5.2 Light Delays

A radar signal sent across the solar system past the Sun to a planet or satellite and returned to Earth suffers a non-Newtonian additional delay in its round-trip time of flight. For a ray which passes close to the Sun this delay, $\delta t(S)$, is given by

$$\delta t(S) \simeq 0.5(1+\gamma)[240 - 20\ln(d^2/r)]$$

where the units are in microseconds, d is the distance of closest approach to the Sun in units of solar radii, and r is the distance of the planet or probe from the Sun in astronomical units. This effect, called the Shapiro effect, was derived as a consequence of GR by Shapiro et al. (1977).

  In practice there is no "relativity free" way of obtaining the light travel times, so the observer allows the reflecting body to approach the Sun, pass close to or behind it, and emerge, plotting the effect on the travel time, fitted to a model. The precise orbital elements needed to do this are obtained by radar ranging the body in question when it is far from superior conjunction, i.e., when the Shapiro delay is negligible. The predicted extra travel time close to superior conjunction is modelled as a function of $(\gamma+1)/2$ and the best least-squares fit of the predicted curve to the observed curve gives the value of this parameter. The method has been performed using planets (e.g., Mercury and Venus) used as passive reflectors of the radar signal, or probes such as Mariner 6 and 7, Voyager 2, and the Viking Mars landers and orbiters which give "active reflection" and hence boost the signal to noise.

  However, the most accurate limit to the departure of gamma from unity via solar system measurements was made using the Cassini probe on its way to Saturn as it made conjunction with the Sun (Bertotti et al. 2003). These authors give the expression for the difference δ in light travel time: ground antenna–spacecraft–ground antenna, produced by the Sun's (as M_s and radius R_s) gravitational field at distances r_1 and r_2 from the spacecraft as

$$\delta t = 2(1+\gamma)GM_s\ln(4r_1r_2/b^2)/c^3 \tag{11}$$

where G is the gravitational constant, b is the impact parameter of the ray (distance of the asymptotic projected ray from the solar centre, assumed to be much smaller than r_1 and r_2, and this condition is true for the Cassini probe), and c is the velocity of light. This can be converted into a numerical Doppler shift in the frequency of the signal, expressed fractionally by $y(gr)$ where

$$y(gr) = d(\delta t)/dt = -2(1+\gamma)GM_s.db/dt/c^3b$$
$$= -(1\times 10^{-5}\,\mathrm{s})(1+\gamma)(db/dt)/b. \tag{12}$$

For a spacecraft much farther away from the Sun than the Earth, db/dt is close in value to the Earth's orbital velocity of $\sim 30\,\mathrm{km\,s^{-1}}$. In the Cassini measurement the maximum measured value for $y(gr)$ was 6×10^{-6} (see Figure 13.1). As always, the practical experimental details of the system required to counteract solar plasma noise and to make the precision observations required are complex and we cannot discuss

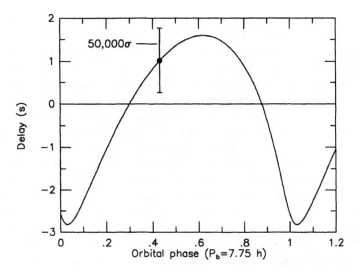

Fig. 13.1. Orbital delays measured for the "original" binary pulsar PSR 1913 + 16, during the month of July 1988. The error bar shown is 50,000 times bigger than the true error bar for an individual five-minute integrated measurement. This impressive accuracy underlies the use of pulsars in binary systems for quantitative testing of GR predictions. (Figure from Taylor, 1994, Fig. 5.)

them here. Also a complex dynamical model was needed to take into account the anisotropic thermal emission of the spacecraft, although this is in fact four orders of magnitude smaller than the relativistic signal. Finally a result of

$$\gamma = 1 + (2.1 \pm 2.3) \times 10^{-5} \tag{13}$$

was obtained, in excellent agreement with GR, though not excluding lower-level effects consistent with superstring theory. This is a remarkably precise result obtained taking the best possible advantage of long-trajectory solar system space probes.

13.6 Binary Pulsars as Probes of GR

13.6.1 Pulsars and Binary Pulsars

As an astronomer I find the most fascinating method developed to test GR to be the use of pulsars in binary orbits around other compact stars, either neutron stars or white dwarfs. Pulsars are remarkable objects in themselves. They are rapidly rotating stars made of matter at nuclear densities, a solar mass compacted into a volume of diameter less than 10 km, whose highly intense magnetic fields cause beamed bursts of energy, which sweep around the sky like a lighthouse. We observe these bursts as the beam sweeps past the Earth. They are exceptional laboratories for high-energy physics, and the stability of their pulse frequency makes them among the best clocks in the universe. This is particularly true for the sub-class of pulsars known as millisecond pulsars, formed by a whip-and-top like spin-up process as the dense neutron star accretes material from a larger less-dense partner. This is not the place to go into a discussion of how pulsars form (as supernova remnants) or even how they form in binaries. We will take it as given that pulsars in binary systems exist. A little history is relevant, particularly in the present context. The first binary pulsar was found by Hulse and Taylor (1975) using the 300 m Arecibo radio telescope, a discovery which later led these researchers to receive the Nobel prize (see Hulse 1994; Taylor 1994). The formidable technical prowess required to detect a pulsing source, and to show how the frequency dependence of the arrival time of the pulses can be used to determine the characteristics of the interstellar medium between us and the emitter, and hence to estimate its distance, is an important chapter in the history of astrophysics. The extreme stability of the pulse frequency enables pulses to be incorporated into a timing scheme over periods of months and even years, in the context of an irregular distribution of observations influenced by observing conditions. The pulse sequence number can usually be unambiguously established, even though the pulsar has rotated between 10^7 and 10^{10} times between one observation run and the next.

From the TOAs (times of arrival) of the pulses, obtained by comparison with a stable reference clock, a wealth of information can be obtained about a pulsar's spin, location in space, and orbital motion. The techniques now used allow pulse timing to an accuracy of order 10 microseconds. It took both technical expertise and perception for Hulse to realise that one of the pulsars he was observing had variations in its pulse timing sequence that could only correspond to its being in orbit around another massive body, while Taylor, his thesis supervisor, was quick to realise the importance of this for testing GR.

13.6.2 Model Fits to Pulsar TOA Sequences

A sequence of TOAs for a binary pulsar can be represented by an equation which first transforms the observed times of arrival to the times at the solar system barycentre (i.e., taking out all secular motions of the observer around this point) and then expresses these times of arrival in terms of the times of emission by the pulsar. The transformation equation must be in terms of space-time 4-vectors, and must take into account terms which depend on (1) the delay due to the interstellar medium between the pulsar and the observing point, (2) propagation delays and relativistic time adjustments within the solar system, and (3) propagation delays and relativistic time adjustments within the binary pulsar's orbit. The time delays under headings (2) and (3) are both separable into three components.

The first is the "Roemer" delay which is essentially the time taken for light to cross the orbit (the Earth's orbit in case (2) and the pulsar's orbit in case (3)). The Roemer terms have amplitude of order the orbital period times v/c where v is the orbital velocity and c the speed of light. The second is the "Einstein" delay which is the integrated effect of the gravitational redshift of the emitting or receiving body and time dilation (these are both important for the pulsar and negligible for the

Earth); these terms are smaller than the Roemer terms by a factor of $\sim ev/c$ where e is the orbital eccentricity. The third delay term is the Shapiro delay, described previously for solar system experiments, which represents the reduced velocities as the light is bent by the body around which the Earth or the pulsar is orbiting. Within the solar system this has a magnitude of up to 120 microseconds for lines of sight grazing the Sun, and varies with the logarithm of the impact parameter (see eqn. (11) above). For a binary pulsar the Shapiro delay depends on the mass of the companion star, the orbital phase, and the angle of inclination i between the orbital angular momentum and the line of sight. Figure 13.1 shows the orbital delay as a function of phase for the binary pulsar PSR $1913 + 16$ for observations during a given month, July 1988. This plot, which is clearly dominated by the Roemer delay, shows also the extremely high S:N possible for the observations which are integrated over five-minute intervals. This should in principle allow both the Einstein and Shapiro delays to be derived.

In order to analyse this type of data a first-order approximation uses the five Keplerian orbital elements: the projected semi-major axis $x = a\sin i/c$ in time units, the eccentricity e, the binary period Pb, the longitude of the periastron point Ω, and the time of periastron T_o. In addition, with accurate enough TOAs and sufficiently long sequences there are five further "post-Keplerian" parameters which can be measured, and which are due to relativistic effects (or their equivalents in corresponding post-Newtonian theories): the time derivatives of ω and Pb, i.e., $\dot{\Omega}$ and $\dot{Pb}$, respectively, the Einstein parameter γ which has been explained in the solar system context above, and the amplitude ("range") and the shape of the Shapiro delay, r and $s = \sin i$, respectively.

Tens of binary pulsars have been observed well enough to establish their five Keplerian parameters, and for these the orbital period and projected semi-major axis can be used to derive the "mass function," f, given by

$$f(m_1, m_2, s) = (m_2 s)^3/(m_1 + m_2)^2 = x^3/T_o(Pb/2\pi^2) \tag{14}$$

In (14) m_1 and m_2 are the masses of the pulsar and its companion, respectively, in units of the solar mass, s refers to $\sin i$, and $T_o = GM_o/c^3 = 4.925490947 \times 10^{-6}$ s where G is the gravitational constant and M_o the mass of the Sun. Although the individual masses of the pulsars or their companions cannot be derived using (15) assuming that the pulsar has a mass not too far from 1.4 M_o (the Chandrasekhar limit for white dwarfs) and that a median value for $\cos i$ is 0.5, i.e., $s = 0.87$, we can make estimates of the companion masses, and a set of these is plotted on Figure 13.2.

Most of them have low eccentricity orbits, and low masses, but five of those plotted in the figure have high eccentricities and probable companion masses greater than $0.8 M_o$. These are almost certainly neutron stars, and cannot be seen themselves as pulsars because the beams do not sweep the Earth. The large orbital eccentricities are very probably due to the "kick" as the second star is converted to a neutron star by way of a supernova. These are ideal GR laboratories, because their compactness ensures the virtual absence of tidal effects which could perturb direct comparison of

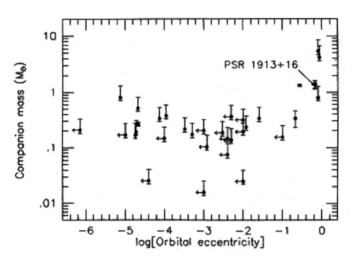

Fig. 13.2. Masses (in units of solar mass) of companion stars to observed binary pulsars derived from the orbital signal delays of the pulsars' emission, as a function of derived orbital eccentricity. The companions at the extreme upper right are on the main sequence, the two companions near the marked companion of PSR 1913 + 16 are (as are all the pulsars themselves) neutron stars, and the remainder are thought to be white dwarfs.

GR prediction with observations (the two much higher mass companions are probably main sequence stars, which complicate computations due to tidal and mass loss effects).

13.6.3 The Post-Keplerian Parameters (PKP)

In order to interpret the departures from the Keplerian parameters, Damour and Deruelle (1986) developed an analytical framework in which each measured post-Keplerian parameter defines a curve in the (m_1, m_2) plane, valid for a specified theory of gravity, so that the formalism allows verifications of GR in the first instance, and then tests of any possible departures as the observational results become more precise. For binary pulsars this precision grows almost linearly with time, which is one reason why they are such powerful test beds for GR. The PKP are:

1. The mean rate of periastron advance $\dot{\Omega}$. This is the equivalent of the rotation rate of the apse line of Mercury's orbit, which as we know was the earliest test of GR (and historically put paid to the "Vulcan" hypothesis). The striking fact is that for the two-neutron-star binary pulsars this is measured in units of degrees of arc per year, while for Mercury's orbit the units are seconds of arc per century, in other words the magnitude of the effect in the strong gravity of the neutron star fields is over four orders of magnitude greater than for the Mercury orbit. (It is interesting to comment that the "normal" eclipsing binary V541 Cygni shows a post-Newtonian component of the rotation of its orbital apse-line of 0.82 degree

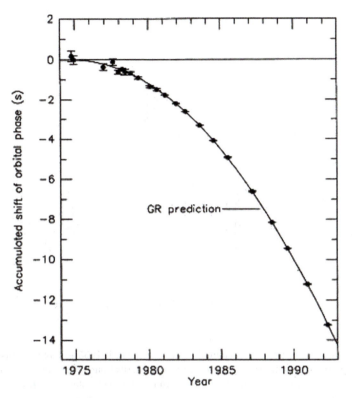

Fig. 13.3. Accumulated shift of the times of periastron in the PSR 1913 + 16 binary system, relative to an assumed orbit with constant period. The parabolic curve represents the GR prediction for energy losses due to gravitational radiation. Only GR or possibly Brans–Dicke would give a fit as good as this. (Figure from Taylor, 1994, following Taylor and Weisberg, 1982.)

per century, in excellent agreement with the GR prediction (Lines et al. 1989), and that this is one of the largest values measured for non-pulsars.)

2. The gravitational redshift/time dilation parameter γ, which for the pulsar PSR 1913 + 16 takes the value of 4.295(2) milliseconds.

3. The derivative of the orbital period $\dot{P}$. If this takes a measurable value, it gives a proof that orbital decay is taking place, and this in turn is an indirect but clear proof of the existence of gravitational radiation. The measurement by Taylor and Weisberg (1982) of this decay rate for PSR 1913 + 16 provided the first clear proof that energy is being radiated from the binary pulsar system by gravitational waves, and their figure showing the measured orbital decay, and giving excellent agreement with that predicted from GR, is shown in Figure 13.3.

4. (including point 5) The Shapiro delay which has two parameters: r, the delay, and s, related to the width, or "shape," of the delay. Binaries with orbits nearly perpendicular to the plane of the sky will show a maximum Shapiro delay. This

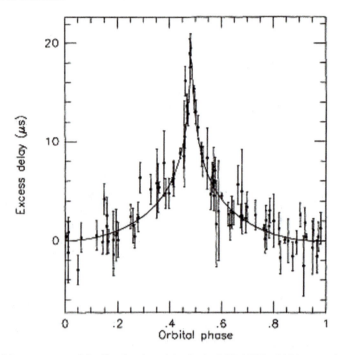

Fig. 13.4. Measurements of the Shapiro time delay in the PSR 1855+09 binary pulsar system, whose edge-on presentation allows us to optimize the derivation of the r and s parameters. The fitted curve is given in eqn. (15). (Data from Kaspi et al. (1994), plotted by Taylor (1994).)

is found to be the case for the binary pulsar PSR 1855 + 09, and the observations for this pulsar are shown in Figure 13.4 (data from Kaspi et al. 1994). A fitted curve with the form

$$\delta(s) = -2r \log(1 - s \cos(2\pi\phi - \phi(o)))$$ (15)

is shown. With measurements of both r and s one can derive the individual masses of the binary pair, which for this pulsar are $m_1 = 1.50(+0.26, -0.14)$ and $m_2 = 0.258(+0.028, -0.016)$ in units of the solar mass. In fact, the majority of measured neutron star masses are obtained in this way.

13.6.4 Using the PKP to Test GR

As mentioned above, with any two of the PKP determined from observations (and assuming that the five Keplerian orbital parameters are established, which they must be prior to deriving the PKP from observation) one can determine the masses of the two binary components. The formulae relating the PKP to the GR dynamical predictions then allow the prediction of the other three PKP, i.e., the measurements are overdetermined, providing that the assumptions of GR are in fact satisfied. This

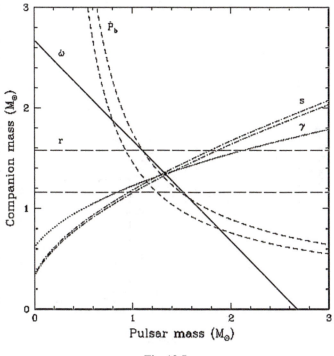

Fig. 13.5.

allows us to use any of the three remaining parameters as direct tests of GR predictions. The difficulty lies in finding systems where more than two of the PKP can in practice be measured accurately. Several binary pulsars satisfy the criteria for this. In his Nobel lecture, Taylor (1994) described the results for two of these, PSR 1913 + 16 and PSR B1534 + 12, at that date. In her more recent review Stairs (2003) gave the measured values for three PKP for PSR 1913 + 16 and for all five of these parameters for PSR B1534 + 12. The best graphical way to treat the results is in a "mass-mass" diagram. A given value for a given PK parameter defines a curve in the mass-mass plane for the two stars. In Figure 13.5 we can see Taylor's results for the three PKP $\dot{\Omega}$, γ, and $\dot{Pb}$. For the originally discovered binary pulsar PST 1316 + 16. Any pair of these enables us to determine the masses of the binary components; intersection of the third parameter at the point of intersection of the other two is a quantitative confirmation of the predictions of GR. In this case the prediction is that of gravitational wave emission by the orbiting stars in the strong fields of their companions.

The pulsar PSR B1534 + 12 allowed Stairs et al. (2002) to measure all five PKP since the angular momentum vector of the pulsar orbit lies close to the plane of the sky. The mass-mass diagram for this pulsar is shown in Figure 13.6 and as we see the two Shapiro delay parameters s and r are included. The thickness of the tracks is a measure of the uncertainties in the respective PK parameters. The agreement with GR here is not so impressive as for PSR 1913 + 16. In particular, $\dot{Pb}$

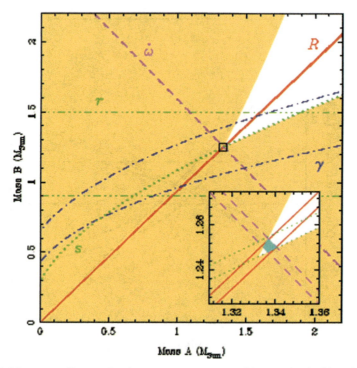

Fig. 13.6. Mass-mass diagram for the two neutron stars making up the double-pulsar PSR J0737–3039A/B, from Possenti et al. (2004). The grey area of the figure is the zone excluded by the Keplerian mass functions of the two pulsars. Pairs of lines enclose regions permitted by GR, which are: (a) advance of the periastron. $\dot{\omega}$ (dashed line, which is really two lines too close to separate on the main figure), (b) direct mass ratio R, from the projected semi-axis ratio (single solid line), (c) gravitational redshift γ (dashed dotted lines), (d) Shapiro parameter r, (dot-dot-dot-dashed horizontal lines), (e) Shapiro parameter s (dotted line which is really two lines too close to separate in the main figure). The inset shows enlarged (by a factor of 16) the region where the constraints overlap, showing that the GR predictions are precisely obeyed. An updated version of this diagram, with improved constraints due to a data base over a longer timescale, can be found in Kramer et al. (2006).

appears to give a disagreement with predictions, though the other four parameters agree well (of course the uncertainties in r are clearly higher than those for the other parameters). The problem with $\dot{P}b$ is that to derive it the observers need a distance-dependent kinematic correction, and the error in determining the distance to the pulsar dominates this correction. A slightly larger assumed distance is required to correct the small discrepancy here, and this systematic error is not at all excluded.

Quite a number of pulsars are now being used to probe the validity of GR, but here I will mention only one more example. This is the double pulsar PST J0737–3039A/B, discovered by Burgay et al. (2003), and reported on by Possenti et al. (2004). Apart from satisfying all the requirements of the single pulsars orbiting compact stars, this double pulsar can be used to measure directly the projected semi-major axis ratio, R, of the two components A and B, which in turn gives the ratio of their masses. In a mass-mass diagram this gives a simple linear plot, which is independent of the gravitational theory used to interpret the data, and in particular independent of strong field effects, and is of course an extra curve for the mass-mass diagram, as seen in Figure 13.6. In the publication by Possenti et al. accurate values for $\dot{\Omega}$ and s are presented, which together with the value of R give a very good fit to GR predictions as seen in the figure. Good values for γ and for $\dot{P}b$ were not then available. Although a moderately accurate value for γ leads to an error band which does include the common zone bounded by the error bars for R, $\dot{\Omega}$, and s. I have recently had a private communication from Possenti who tells me that there is now a good determination of $\dot{P}b$ which takes the value of $-1.2(\pm 0.30) \times 10^{-12}(ss^{-1})$ with corrections still to be applied for galactic acceleration and transverse velocity. Estimates do show that this prediction is in good agreement with the GR predictions from the other PK parameters. In fact, the coincidence area for the measured PK parameters keeps shrinking with observation time, but so far shows no deviation from GR.

13.7 Refined Tests for the Near Future

Tests of GR are always of considerable importance to the physics community, and a number of sensitive tests using satellites have been proposed, the proposals funded (by NASA and ESA), and elegant techniques of high sensitivity developed. Some have been proposed and indeed some of these are in orbit. In addition, there are powerful ground-based searches, some already under way, whose aim is to detect gravitational waves essentially from powerful astronomical explosions. However, it is beyond the scope of this review to describe what is a burgeoning field, and I have confined this article to what are the most outstanding historical tests. The informed reader will certainly be able to keep abreast of quantitative refinements, and one may only speculate on whether, or perhaps when, any discrepancies between the observations and Einstein's theory will be found.

13.8 Conclusions

During my active carreer as a physicist GR has evolved from being theoreticians' territory and has entered the realms of mature science. It is a remarkable tribute to the geniality of Einstein that all the experimental and observational tests so far performed are in better agreement with GR than with any of its rivals, and that so far no discrepancy with GR predictions has been established, at least within the limits of accuracy of the measurements described. To be sure the rival theories have tended to come from relativists who are keen to offer rival models in the abstract (e.g., the Brans–Dicke scalar-tensor theory which is one of the most sophisticated and successful). However, theorists have not yet directed very much attention to quantum gravity, or to the implications of particle theory for gravitation in general. This is surely because our techniques still seem insufficiently precise to test any small deviations from GR which may be implied. As the precision of the pulsar tests grow with time or perhaps with the arrival of the space experiments, observational deviations from GR at some level will become apparent. In that case the push for new theory will have come from observation, as has so often been the case in astrophysics and cosmology.

References

Bertotti B., Iess L. and Tortora P., 2003, "A test of general relativity using radio links with the Cassini spacecraft", *Nature* 425, 374.

Braginsky V. B. and Panov V. I., 1972, "Verification of the equivalence of inertial and gravitational mass", *Sov. Phys. JETP* 34, 463.

Brans C. H. and Dicke R. H., 1961, "Mach's principle and a relativistic theory of gravitation", *Phys. Rev.* 124, 925.

Burgay M., D'Amico N., Possenti A., Manchester R. N., Lyne A. G., Joshi B. C., McLaughlin M. A., Kreamer M., Sarkissian J. M., Camilo F., Kalogera V., Kim C. and Lorimer D. R., 2003, "An increased estimate of the merger rate of double neutron stars from observations of a highly relativistic system", *Nature* 426, 531.

Burgay M., D'Amico N., Possenti A., Manchester R. N., Lyne A. G., Kramer M., McLaughlin M. A., Lorimer D. R., Camilo F., Stairs I. H., Freire P. C. C. and Joshi B. C., 2006, "The double pulsar system J0737-3039", *Mem. Soc. Astron. Ital. Suppl.* 9, 345.

Damour T. and Deruelle B., 1986, "General relativistic celestial mechanics of binary systems, II. The post-Newtonian timing formula", *Ann. Inst. Henri Poincare (Physique Te'orique)* 44, 263.

Eddington A. S., 1919, "The total eclipse of 1919 May 29 and the influence of gravitation on light", *Observatory* 42, 119.

Eötvös R. von, 1890, *Mathematische und Naturwissenschaftliche Berichte aus Ungarn* 8, 65.

Freedman Wendy L., Madore Barry F., Gibson Brad K., Ferrarese Laura, Kelson Daniel D., Sakai Shoko, Mould Jeremy R., Kennicutt Robert C., Jr., Ford Holland C., Graham John A., Huchra John P., Hughes Shaun M. G., Illingworth Garth D., Macri Lucas M. and Stetson Peter B., 2001, "Final Results from the Hubble Space Telescope Key Project to Measure the Hubble Constant", *ApJ* 553, 47.

Hulse R. A., 1994, "Nobel Lecture, The discovery of the binary pulsar", *Rev. Mod. Phys.* 66, 699.

Hulse R. A. and Taylor J. H., 1975, "The discovery of a pulsar in a binary system", *ApJ* 195, L51.

Kaspi V. M., Taylor J. H. and Ryba M. F., 1994, "High precision timing of millisecond pulsars III. Long-term monitoring of PSRs B1855+09 and B1937+21", *ApJ* 428, 713.

Kramer M., Staris I. H., Manchester R. N., McLaughlin M. A., Lyne A. G., Ferdman R. D., Burgay M., Lorimer D. R., Possenti A., D'Amico N., Sarkissian J. M., Hobbs G. B., Reynolds J. E., Freire P. C. C. and Camilo F., 2006, "Tests of general relativity from timing the double pulsar", *Science* 314, 97.

Laplace P. S. de, 1774, Memoirs de Mathématique et Physique présente's a l'Académie Royale de Science, y lus dans ses Assemblées.

Lebach D. E., Corey B. E., Shapiro I. I., Ratner M. I., Webber J. C., Rogers A. E. E., Davis J. L. and Herring, T. A., 1995, "Measurement of the solar gravitational deflection of radio waves using very long baseline interferometry", *Phys. Rev. Lett.* 75, 1439.

Lines R. D., Lines H., Guinan E. F. and Carroll, S. M., 1989, "Time of minimum determination of the eclipsing binary V541 Cygni", *Informational Bulletin on Variable Stars of the IAU*, No. 3286.

Newton I., 1687, *Principia*, Book III, Proposition VI, Theory VI.

Nordtvedt K., 1995, "The relativistic orbit observables in lunar laser ranging", *Icarus* 114, 51.

Possenti A., Burgay M., D'Amico N., Lyne A. G., Kramer M., Manchester R. N., Camilo F., McLaughlin M. A., Lorimer D., Joshi B. C., Sarkissian J. M. and Freire P. C. C., 2004, "The double-pulsar PSR J0737-3039A/B", *Mem. Soc. Astron. Ital. Suppl.* 5, 142.

Potter H. H., 1923, *Proc. R. Soc. A* 104, 588.

Roll P. G., Krotkov R. and Dicke R. H., 1964, "The equivalence of inertial and passive gravitational mass", *Ann. Phys.* (N.Y.) 26, 442.

Shapiro I. I., Reasenberg R. D., MacNeil P. E., Goldstein R. B., Brenkle J. P., Cain D. L., Komarek T., Zygielbaum A. I., Cuddihy W. F. and Michael W. H., Jr., 1977, "The Viking relativity experiment", *JGR* 82, 4329.

Shapiro S. S., Davis J. L., Lebach D. E. and Gregory J. S., 2004, "Measurement of the solar gravitational deflection of radio waves using geodetic very long baseline interferometry data, 1979–1999", *Phys. Rev. Lett.* 92, 121101.

Spergel D. N., Verde L., Peiris H. V., Komatsu E., Nolta M. R., Bennett C. L., Halpern M., Hinshaw G., Jarosik N., Kogut A., Limon M., Meyer S. S., Page L., Tucker G. S., Weiland J. L., Wollack E. and Wright E. L., 2003, "First-year Wilkinson Microwave Anisotropy Probe (WMAP) observations: Determination of cosmological parameters", *ApJS* 148, 175.

Stairs I. H., 2003, "Testing general relativity with pulsar timing", *Living Reviews in Relativity* 6, Irr-2003-5.

Stairs I. H., Thorsett S. E., Taylor J. H. and Woclzan A., 2002, *ApJ* 581, 501.

Su Y., Heckel B. R., Adelberger E. G., Gundlach J. H., Harris M., Smith G. L. and Swoanson H. E., 1994, "New tests of the universality of free fall", *Phys. Rev. D* 50, 3614.

Taylor J. H., 1994, "Nobel Lecture, Binary pulsars and relativistic gravity", *Rev. Mod. Phys.* 66, 711.

Taylor J. H. and Weisberg J. M., 1982, "A new test of general relativity— Gravitational radiation and the binary pulsar PSR 1913 + 16", *ApJ* 253, 908.

Will C., 2006, "The confrontation between general relativity and experiment", *Living Reviews in Relativity* 9, Irr-2006-3.

Williams J. G., Turyshev S. G. and Boggs D. H., 2004, "Progress in lunar laser ranging tests of relativistic gravity", *Phys. Rev. Lett.* 93, 261101-1-4.

14

Primordial Magnetic Fields and Cosmic Microwave Background

Eduardo Battaner and Estrella Florido

Departamento de Física Teórica y del Cosmos, Universidad de Granada, Spain

14.1 Introduction

Magnetic fields could play an important role in determining the parameters of the CMB (Cosmic Microwave Background Radiation). If there were significant primordial fields our interpretation of the spectrum of anisotropies of temperature and polarization which we use to estimate the parameter defining a cosmological model would require modification. For such an influence to be detected primordial commoving magnetic field strengths in the range 10^{-9}–10^{-8} Gauss are needed, values compatible with at least some of the astrophysical restrictions. Magnetic fields contribute to the energy-momentum tensor and are therefore a source of curvature. The LSS (Last Scattering Surface) could be crossed by radiation energy density filaments, inheritors of primordial magnetic flux tubes. The possibilities of measuring magnetic field within the space mission Planck are examined here.

Magnetic fields are present in all astrophysical subsystems and in general classical magnetohydrodynamics (MHD) provides a satisfactory physical basis for their interpretation. However, cosmology requires the full use of general relativity (GR), and as a consequence the treatment of primordial magnetic fields before recombination ($z \geq 1000$) requires the application of general relativistic MHD. Observations of primordial fields should be possible only via the CMB. However, no measurements to date have succeeded in detecting primordial magnetic fields. The best opportunity in the near future will be offered by the Planck Satellite, with which the small-scale anisotropies detected could be shown to exhibit Faraday rotation. In spite of the current absence of observational evidence for magnetic fields from CMB measurements, GR fluid dynamicists are making a considerable effort as the incorporation of magnetic fields into cosmological models may change our interpretation of the observed parameters and hence our understanding of the Universe.

Cosmological MHD is an unexpected but beautiful extension of GR which Einstein himself did not explore. In common with any piece of useful theory it can have an ever-extending scope, well beyond the field for which it was conceived.

As for any type of energy density, magnetic energy density can produce curvature of space-time. However, the specific action of magnetic fields is anisotropic, which is different from other sources of curvature such as those caused by matter or radiation, or any other phenomenon represented by an energy-momentum tensor.

The Cosmological Principle has shown itself to be an effective guiding principle for our understanding of the Universe, and according to this we would not expect a net cosmological magnetic field $\vec{B}$ to exist. However, there could be an effective magnetic energy density on smaller scales which would affect the Universe giving rise to anisotropies in the CMB and possibly giving rise to density fluctuations in the Universe as a whole. These should not have a significant effect on either the Hubble expansion or the cooling rate of the Universe. In other words the mean field $\langle \vec{B} \rangle$ should vanish but the mean energy density $\langle B^2/8\pi \rangle \neq 0$, as $\vec{B}$ could have arbitrarily large values on certain scales. $\langle B^2/8\pi \rangle$ could be especially large at specific points, or along filaments or sheets in the early Universe.

There are four aspects of the subject which need to be studied: a) Magnetogenesis, when and how were magnetic fields created? b) What are the physical, and the mathematic descriptions of cosmological magnetic fields? c) What is the effect of magnetic fields on the CMB radiation? d) How can we measure these in the CMB time window?

14.2 History of the Discovery of Cosmic Magnetism

A galaxy such as ours has magnetic fields of order $10^{-6} - 10^{-5}$ G. The theory of the $\alpha - \Omega$ dynamo postulates that this range of values is achieved via turbulent motions combined with differential rotation. The $\alpha - \Omega$ dynamo model predicts that for each rotation of a galaxy the magnetic field should increase by a factor e. As our Galaxy has made about twenty turns since it was formed, this amplification factor should be e^{20}, i.e., 5×10^8. We can infer that when the Galaxy formed, its initial field must have been of order 10^{-13} G. A rough calculation suggests that during the pre-galactic collapse phase the field should have increased by a factor of $10^2 - 10^4$, so the pre-galactic field would have been of order 10^{-17} G. The expansion of the universe implies a decline in field strength proportional to a^2 (where a is the cosmic scale factor). At Recombination (when $z \approx a \approx 1000$) the field strength should have been of the order of $B_R \equiv 10^6 \times 10^{-17}$ G $\approx 10^{-11}$ G. A field of this magnitude would have had no influence on the CMB we observe today.

This rough calculation gives an idea of the accepted scenario some ten years ago, when it was regarded that magnetic fields were unimportant during the CMB epoch. However, our ideas about the origin of magnetic fields in galaxies probably need revision. In the first place Kulsrud and Anderson (1992) noted that the dynamo calculations had to be revised because the influence of small-scale processes on larger scales had not been taken into account. Of key importance here is that we can now measure directly the magnetic field strength prior to galaxy formation. The Ly-α forest clouds are pre-galactic structures at high redshift, and field strengths as large as those in the Milky Way have been found in them (Kronberg and Perry 1982; Wolfe, Lanzetta, and Oren 1992). This means that in pre-galactic structures we can set a canonical value for the field strength in pre-galactic structures of about 10^{-6} G.

Assuming that in the contraction phase of the Ly-α forest clouds there is a concentration factor of $10^2 - 10^3$ over the background, we would calculate a value of the recombination field B_o of between 10^{-9} and 10^{-8} G where B_o is the commoving field strength, $B_o = a^2 B$. Values as high as this can seriously affect our interpretation of the CMB. Taking into account the expansion of the universe this would correspond, at $z = 1000$, the recombination epoch, to $10^{-2} - 10^{-3}$ G.

It is interesting to point out that this is not the first time that this scenario has been proposed. Fermi (1949) became interested again in cosmic rays after his participation in the Manhattan Project, after Alfvèn in 1943 had shown that magnetic fields are long lived in fluids with infinite conductivity, a condition to which conditions in the ionized interstellar medium can be reasonably approximated. Fermi suggested a value of 10^{-6} G for the Galactic field assuming energy equipartition with cosmic rays. He also assumed that the origin of galactic magnetic fields was primordial (see Giovannini 2001, 2003). Hoyle in 1958 assumed that magnetic fields existed before galaxies, rather than created by the galactic dynamo as described by Parker (1979) and others. The hypothesis of the primordial origin of the Galactic field was used by Chandrasekhar and Fermi (1953) in a model of gravitational instabilities in the presence of large magnetic fields. At that time there was already some observational evidence of galactic magnetic fields from observations of polarized optical emission, even though it is very difficult, even today, to deduce the field strength from this type of observation. The possibility of good measurements of magnetic fields in the Galaxy was opened when Wielebinski, Shakeshaft, and Pauliny-Toth (1962) and Wielebinski and Shakeshaft (1964) discovered polarized radio emission in the cm-range synchrotron continuum. A variety of techniques were then developed, useful not only for measuring Galactic, but also extragalactic fields.

Wasserman (1978) and Peebles (1980) developed interesting models for explaining the formation of cosmic structures from primordial magnetic fields, and special mention should be made of the work of Zel'dovich and Novikov (1970) who assumed that all the CMB anisotropies must be produced by magnetic fields.

The fact that fields of order microgauss seem to be ubiquitous, as they are found in galaxies, in clusters, and even in superclusters, as well as in protogalaxies, led Kronberg (1994) to suggest the existence of a background magnetic field. Considering that the magnetic energy density may be similar to that in the CMB (assuming a value of 3 K for its temperature) he suggested an explicit equipartition between the magnetic energy density $(B^2/8\pi)$ and the CMB energy density (aT^4). It is interesting to note that both of these energy densities decrease with the universal expansion at the same rate, i.e., proportional to a^4. This would mean that the radiation-dominated epoch could have been co-dominated by magnetic fields. This suggestion, though intriguing, should be taken with caution. Firstly no measurement of magnetic fields could well be as inhomogeneous as that of matter. In addition, a ubiquitous field of this magnitude would be in contradiction with some existing measured astrophysical upper limits (Vallée 2004).

Most probably, the earliest work in which primordial magnetic fields were considered was that of Lemaître, as cited by Peebles (1993). Considering the oscillating universe as a series of collapses he proposed the possibility that in the bounce, magnetic flux preserved from the fields of the previous generation of galaxies could provide nucleation sites for the next generation. Even though there was very little justification for this hypothesis it was the first claim that magnetic fields can create large-scale structures and hence give rise to radiation anisotropies. Another interesting curiosity was a model by Khalatnikov who considered a universe with a homogeneous magnetic field in a homogeneous but clearly anisotropic universe, which then adopted an elongated shape. This model, which did not assume the Cosmological Principle, was soon rejected by Khalatnikov himself (private communication).

14.3 The History of Cosmic Magnetism

Idealized cosmic plasmas have infinite conductivity. There are two important consequences. Firstly they are long-lived, with a diffusion time typically longer than the age of the Universe. Secondly they can be approximated by an intuitive description using the condition of "frozen-in magnetic field lines," since in a plasma with infinite conductivity field lines evolve as if they were fixed to the fluid, and follow the fluid motion, including expansion and compression. There are effects such as reconnection, diffusion, turbulent magnetic diffusion, dynamos, and instabilities which cause the plasma to depart from ideal conditions, but to a first approximation field lines are neither created nor eliminated. This is only a zero-order description, but just as we can intuitively represent particles as points, we can conceive of field lines as long-lived one-dimensional lines.

It is not unreasonable, using this approximation, to propose early origins for cosmic magnetic fields. They may be primordial, created during inflation, or in a cosmological phase transition, all of these before photon decoupling. The fact that magnetic fields have been found in many types of cosmic subsystems (also in the extra-cluster intergalactic medium) does point to a primordial origin. There are a variety of magnetogenesis theories. Among reviews we can cite Rees (1987), Coles (1992), Kronberg (1994), Grasso and Rubinstein (1996), Olesen (1997), Battaner and Florido (2000, 2009), Battaner and Lesch (2000), Giovannini (2001, 2006), and Banerjee and Jedamzik (2004). Some authors (e.g. Lesch and Chiva 1997, Kronberg 2009) propose post-Recombination magnetogenesis. Matsude, Sato, and Takeda (1971) and Harrison (1973) proposed turbulent generation in the relativistic radiative medium before decoupling, but turbulence in this fluid has special characteristics, and its existence is in fact doubtful (Rees 1987). The Jeans mass is very low, and a concentration introduced by turbulent flow will be followed by collapse, rendering subsequent expansion impossible.

Cosmological phase transitions are at present one of the favoured processes of magnetogenesis. This was first proposed by Hogan (1983). The electroweak, QCD and GUT phase transitions have been considered by a number of authors, among them Vachaspati (1991), Enqvist and Olesen (1993), Quashnock, Loeb, and Spergel (1989), Sigl, Olinto, and Jedamzik (1996). These models, however, yield low field strengths and very short coherence lengths. Small-scale fields also have problems in surviving to reach decoupling. The main ones are damping (Jedamzik, Katalinič, and Olinto 1998) and diffusion due to finite conductivity (Lesch and Birk 1998). Brandenburg, Enqvist, and Olesen (1996) theorized an inverse cascade to alleviate the problem on small scales. However, even if this were possible, it would never be feasible to obtain cells larger than the horizon scale.

Even though some of these magnetogenetic models are competitive and attractive some authors prefer to set the origin of cosmic fields at inflation. This possibility was first put forward by Turner and Widrow (1988). Just as with large inhomogeneities, a coherent magnetic cell could be on a subhorizon scale when produced during inflation, become superhorizon until just before photon decoupling, and subhorizon again within the CMB. This avoids problems with causal processes (resistivity, damping) and fields on any scale could be created.

Any small-scale field, e.g., $\vec{E}$ and $\vec{B}$ in an electromagnetic wave, would have its scale amplified during inflation by a factor of at least 10^{21}. The problem which must be overcome in these inflationary models is the rapid decrease of the field strength during inflation. Under conformal invariance the strength would be reduced by a very large factor. Among the possibilities which have been proposed to overcome this problem are gravitational coupling of the photon field (Turner and Widrow 1988), and coupling of the inflation and the Maxwell fields (Ratra 1992). Superstring theories (Veneziano 1991; Gasperini, Giovannini, and Veneziano 1995) show promise here, producing fields on scales larger than 100 Mpc and commoving field strengths close to the desired magnitude of 10^{-9}G.

A "Melvin" (1964) and dilatonic Melvin universes with 10 dimensions are a very interesting possibility for primordial field production as some dilatonic versions can be obtained via a dimensional reduction of flat spacetime (Gibbons and Wiltshire 1987; Gibbons and Maeda 1988; Dowker et al. 1994; Gheerardyn and Janssen 2003), such that the lower dimensional solution describes a flux tube, with the vector potential the Kaluza–Klein vector which results from the reduction. Kaluza–Klein and superstring theories, whose aims are the unification of gravitational and gauge interactions within a framework of higher dimension, are both to give rise to cosmological magnetic fields.

Inflationary models are as capable of producing finite primordial magnetic fields as they are at producing primordial inhomogeneities. Both could be driven by similar mechanisms (Giovannini 2003). Superstring theories and theories with a higher number of dimensions incorporate the necessary interplay of gravitational and gauge interactions. Quantum fluctuations before inflation could be the source of the primordial coherence cells, which would be anisotropic and probably give rise to filaments. A coherence cell with the field well oriented inside it is a filament or flux tube (see Battaner, Florido, and Jimenez-Vicente 1997; Florido and Battaner 1997). Because of the condition $\nabla \cdot \vec{B} = 0$, which as we will see below has to hold even at very early times, these filaments must produce loops. (Infinite flux tubes are not easy to reconcile with isotropy within the cosmological principle.) It has been suggested that these loops would in turn form a network (Battaner, Florido, and Garcia-Ruiz 1997; Battaner and Florido 1998; Battaner 1998), but observational support for the existence of such a network is required.

In the inflationary scenario of magnetogenesis, primordial magnetic flux tubes produce filamentary radiative structures (including dark matter and baryons as minor components) which would still be observable in the CMB. After decoupling, filaments of dark matter and baryons remain as potential seeds for the formation of large-scale structure. Of particular interest here would be filaments in the Sachs–Wolfe region of the spectrum of the anisotropies, as these would come from unperturbed primordial magnetic fields, unaffected by the non-linear post-Recombination effects which would affect small-scale fields.

This primordial field would be present before galaxy formation. It would be concentrated by factors between 10^2 and 10^4 during gravitational collapse, so that galaxies could have been formed from a magnetized medium. These initial galactic fields could have been oriented on the scale of a whole galaxy by differential rotation, and maintained by dynamo effects, as well as by turbulent magnetic diffusion, which could ensure permanent equilibrium between galactic and extragalactic magnetic fields (Battaner, Lesch, and Florido 1998).

14.4 Magnetic Fields and General Relativity

We will consider a pre-Recombination situation before the epoch of equality of dark matter and radiation. We will assume that $\langle \vec{B} \rangle = 0$ over the whole universe, but that $\langle B^2 \rangle \neq 0$. We will take a speculative look at the possible influence of magnetic fields on the overall expansion of the Universe. We obtain a similar expression for the behaviour of the scale parameter a with time as for a radiation-dominated universe, i.e.,

$$a^2 = 2 \left(\frac{8\pi}{3} [\epsilon R^4] \right)^{1/2} t \tag{1}$$

where $[\epsilon R^4]$ is a constant during the expansion, with $\epsilon = \epsilon_R + \epsilon_M$, where the first term refers to the radiation, and the second to the magnetic field. We can see from this that magnetic fields would not alter the fundamental behaviour of the scale parameter $a \propto t^{1/2}$ during the radiation-dominated era, but the expansion may be faster. This increased expansion rate would yield changes in the parameters of Big Bang nucleosynthesis, indirectly by modifying the density, or more directly by modifying the reaction rates leading to the production of D, He, and Li (even Be and B). The fact that Big Bang nucleosynthesis gives such a good quantitative account of the abundances of these elements suggests that $\epsilon_M \ll \epsilon_R$, which would imply that in all probability, magnetic fields do not make a significant contribution to the general expansion rate of the Universe. By the same token they probably have contributed little to the general cooling. In the unlikely extreme case of equipartition, $\epsilon_M = \epsilon_R$, and the expansion would be faster by $\sqrt{2}$ than in the equivalent conventional model.

If the influence of magnetic fields on the expansion and the cooling of the Universe is probably slight, nevertheless they could have a more important influence on the perturbations of the energy densities and the tensor metrics, i.e., on the CMB anisotropies and on the formation of large-scale structure.

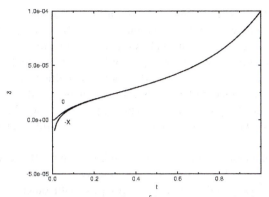

Fig. 14.1. Time evolution of the value of $\delta = \frac{\delta \epsilon}{\epsilon}$ at the centre of a filament. Curve 0 for boundary conditions of homogeneity. Curve-X for boundary conditions of isocurvature. Unit time is Equality epoch. From Florido and Battaner (1997).

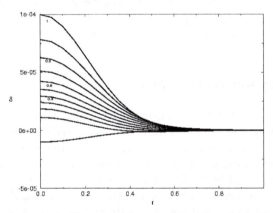

Fig. 14.2. Time evolution of the filamentary Gaussian inhomogeneity profile for the boundary condition of isocurvature. The parameter characterizing the different curves is a time parameter (equality epoch as unity).

We show a selected example in Figure 14.1 in which the effect of a magnetic field of commoving field strength $B_o \sim 10^{-8}$ G on the development of a filament is demonstrated. Figure 14.1 shows that an incremental overdensity δ of 10^{-4} times, the mean density is developed by the epoch of equality between matter and energy densities. In Figure 14.2 we show the evolution of the transverse profile of the same filament, for the same value of the magnetic field.

This confirms earlier studies by Wasserman (1978) and Peebles (1980) and demonstrates the evolution of a radiation filament created via a magnetic flux tube. The relevant equations have been generalized from those from Weinberg (1972), in which no magnetic fields were included, following a set of pioneering studies based on initial work by Lifschitz. This computation shows that magnetic inhomogeneities can be an extra source of CMB anisotropies, especially on large scales.

It is more difficult for small-scale structures to remain until the present time. The physical parameters of the initial primordial filaments can be affected by finite conductivity, viscosity, and other distorting non-linear effects. Large-scale structures are essentially unaffected by these processes and they would become observable at decoupling and at the present epoch. This implies that the filamentary structures easiest to observe will be the largest ones, in the Sachs–Wolfe region of the spectrum of anisotropies.

Commoving magnetic fields of larger amplitudes than 10^{-8} G would have led to formation of galaxies too early in the universe. On the other hand, field strengths lower than 10^{-9} G would be unable to compete with other sources of inhomogeneities, and the models would essentially reproduce the standard non-magnetic genesis of large-scale structures.

In this article, in order to keep the mathematical treatment simple we have not presented the growth of the magnetically driven inhomogeneities between the Equality epoch and Recombination, but we would not expect significant changes in filament properties in this relatively short time interval. However, strictly speaking these equations would be valid from $z = 10^8$ to $z = 10^5$, approximately.

14.5 Faraday Rotation in CMB

In this section we will show that magnetic fields may indeed play a role in determining the observed properties of the CMB, and therefore including them in cosmological models may change our interpretation of the basic parameters of the Universe. In the previous section, based on published work by Battaner, Florido, and Jiménez-Vicente (1997) and by Florido and Battaner (1997) we have shown that magnetic fields may produce CMB anisotropies. A similar conclusion was reached by Gasperini, Giovannini, and Veneziano (1995). It has also been shown that magnetic fields can distort the Planck spectrum of the CMB (Jedamzik, Katalinić, and Olinto 2000), can distort the acoustic anisotropy peaks (Adams et al. 1996), can be a source of non-Gaussianity (Lewis 2004; Naselsky et al. 2004), and can produce depolarization (Harari, Hayward, and Zaldarriaga 1997).

A further possible effect of magnetic fields on the CMB is Faraday rotation of the polarization plane when the polarization is due to Thomson scattering (Rees 1987). This effect is particularly important because it could permit us to measure primordial effects. The other postulated changes, such as the need to reinterpret the anisotropy spectrum of the CMB, must remain purely speculative if the magnetic fields cannot be observed directly. Unfortunately, the degree of polarization of the CMB is not high (around 10%), and the expected Faraday rotation is therefore low. However, the high sensitivity of the ESA Planck mission (successfully launched in May 2009) could reveal this effect observationally, or in the worst case set upper limits to the primordial magnetic fields.

Faraday rotation produces a change in the plane of polarization through an angle given by α:

$$\alpha = K(RM)\lambda^2 \tag{2}$$

where RM is the rotation measure given by

$$RM = K \int_L \left(\vec{B} \cdot \hat{u}_x \right) dx n_e \approx B_{\parallel} L n_e. \tag{3}$$

In the above, K is a constant, $\hat{u}_x$ is a unit vector along the line of sight (so that only the component of $\vec{B}$ in this direction, $B_{\parallel}$ produces Faraday rotation), x is the coordinate along the line of sight, the integral is performed along a path length L through the medium causing the Faraday rotation, n_e is the electron number density, and λ is the wavelength. By fitting the curve $\left[\alpha, \lambda^2\right]$ we can deduce RM and can infer $B_{\parallel}$ provided that we can find another way to determine n_e and L.

As Faraday rotation (FR) depends on λ, it is normally observed in galaxies, at long wavelengths, i.e., in the centimetre range. The CMB has its intensity peak at shorter wavelengths. However, the expected field strength should be higher, as $B = B_o/a^2 = 10^{-9} \times 10^6 \sim 10^{-3}$ G. The expected values for the electron density are not so different from those in galaxies, and can be derived using $n_e = n_{eo}/a^3$. Taking the present-day baryon density as 4% of the closure density we have for the mean electron density in the present universe $n_e \sim 10^{-8}$ cm^{-3}, from which the number density at Recombination would be of the order of $10^{-8} \times (10^3)^3$, i.e., 10 cm^{-3}. This is higher than a mean value for present-day galaxies, of order 10^{-2} cm^{-3} but not extremely different. However, L is more difficult to estimate, as it does not correspond fully to Δz for the CMB, i.e., the thickness of the surface of last scattering, ($\Delta z \sim 80$).

Kosowsky and Loeb (1996) and Kosowsky et al. (2005) estimated the order of magnitude of the predicted angle for Faraday rotation, finding a value of some $3°$ at 30 GHz, assuming stochastic fields of order 10^{-9} G. However, in specific filaments the field could be as much as an order of magnitude higher, so that the possibility of measuring Faraday rotation using Planck is not negligible, and this would reveal, in principle, the presence of magnetic fields in the Universe when it was no more than some 400,000 years old.

At millimetre wavelengths low field strengths produce negligible values of RM, which means that correcting any cosmic FR for Galactic FR should be feasible. On the other hand, galaxy clusters have been observed to induce significant rotation in the plane of polarization of the CMB (Clarke, Kronberg, and Böhringer 2001; Ohno et al. 2003) showing measured increases in RM as the centre of the cluster is approached, where values as high as 200 rad m^{-2} have been found. This effect would be straightforward to correct by avoiding observations around clusters.

Another interesting possibility related to FR is the fact that it will induce some B-mode polarization. A strength of 1 nG would rotate E-modes into B-modes with amplitude enough to be detected.

14.6 Conclusions

Magnetic fields are found in all astrophysical subsystems: planets, the interplanetary medium, stars, the interstellar medium, galaxies, and the intracluster and intercluster extragalactic media. There is no reason not to assume that primordial magnetic fields existed, and were present before the epoch of decoupling (or Recombination as it is conventionally called). The universal energy-momentum tensor should then include the contribution due to magnetic fields, which implies that primordial magnetic fields might contribute to create curvature. This contribution would differ from other sources of curvature, as it would be anisotropic. It should be feasible to detect the effects of primordial fields on the production and properties of the CMB anisotropies. Even though the predicted degrees of polarization and of Faraday rotation are low, the FR in particular could be detected in the context of the forthcoming Planck mission, either directly or statistically. If so, our standard models for the Universe would be altered, to a degree which at this stage is not possible to estimate.

It is interesting to quote here a statement by Peebles (1993): "If this proves impossible [referring to a dynamo in the Ly-α forest] indicating that magnetic fields existed before galaxies, it will be a fascinating hint of what happened in the early universe, and a considerable challenge to conventional ideas." Nearly fifteen years later Peebles' statement still holds.

References

Adams, Jenni, Danielsson, Ulf H., Grasso, Dario and Rubinstein, Héctor (1996). "Distortion of the acoustic peaks in the CMBR due to a primordial magnetic field". *Physics Letters B* 338, 253–258.

Banerjee, Robi and Jedamzik, Karsten (2004). "The evolution of cosmic magnetic fields: from the very early Universe, to recombination, to the present". *Physical Review D* 70, id. 123003.

Battaner, Eduardo (1998). "The fractal octahedron network of the large scale structure". *Astronomy and Astrophysics* 334, 770–771.

Battaner, Eduardo and Florido, Estrella (2000). "The rotation curve of spiral galaxies and its cosmological implications". *Fundamentals of Cosmic Physics* 21, 1–154.

——. (1998). "Magnetic fields and large scale structure in a hot Universe. IV. The egg-carton Universe". *Astronomy and Astrophysics* 338, 383–385.

——. (2009). "Magnetic Fields in the Early Universe" in Cosmic Magnetic Fields: from Planets, to Stars and Galaxies. Proc. of the International Astronomical Union, IAU Symp. 259, p. 529–538.

Battaner, Eduardo, Florido, Estrella and Garcia-Ruiz, Juan M. (1997). "Magnetic fields and large scale structure in a hot Universe. III. The polyhedric network". *Astronomy and Astrophysics* 327, 8–10.

Battaner, Eduardo, Florido, Estrella and Jimenez-Vicente, Jorge (1997). "Magnetic fields and large scale structure in a hot universe. I. General equations". *Astronomy and Astrophysics* 326, 13–22.

Battaner, Eduardo and Lesch, Harald (2000). "On the physics of primordial magnetic fields". *Anales de Física* 95, 213–225.

Battaner, Eduardo, Lesch, Harald and Florido, Estrella (1998). "Magnetic fields and rotation of spiral galaxies". *Anales de Física* 94, 98–102.

Brandenburg, Axel, Enqvist, Kari and Olesen, Poul (1996). "Large-scale magnetic fields from hydromagnetic turbulence in the very early universe". *Physical Review D* 54, 1291–1300.

Chandrasekhar, Subrahmanyan and Fermi, Enrico (1953). "Problems of gravitational stability in the presence of a magnetic field". *Astrophysical Journal* 118, 116.

Clarke, Tracy E., Kronberg, Phil P. and Böhringer, Hans (2001). "A new radio-X-ray probe of galaxy cluster magnetic fields". *Astrophysical Journal* 547, L111–L114.

Coles, Peter (1992). "Primordial magnetic fields and the large-scale structure of the Universe". *Comments on Astrophysics* 16, 45.

Dowker, Fay, Gauntlett, Jerome P., Giddings, Steven B. and Horowitz, Gary T. (1994). "Pair creation of extremal black holes and Kaluza-Klein monopoles". *Physical Review D* 50, 2662–2679.

Enqvist, K. and Olesen, Poul (1993). "On primordial magnetic fields of electroweak origin". *Physics Letters B* 319, 178–185.

Fermi, Enrico (1949). "On the origin of the cosmic radiation". *Physical Review* 75, 1169–1174.

Florido, Estrella and Battaner, Eduardo (1997). "Magnetic fields and large-scale structure in a hot universe. II. Magnetic flux tubes and filamentary structure". *Astronomy and Astrophysics* 327, 1–7.

Gasperini, Maurizio, Giovannini, Massimo and Veneziano, Gabriele (1995). "Primordial magnetic fields from string cosmology". *Physical Review Letters* 75, 3796–3799.

Gheerardyn, Jos and Janssen, Bert (2003). "Probes in fluxbranes and supersymmetry breaking through Hodge-duality". *Physics Letters B* 577, 263–272.

Gibbons, Gary W. and Maeda, Kei-Ichi (1988). "Black holes and membranes in higher-dimensional theories with dilation fields". *Nuclear Physics B* 298, 741–775.

Gibbons, Gary W. and Wiltshire, David L. (1987). "Spacetime as a membrane in higher dimensions". *Nuclear Physics B* 287, 717–742.

Giovannini, Massimo (2001). "Thick branes and Gauss-Bonnet self-interactions". *Physical Review D* 64, 124004.

——. (2003). "Assigning quantum-mechanical initial conditions to cosmological perturbations". *Classical and Quantum Gravity* 20, 5455–5473.

——. (2006). "Magnetized CMB anisotropies". *Classical and Quantum Gravity* 23, R1–R44.

Grasso, Dario and Rubinstein, Hector R. (1996). "Revisiting nucleosynthesis constraints on primordial magnetic fields". *Physics Letters B* 379, 73–79.

Harari, Diego D and Hayward, Justin D. and Zaldarriaga, Matias (1997). "Depolarization of the cosmic microwave background by a primordial magnetic field and its effect upon temperature anisotropy". *Physical Review D* 55, 1841–1850.

Harrison, Edward H. (1973). "Magnetic fields in the early Universe". *Monthly Notices of Royal Astronomical Society* 165, 185.

Hogan, Craig J. (1983). "Magnetohydrodynamic effects of a first-order cosmological phase transition". *Physical Review Letters* 51, 1488–1491.

Jedamzik, Karsten, Katalinić, Visnja and Olinto, Angela V. (1998). "Damping of cosmic magnetic fields". *Physical Review D* 57, 3264–3284.

———. (2000). "Limit on primordial small-scale magnetic fields from cosmic microwave background". *Physical Review Letters* 85, 700–703.

Kosowsky, Arthur and Loeb, Abraham (1996). "Faraday rotation of microwave background polarization by a primordial magnetic field". *Astrophysical Journal* 461, 1.

Kosowsky, Arthur, Kahniashvili, Tina, Laurelashvili, George and Ratra, Bharat (2005). "Faraday rotation of the cosmic microwave background polarization by a stochastic magnetic field". *Physical Review D* 71, 043006.

Kronberg, Philipp P. (1994). "Extragalactic magnetic fields". *Reports on Progress in Physics* 57, 325–382.

———. (2009). "Magnetic Field Transport from AGN Cores to Jets, Lobes, and the IGM" in *Cosmic Magnetic Fields: from Planets, to Stars and Galaxies. Proc. of the International Astronomical Union, IAU Symp.* 259, p. 499–508.

Kronberg, Philipp P. and Perry, Jason J. (1982). "Absorption lines, Faraday rotation, and magnetic field estimates for QSO absorption–line clouds". *Astrophysical Journal* 263, 518–532.

Kulsrud, Russell M. and Anderson, Stephen W. (1992). "The spectrum of random magnetic fields in the mean field dynamo theory of the Galactic magnetic fields". *Astrophysical Journal* 396, 606–630.

Lesch, Harald and Birk, Guido T. (1998). "Can large-scale magnetic fields survive during the pre-recombination era of the universe?" *Physics of Plasmas* 5, 2773–2776.

Lesch, Harald and Chiba, Masashi (1997). "On the origin and evolution of Galactic magnetic fields". *Fundamentals of Cosmic Physics* 18, 273–368.

Lewis, Antony (2004). "CMB anisotropies from primordial inhomogeneous magnetic fields". *Physical Review D* 70, id. 043011.

Matsuda, Takuya, Sato, H. and Takeda, Hiroshi (1971). "Pre-Galactic magnetic fields and cosmic rays in the expanding universe". *Publications of the Astronomical Society of Japan* 23, 1.

Melvin, M. (1964). *Physical Letters* 8, 65. "B polarization of the CMB from Faraday rotation". *Physical Review D* 70, 063003.

Naselsky, Pavel D., Chiang, Lung-Yih, Olesen, Poul and Verkhodanov, Oleg V. (2004). "Primordial magnetic field and non-Gaussianity of the one-year Wilkinson microwave anisotropy probe data". *Astrophysical Journal* 615, 45–54.

Ohno, Hiroshi, Takada, Masahiro, Dolag, Klaus, Bartelmann, Matthias and Sugiyama, Naoshi (2003). "Probing intracluster magnetic fields with cosmic microwave background polarization". *Astrophysical Journal* 584, 599–607.

Olesen, Poul (1997). "Inverse cascades and primordial magnetic fields". *Physics Letters B* 398, 321–325.

Parker, Eugene N. (1979). *Cosmical Magnetic Fields: Their Origin and Their Activity*. Clarendon Press, Oxford.

Peebles, Phillip (1980). *The Large Scale Structure of the Universe*. Princeton University Press, Princeton, New Jersey.

——. (1993). *Principles of Physical Cosmology*. Princeton University Press, Princeton, New Jersey.

Quashnock, Jean M., Loeb, Abraham and Spergel, David N. (1989). "Magnetic field generation during the cosmological QCD phase transition". *Astrophysical Journal Letters* 344, L49–L51.

Ratra, Bharat (1992). "Cosmological seed magnetic field from inflation". *Astrophysical Journal Letters* 391, L1–L4.

Rees, Martin (1987). "The origin and cosmogonic implications of seed magnetic fields". *Royal Astronomical Society, Quarterly Journal* 28, 197–206.

Sigl, Gunter, Olinto, Angela V. and Jedamzik, Karsten (1996). "Primordial magnetic fields from cosmological first order phase transitions". *Physical Review D* 55, 4852–4590.

Turner, Michael S. and Widrow, Lawrence M. (1988). "Inflation-produced, large-scale magnetic fields". *Physical Review D* 37, 2743–2754.

Vachaspati, Tanmay (1991). "Magnetic fields from cosmological phase transitions". *Physics Letters B* 265, 258–261.

Vallée, Jacques P. (2004). "Cosmic magnetic fields—as observed in the Universe, in galactic dynamos, and in the Milky Way". *New Astronomy Reviews* 48, 763–841.

Veneziano, Gabriele (1991). "Scale factor duality for classical and quantum strings". *Physics Letters B* 265, 287–294.

Wasserman, Ira (1978). "On the origins of galaxies, galactic angular momenta, and galactic magnetic fields". *Astrophysical Journal* 224, 337–343.

Weinberg, Steven (1972). *Gravitation and Cosmology: Principles and Applications of the General Theory of Relativity*. Wiley-VCH, New York.

Wielebinski, Richard and Shakeshaft, John R. (1964). "A survey of the linearly polarized component of galactic radio emission at 408 Mc/s". *Monthly Notices of the Royal Astronomical Society* 128, 19.

Wielebinski, Richard, Shakeshaft, John R. and Pauliny-Toth, Ivan I. K. (1962). "A search for a linearly polarized component of the galactic radio emission at 4087 Mc/s". *The Observatory* 82, 158–164.

Wolfe, Arthur M., Lanzetta, Kenneth M. and Oren, Aharon L. (1992). "Magnetic fields in damped Ly-alpha systems". *Astrophysical Journal* 388, 17–22.

Zel'dovich, Yakov B., Novikov, Igor D. (1970). "A hypothesis for the initial spectrum of perturbations in the metric of the Friedmann model universe". *Soviet Astronomy* 13, 754.

15

Singularity Theorems in General Relativity: Achievements and Open Questions

José M. M. Senovilla

Universidad del País Vasco, Spain

15.1 Introduction

In this short note, written by a theoretical physicist, not a historian, I would like to present a brief overview of the acclaimed singularity theorems, which are often quoted as one of the greatest theoretical accomplishments in general relativity and mathematical physics.

Arguably, the singularity theorems are one of the few consequences, if not the only one, of Einstein's greatest theory which were not foreseen, and probably not even suspected, by its founder. Many other consequences also came to a scientifically firm basis only after Einstein's demise. To mention a few outstanding examples, let me cite gravitational lensing, gravitational radiation in binary systems (as in the PSR B1913+16 system), or the variation of frequency in electromagnetic radiation travelling in a gravity field. *All of them*, however, had been explicitly predicted in one way or another by Einstein.

On the contrary, the singularity theorems such as we understand them now (the result in Einstein (1941) is concerned with quite another type of singularity), the global developments needed for them, and the derived inferences were not mentioned, either directly or indirectly, in any of his writings. This was so despite the various clear indications that the appearance of some kind of "singularity," or catastrophical behaviour, was a serious possibility within the orthodox theory. For instance, the Friedman–Lemaître models (Friedman 1922, 1924; Lemaître 1927), by the 1930s generally accepted as providing an explanation for the observed galactic (or nebular) redshifts, contain the famous "creation time" (Friedman's wording) identified by Friedman and Lemaître themselves, at which the space-time is simply created. An even more basic example is the proof by Oppenheimer and Snyder (1939) that the "Schwarzschild surface"—which, as we know, is *not* a singularity, but a horizon, see for instance the well-documented historical review of this subject presented in Tipler, Clarke, and Ellis (1980) and the more recent discussion in Senovilla (2007a) — could be reached and crossed by matter in the form of dust in quite simple solutions to the field equations. (By the way, this collapsing dust eventually ends in a catastrophic (real) future singularity.) We also mention the impressive result found by Chandrasekhar—to which Eddington was furiously opposed—of an upper mass limit for stars in equilibrium, even when taking into account the quantum effects, implying that stars with a larger mass must inevitably collapse (Chandrasekhar 1931).

In spite of all these achievements, as I was saying, Einstein and orthodoxy simply dismissed these catastrophic behaviours (singularities) as either a mathematical artifact due to the presence of (impossible) exact spherical symmetry, or as utterly impossible effects, scientifically untenable, obviously unattainable, beyond feasibility in the physical world—see, e.g., Einstein (1939). Of course, this probably is correct in a deep sense, since infinite values of physical observables must not be accepted, and every sensible scientist would defend similar assertions. However, one must be prepared to probe the limits of any particular theory, and this was simply not done with general relativity at the time. It was necessary to wait for a new generation of physicists and mathematicians, without the old prejudices and less inhibited, and probably also less busy with the quantum revolution, who finally could take seriously the questions of singularities, gravitational collapse, and the past of the Universe within the theory.

Thus, one can say that the singularity theorems are the most genuine post-Einsteinian content of general relativity.

15.1.1 The Raychaudhuri Equation

Curiosuly enough, the first result concerning the prediction of singularities under reasonable physical conditions, due to Raychaudhuri, came to light in exactly the same year as Einstein's death. In 1955 Raychaudhuri published what is considered the first singularity theorem, including a version of the equation named after him which is the basis of *all* later developments and of the singularity theorems (Raychaudhuri 1955). The Raychaudhuri equation can easily be derived (see, for instance, Dadhich (2007) in a recent volume in honour of Raychaudhuri), as it has a very simple geometrical interpretation: take the well-known Ricci identity

$$(\nabla_\mu \nabla_\nu - \nabla_\nu \nabla_\mu) u^\alpha = R^\alpha_{\rho\mu\nu} u^\rho \tag{15.1}$$

which is mathematically equivalent to the standard definition of the Riemann tensor

$$(\nabla_{\vec{X}} \nabla_{\vec{Y}} - \nabla_{\vec{Y}} \nabla_{\vec{X}} - \nabla_{[\vec{X},\vec{Y}]}) \vec{Z} = R(\vec{X},\vec{Y}) \vec{Z} \qquad \forall \vec{X}, \vec{Y}, \vec{Z}$$

and contract α with μ in (15.1), and then with u^ν, to obtain

$$u^\nu \nabla_\mu \nabla_\nu u^\mu - u^\nu \nabla_\nu \nabla_\mu u^\mu = R_{\rho\nu} u^\rho u^\nu$$

where $R_{\rho\nu}$ is the Ricci tensor. Reorganizing by parts the first summand on the left-hand side, one obtains

$$u^\nu \nabla_\nu \nabla_\mu u^\mu + \nabla_\mu u_\nu \nabla^\nu u^\mu - \nabla_\mu (u^\nu \nabla_\nu u^\mu) + R_{\rho\nu} u^\rho u^\nu = 0 \tag{15.2}$$

which is the Raychaudhuri equation. Raychaudhuri's important contribution amounts to understanding the physical implications of this relation. Observe that, in the case in which u^μ defines an (affinely-parametrized) *geodesic* vector field, then $u^\nu \nabla_\nu u^\mu = 0$ and the third term vanishes. The second term can be rewritten by splitting

$$\nabla_\mu u_\nu = S_{\mu\nu} + A_{\mu\nu}$$

into its symmetric $S_{\mu\nu}$ and antisymmetric $A_{\mu\nu}$ parts, so that

$$\nabla_\mu u_\nu \nabla^\nu u^\mu = S_{\mu\nu} S^{\mu\nu} - A_{\mu\nu} A^{\mu\nu}.$$

Now the point is to realize that (i) if u^μ is time-like (and normalized) or null, then both $S_{\mu\nu} S^{\mu\nu}$ and $A_{\mu\nu} A^{\mu\nu}$ are non-negative; and (ii) u_μ is proportional to a gradient (therefore defines orthogonal hypersurfaces) if and only if $A_{\mu\nu} = 0$. In summary, for hypersurface-orthogonal time-like or null geodesic vector fields u^μ, one has

$$u^\nu \nabla_\nu \nabla_\mu u^\mu = -S_{\mu\nu} S^{\mu\nu} - R_{\rho\nu} u^\rho u^\nu$$

so that the sign of the derivative of the divergence $\nabla_\mu u^\mu$ along these geodesics is governed by the sign of $R_{\rho\nu} u^\rho u^\nu$. If the latter is non-negative, then the former is non-positive. In particular, if the divergence is negative at some point and $R_{\rho\nu} u^\rho u^\nu \geq 0$, then necessarily the divergence will reach an infinite negative value in a finite affine parameter interval (unless all the quantities are zero everywhere).

If there are physical particles moving along these geodesics, then clearly a physical singularity is obtained, as the average volume decreases and the density of particles will be unbounded. This was the situation treated by Raychaudhuri for the case of irrotational dust. In general, no singularity is predicted, though, and one only gets a typical *caustic* along the flow lines of the congruence defined by u^μ. This generic property is usually called the *focusing effect* on causal geodesics. For this to take place, of course, one needs the condition

$$R_{\rho\nu} u^\rho u^\nu \geq 0 \tag{15.3}$$

which is a *geometric* condition and independent of the particular theory. However, in general relativity, one can relate the Ricci tensor to the energy-momentum tensor $T_{\mu\nu}$ via Einstein's field equations

$$R_{\mu\nu} - \frac{1}{2} g_{\mu\nu} R + \Lambda g_{\mu\nu} = \frac{8\pi G}{c^4} T_{\mu\nu} \tag{15.4}$$

where R is the scalar curvature, G is Newton's gravitational constant, c is the speed of light in a vacuum, and Λ is the cosmological constant. Thereby, the condition (15.3) can be rewritten in terms of physical quantities. This is why sometimes (15.3), when valid for all time-like u^μ, is called the *strong energy condition* (Hawking and Ellis 1973). One should bear in mind, however, that this is a condition on the Ricci tensor (a geometrical object), and therefore it will not always hold: see the discussion in Section 15.3 and (Senovilla 1998a, sect. 6.2).

An important remark of historical importance is that, before 1955, Gödel wrote his famous paper (Gödel 1949) in a volume dedicated to Einstein's 70th anniversary.

This paper is considered (see Tipler, Clarke, and Ellis 1980, sect. 3) to be the genesis of many of the necessary techniques and some of the global ideas which were used on the path to the singularity theorems, especially concerning causality theory. For further information the reader is referred to (Ellis 1998). However, the subject was not ripe and had to wait, first, for the contribution by Raychaudhuri, and then for the imaginative and fruitful ideas put forward by Roger Penrose in the 1960s.

15.2 Remarks on Singularities and Extensions

The problem of the definition of the concept of singularity in general relativity is very difficult indeed, as can be appreciated by reading about its historical development (Hawking and Ellis 1973; Tipler, Clarke, and Ellis 1980). The intuitive ideas are clear: if any physical or geometrical quantity blows up, this signals a singularity. However, there are problems of two kinds:

- The singular points, by definition, do not belong to the space-time which is only constituted by regular points. Therefore, one cannot say, in principle, "when" or "where" the singularity is.
- Characterizing the singularities is also difficult, because the divergences (say) of the curvature tensor can depend on a bad choice of basis, and even if one uses only curvature invariants, independent of the bases, it can happen that all of them vanish and there still will be singularities.

The second point is a genuine property of Lorentzian geometry, that is, of the existence of one axis of time of a different nature than the space axes.

Therefore, the only sensible definition having a certain consensus within the relativity community is by "signalling" the singularities, i.e., "pointing at them" by means of quantities belonging exclusively to the space-time. And the best and simplest pointers are curves, of course. One can imagine what happens if a brave traveller approaches a singularity: he/she will disappear from our world in a *finite* time. The same thing, but time-reversed, can be imagined for the "creation" of the Universe: things *suddenly* appeared from nowhere a *finite* time ago. All in all, it seems reasonable to diagnose the existence of singularities whenever particles (real or hypothetical) unexpectedly appear from or disappear into some region of space-time.

And this is the basic definition of a singularity (Geroch 1968; Hawking and Ellis 1973), the existence of *incomplete and inextensible* curves; that is, curves which cannot be extended in a regular manner within the space-time and do not take all possible values of their canonical parameter. Usually, only causal (time-like or null) curves are used, but in principle incomplete space-like curves also define singularities. The curves do not need to be geodesic; as a matter of fact, there are known examples of geodesically complete space-times with incomplete time-like curves with everywhere-bounded acceleration (Geroch 1968). It must be remarked, however, that all singularity theorems prove merely the existence of *geodesic* incompleteness, which of course is a *sufficient* condition for the existence of singularities according to the definition.

Some fundamental questions, which are sometimes omitted, arise at this point. How can one give structure and properties to the singularities? What is the relation between geodesic incompleteness and curvature problems, if any? Singularities in the sense defined above clearly reach, or come from, the *edge* of space-time. This is some kind of boundary, or margin, which is not part of the space-time but that, somehow, is accessible from within it. Thus, there arose the necessity of a rigourous definition of the boundary of a space-time. Given that a Lorentzian manifold is not a metric space in the usual sense, one cannot use the traditional completion procedure by means of Cauchy sequences. The most popular approach in this sense has been the attempt to attach a *causal boundary* to the space-time (see García-Parrado and Senovilla 2005 for an up-to-date review). But this approach is not exempt from recurrent problems which have not yet been completely solved.

Furthermore, the existence of incomplete geodesics which are not extensible in a given space-time may indicate *not* a problem with the curve and the geometrical properties along it when it approaches the edge, but rather *incompleteness* of the space-time itself. For instance, flat space-time without the origin has infinite inextensible incomplete curves. These cases, however, can be circumvented, as one can easily see that the problem arises due to the excision of regular points. This is why one usually restricts consideration to inextensible space-times when dealing with singularity theorems (Hawking and Ellis 1973). The physical problem, however, is hidden under the carpet with this attitude: what are we supposed to do with given extensible space-times? The answer may seem simple and easy: just extend it until you cannot extend it anymore. However, this is not that simple for several reasons (see Senovilla 1998a, sects. 3 and 7):

- Extensions are not obvious, and generally not unique. Usually there are infinite inequivalent choices.
- Not even analytical extensions are generally unique, let alone smooth extensions.
- It can happen that there are incomplete curves, no curvature problems, and no possible (regular) extension.
- Sometimes, for a given fixed extensible space-time, there are extensions leading to new extensible space-times, other extensions leading to singular space-times, and still other extensions which are free of singularities and inextensible. It may seem obvious that the last choice would be the preferred one by relativists, but this is simply not the case—if the singularity-free extension violates a physical condition, such as causality or energy positivity, then the other extensions will be chosen.
- What physical criteria are to be used to discriminate between inequivalent extensions?

As a drastic example of the above problems, take the traditional case of the Schwarzschild solution which, as is known, is extensible through the horizon. In textbooks one usually encounters the *unique* maximal analytical vacuum extension due to Kruskal and Szekeres that keeps spherical symmetry. However, if one drops any one of the conditions (vacuum, analyticity), many other maximal extensions are possible; see, e.g., (Senovilla 1998a) where at least eleven inequivalent extensions were explicitly given. This should make it plain that the question of singularities is intimately related to the problem of how, why, and to where a given extensible space-time should be extended.

15.3 Singularity Theorems: Critical Appraisal

The first singularity theorem in the modern sense is due to Penrose (Penrose 1965), who in this seminal paper introduced the fundamental concept of a *closed trapped surface* and opened up a new avenue of fruitful research. His main, certainly innovative idea was to prove null geodesic incompleteness if certain *initial* conditions, reasonable for known physical states of collapse, were given *irrespective of any symmetry or similar restrictive properties.*

Since then, many singularity theorems have been proven (see Hawking and Ellis 1973; Senovilla 1998a), some of them applicable to cosmological situations, some to star or galaxy collapse, and others to the collision of gravitational waves. The culmination was the celebrated Hawking–Penrose theorem (Hawking and Penrose 1970), which since then has been the singularity theorem *par excellence.* However, all of the singularity theorems share a well-defined skeleton, the very same pattern. This is stated succinctly as follows (Senovilla 1998a).

Theorem 1 (Pattern Singularity Theorem). *If a space-time of sufficient differentiability satisfies*

 1. *a condition on the curvature,*
 2. *a causality condition, and*
 3. *an appropriate initial and/or boundary condition,*

then there are null or time-like inextensible incomplete geodesics.

I have started by adding explicitly the condition of sufficient differentiability. This is often ignored, but it is important from both the mathematical and the physical points of view. A breakdown of the differentiability should not be seen as a true singularity, especially if the problem is mild and the geodesic curves are continuous. The theorems are valid if the space-time metric tensor field $g_{\mu\nu}$ is of class C^2 (twice differentiable with continuity), but they have not been proven in general if the first derivatives of $g_{\mu\nu}$ satisfy only the Lipschitz condition. This problem is of physical relevance, because the entire space-time of a star or a galaxy, say, is usually considered to have two different parts matched together at the surface of the body. Then the metric tensor is not C^2 at this surface: it cannot be, for there must exist a jump in the matter density which is directly related to the second derivatives of $g_{\mu\nu}$ via equations (15.4). As an example, the Oppenheimer–Snyder collapsing model does not satisfy the C^2 condition. A list of the very many places where this condition is used in the singularity theorems can be found in (Senovilla 1998a, p. 799).

Then there is the "curvature condition." I have used this name rather than the usual *energy* and *generic* condition to stress the fact that this assumption is of a geometric nature, and it is absolutely indispensable: it enforces the geodesic focusing via the Raychaudhuri equation and other similar identities. The majority of the theorems, specially the stronger ones, use the condition (15.3), which is usually called the *strong energy condition* if it is valid for all time-like u^μ, and the *null convergence condition* if it is valid only for null vectors. The former name is due to the equivalent relation, via the field equations (15.4)

$$T_{\rho\nu}u^\rho u^\nu \geq \frac{c^4 \Lambda}{8\pi G} - \frac{1}{2}T^\rho_\rho$$

for unit time-like u^μ. This involves energy-matter variables. However, this condition does *not* have to be satisfied in general by realistic physical fields. To start with, it depends on the sign of Λ. But furthermore, even if $\Lambda = 0$, the previous condition does not hold in many physical systems, such as scalar fields (Hawking and Ellis 1973, p. 95). As a matter of fact, most of the inflationary cosmological models violate the above condition. Let us stress that the physically compelling energy condition is the *dominant energy condition* (Hawking and Ellis 1973), but this has, in principle, nothing to do with the assumptions of the singularity theorems. In particular, there are many examples of reasonable singularity-free space-times satisfying the dominant, but not the strong, energy condition (see, e.g., Senovilla 1998a, sect. 7.3). In my opinion, this is one of the weak points of the singularity theorems.

The "causality condition" is probably the most reasonable and well founded, and perhaps also the least restrictive, condition. There are several examples of singularity theorems without any causality condition (Hawking and Ellis 1973, Theorem 4, p. 272; Maeda and Ishibashi 1996). The causality condition is assumed for two types of reasons. Firstly, it is used to prevent the possibility of avoiding the future, that is, to provide a well-defined global time-arrow. This may seem superfluous, but it has been known since the results in (Gödel 1949) that there may be closed time-like lines, that is, curves for which the time passes to the future permanently and that nevertheless reach their own past. Secondly, it is used to ensure the existence of geodesics of maximal proper time between two events, and therefore geodesics without focal points (Hawking and Ellis 1973, Prop. 4.5.8).

Recapitulating, the first two conditions in the theorems imply

- the focusing of *all* geodesics, ergo the existence of caustics and focal points, due to the curvature condition;
- the existence of geodesics of maximal proper time, and therefore necessarily *without focal points*, joining any two events of the space-time.

Obviously, a contradiction starts to glimmer if all geodesics are complete. However, there is no such contradiction yet because, at this stage, there is no finite *upper bound* for the proper time of selected families of time-like geodesics (and analogously for null geodesics).

To finally get the contradiction, one needs to add the third condition. And this is why the initial/boundary condition is absolutely essential in the theorems. There are several possibilities for this condition, among them (i) an instant of time at which the whole Universe is strictly expanding; (ii) closed universes so that they contain space-like compact slices; (iii) the existence of closed (that is, compact without boundary) trapped surfaces.

This last concept, due to Penrose (1965), has become the most popular and probably the most important one, especially due to its relevance and applicability in many other branches of relativity. Trapped surfaces are two-dimensional differentiable surfaces (such as a donut's skin or a sphere) with a mean curvature vector which is time-like everywhere, or in simpler words, such that the initial variation of area along *any future direction* is always decreasing (or always increasing). An example of a non-compact trapped surface in flat space-time (in Cartesian coordinates) is given by

$$e^t = \cosh z, \quad x = \text{const.}$$

However, there cannot be compact trapped surfaces in flat space-time. An example of a compact trapped surface is given by any 2-sphere $t = \text{const}$, $x^2 + y^2 + z^2 = r^2 = \text{const}$, in a Friedman model

$$ds^2 = -dt^2 + a^2(t)\left(dx^2 + dy^2 + dz^2\right)$$

as long as $r > 1/|\dot{a}|$, which is always possible.

Whether or not the initial or boundary condition is realistic, or satisfied by actual physical systems, is debatable. We will probably find it very difficult to ascertain if the Universe is spatially finite, or if the *whole* Universe is strictly expanding now. There is a wide agreement, however, that it is at least possible that in some situations the third condition in the theorems will hold. For example, given the extremely high degree of isotropy observed in the cosmic microwave background radiation, one can try to infer that the Universe resembles a Friedman–Lemaître model, thus containing trapped spheres like the one shown above. Nevertheless, there are several ways out of this scheme of reasoning; for example, a cosmological constant, or deviations from the model at distances (or redshifts) of the order of or higher than the visible horizon. For a discussion see (Senovilla 1998a, sects. 6.4 and 7.2).

Let us finally come to the conclusion of the theorems. In most singularity theorems, this is just the existence of at least one incomplete causal geodesic. This is very mild, as it can be a mere localized singularity. This leaves one door open (extensions), and furthermore it may be a very mild singularity. In addition, the theorems do not say anything, in general, about the situation and the character of the singularity. We cannot know whether it is in the future or past, or whether it will lead to a blow-up of the curvature or not.

In the next section, I am going to present a very simple solution of Einstein's field equations which shows explicitly the need for the "small print" in the theorems, and that sometimes their assumptions are more demanding than generally thought.

15.4 An Illustrative Singularity-Free Space-Time

Nowadays there are many known singularity-free space-times. Some of them are spatially inhomogeneous universes, others contain a cosmological constant, and there are a wide variety of other types (see Senovilla 1998a, and references therein). Perhaps the most famous singularity-free "cosmological model" was presented in (Senovilla 1990), because it had some impact on the scientific community (see, e.g., Maddox 1990) and opened up the door for many of the rest. Its impact was probably due to the general belief that such a space-time was forbidden by the singularity theorems. But, of course, this was simply a misunderstanding, and only meant that we had thought that the singularity theorems were implying more than they were actually saying.

The space-time has cylindrical symmetry, and in typical cylindrical coordinates $\{t, \rho, \varphi, z\}$ its line element takes the form

$$ds^2 = \cosh^4(act)\cosh^2(3a\rho)(-c^2 dt^2 + d\rho^2)$$
$$+ \frac{1}{9a^2}\cosh^4(act)\cosh^{-2/3}(3a\rho)\sinh^2(3a\rho)d\varphi^2$$
$$+ \cosh^{-2}(act)\cosh^{-2/3}(3a\rho)dz^2$$

where $a > 0$ is a constant. This is a solution of the field equations (15.4) for $\Lambda = 0$ and an energy-momentum tensor of perfect-fluid type

$$T_{\mu\nu} = \rho u_\mu u_\nu + p(g_{\mu\nu} + u_\mu u_\nu)$$

where ρ is the energy density of the fluid, given by

$$\frac{8\pi G}{c^4}\varrho = 15a^2 \cosh^{-4}(act)\cosh^{-4}(3a\rho)$$

p is its isotropic pressure, and

$$u_\mu = \left(-c\cosh^2(act)\cosh(3a\rho), 0, 0, 0\right).$$

defines the unit velocity vector field of the fluid. Observe that u^μ is not geodesic (except at the axis). The fluid has a realistic barotropic equation of state

$$p = \frac{1}{3}\rho.$$

This is the canonical equation of state for radiation-dominated matter and is usually assumed at the early stages of the Universe. Note that the density and the pressure are regular everywhere, and in fact one can prove that the space-time is completely free of singularities and geodesically complete (Chinea, Fernández-Jambrina, and Senovilla 1992).

The space-time satisfies the stronger causality conditions (it is globally hyperbolic), and also all energy conditions (dominant, strict strong). The fluid divergence is given by

$$\nabla_\mu u^\mu = 3a \frac{\sinh(act)}{\cosh^3(act)\cosh(3a\rho)} \tag{15.5}$$

so that this universe is contracting for half of its history ($t < 0$) and expanding for the second half ($t > 0$), having a rebound at $t = 0$ which is driven by the spatial gradient of pressure. Observe that the whole universe is expanding (that is, $\nabla_\mu u^\mu > 0$) *everywhere* if $t > 0$, and recall that this was one of the possibilities we mentioned for the third condition in the singularity theorems: an instant of time with a strictly-expanding whole universe. So, how can this model be geodesically complete and singularity-free?

Well, the precise condition demanded by one of the theorems in globally hyperbolic space-times is that $\nabla_\mu u^\mu > b > 0$ for a constant b. That is, $\nabla_\mu u^\mu$ has to be bounded from below by a positive constant. But this is not the case for (15.5), which is strictly positive everywhere but not bounded from below by a positive constant because $\lim_{\rho\to\infty} \nabla_\mu u^\mu = 0$. This minor, subtle difference allows the model to be singularity-free! In a similar manner, all other possibilities for the initial/boundary condition in the several would-be applicable singularity theorems can be seen to *just* fail in the model. For a complete discussion see (Chinea, Fernández-Jambrina, and Senovilla 1992) and (Senovilla 1998a, sect. 7.6). One can also see that the focusing effect on geodesics takes place fully in this space-time, but nevertheless there is no problem with the existence of maximal geodesics between any pair of points (see Senovilla 1998a, pp. 829–830).

This simple model showed that there exist well-founded, well-behaved classical models expanding everywhere, satisfying all energy and causality conditions, and that are singularity-free. Of course, the model is not realistic in the sense that it cannot describe the actual Universe—for instance, the isotropy of the cosmic background radiation cannot be explained—but the question arises whether or not there is room left by the singularity theorems to construct geodesically-complete *realistic* universes.

It should be stressed that this model does not describe a "cylindrical star," because the pressure of the fluid does not vanish anywhere. Nevertheless, as can be seen from the previous formulae, for example, the energy density is mainly concentrated in an area around the axis of symmetry, dying away from it very quickly as ρ increases. This may somehow raise some doubts about the relevance of this type of model for *cosmological* purposes. In this sense, there was an interesting contribution by Raychaudhuri himself (Raychaudhuri 1998), in which he tried to quantify this property in a mathematical condition. Unfortunately, this time he was not completely right (see Senovilla 1998b), because he used *space-time* averages of the physical variables such as the energy density. But one can easily see that the vanishing of such averages is a property shared by the majority of the models, be they singular or not. However, his work provided crucial inspiration, and one can actually prove that the vanishing of the *spatial* averages of the physical variables allows distinguishing between the singularity-free models permitted by the singularity theorems and the singular ones. This was conjectured in (Senovilla 1998b) and partly proved in (Senovilla 2007b), in a contribution to the volume devoted to a tribute to Raychaudhuri.

All in all, the main conclusion of this contribution is to remind ourselves that it is still important to develop further, understand better, and study carefully the singularity theorems and their consequences for realistic physical systems.

References

Chandrasekhar, Subrahmanyan (1931) "The maximum mass of ideal white dwarfs" *Astrophysical Journal* **74** 81–82.

Chinea, F. Javier, Fernández-Jambrina, Leonardo and Senovilla, José M. M. (1992) "Singularity-free space-time" *Physical Review D* **45** 481–486.

Dadhich, Naresh (2007) "Singularity: Raychaudhuri equation once again" In *Raychaudhuri equation and its role in Modern Cosmology*, Pramana special issue dedicated to A. K. Raychaudhuri, *Pramana* **69** 23–29.

Einstein, Albert (1939) "On a stationary system with spherical symmetry consisting of many gravitating masses" *Annals of Mathematics* **40** 922–936.

—— (1941) "Demonstration of the non-existence of gravitational fields with a non-vanishing total mass free of singularities" *Revista de la Universidad Nacional de Tucumán* **A2** 11–16.

Ellis, George F. R. (1998) "Contributions of K. Gödel to Relativity and Cosmology" In *Gödel '96: Foundations of Mathematics, Computer Science and Physics: Kurt Gödel's Legacy* Lectures Notes in Logic 6, Petr Hajek ed. Berlin: Springer, 34–49.

Friedman, Alexander (1922) "Über die Krümmung des Raumes" *Zeitschrift für Physik* **10** 377–386;
Translated and reproduced in (1999) *General Relativity and Gravitation* **31** 1991–2000.

—— (1924) "Über die Möglichkeit einer Welt mit konstanter negativer Krümmung des Raumes" *Zeitschrift für Physik* **21** 326–332;
Translated and reproduced in (1999) *General Relativity and Gravitation* **31** 2001–2008.

García-Parrado, Alfonso and Senovilla, José M. M. (2005) "Causal structures and causal boundaries" (Topical Review) *Classical and Quantum Gravity* **22** R1–R84.

Geroch, Robert (1968) "What is a singularity in general relativity?" *Annals of Physics (New York)* **48** 526–540.

Gödel, Kurt (1949) "An example of a new type of cosmological solution of Einstein's field equations of gravitation" *Reviews in Modern Physics* **21** 447–450.

Hawking, Stephen W. and Ellis, Georges F. R. (1973) *The Large Scale Structure of Space-time*. Cambridge: Cambridge University Press.

Hawking, Stephen W. and Penrose, Roger (1970) "The singularities of gravitational collapse and cosmology" *Proceedings of the Royal Society London* **A314** 529–548.

Lemaître, Georges (1927) "Un Universe homogène de masse constante et de rayon croissant, rendant compte de la vittese radiale des nébuleuses extragalactiques" *Annales de la Société Scientifique de Bruxelles* **A47** 49–59.

Maddox, John (1990) "Another gravitational solution found" *Nature* **345** 201.

Maeda, Kengo and Ishibashi, Akihiro (1996) "Causality violation and singularities" **13** 2569–2576.

Oppenheimer, Julius Robert and Snyder, Hartland (1939) "On continued gravitational contraction" *Physical Review* **56** 455–459.

Penrose, Roger (1965) "Gravitational collapse and space-time singularities" *Physical Review Letters* **14** 57–59.

Raychaudhuri, Amal Kumar (1955) "Relativistic cosmology I" *Physical Review* **98** 1123–1126.

—— (1998) "Theorem for non-rotating singularity-free universes" *Physical Review Letters* **80** 654–655.

Senovilla, José M. M. (1990) "New class of inhomogeneous cosmological perfect-fluid solutions without big-bang singularity" *Physical Review Letters* **64** 2219–2221.

—— (1998a) "Singularity theorems and their consequences" (Review) *General Relativity and Gravitation* **30** 701–848.

—— (1998b) "Comment on 'Theorem for non-rotating singularity-free universes'" *Physical Review Letters* **81** 5032.

—— (2007a) "The Schwarzschild solution: corrections to the editorial note" *General Relativity and Gravitation* **39** 685–693.

—— (2007b) "A singularity theorem based on spatial averages" In *Raychaudhuri equation and its role in Modern Cosmology*, Pramana special issue dedicated to A. K. Raychaudhuri, *Pramana* **69** 31–47.

Tipler, Frank J., Clarke, Chris J. S. and Ellis, George F. R. (1980) "Singularities and Horizons—A Review Article" In *General Relativity and Gravitation: One Hundred Years After the Birth of Albert Einstein*. Alan Held ed. New York: Plenum Press, 97–206.

16

The History and Present Status of Quantum Field Theory in Curved Spacetime

Robert M. Wald

Enrico Fermi Institute and Department of Physics, The University of Chicago, USA

16.1 Introduction

Quantum field theory in curved spacetime is the theory of quantum fields propagating in a classical curved spacetime. Here the spacetime is described, in accord with general relativity, by a manifold, M, on which is defined a Lorentz metric, g_{ab}. In order to ensure that classical dynamics is well defined on (M, g_{ab}), we restrict attention to the case where (M, g_{ab}) is globally hyperbolic (see, e.g., (Wald 1984)). In the framework of quantum field theory in curved spacetime, back-reaction of the quantum fields on the spacetime geometry can be taken into account by imposing the semi-classical Einstein equation $G_{ab} = 8\pi \langle T_{ab} \rangle$. However, I will not consider issues associated with back-reaction here, so in the following, (M, g_{ab}) may be taken to be an arbitrary, fixed globally-hyperbolic spacetime.

Some of the earliest applications of quantum field theory in curved spacetime were to study particle creation effects in an expanding universe. A major impetus to the theory was then provided by Hawking's calculation of particle creation by black holes, showing that black holes radiate as perfect black bodies. However, as the ramifications of this work were explored, it became clear that major issues of principle regarding the formulation of the theory arise from the lack of Poincaré symmetry, the absence of a preferred vacuum state, and, in general, the absence of asymptotic regions in which particle states can be defined. For free (i.e., non-self-interacting) quantum fields, it was understood by the mid-1980s how all of these difficulties could be overcome by formulating the theory via the algebraic approach and focusing attention on the local field observables rather than a notion of "particles." However, these ideas, by themselves, were not adequate for the formulation of interacting quantum field theory, even at a perturbative level, since standard renormalization prescriptions in Minkowski spacetime rely heavily on Poincaré invariance and the existence of a Poincaré-invariant vacuum state. However, during the past decade, considerable further progress has been made, mainly due to the importation into the theory of the methods of "microlocal analysis."

This article will describe the historical development of the subject and summarize some of the recent progress, focusing primarily on issues concerning the formulation of quantum field theory in curved spacetime. I will describe some aspects of the historical evolution of the subject through the mid-1970s, at which point it had become clear that a proper formulation of the theory could not be based upon a notion of "particles." I will then describe how the major conceptual obstacles to formulating the theory in the case of free quantum fields were overcome by adopting the algebraic approach. Finally, I will describe some of the progress that has been made during the past decade toward the formulation of interacting quantum field theory in curved spacetime.

Much of the quantum theory of a free field follows directly from the analysis of an ordinary quantum mechanical harmonic oscillator, described by the Hamiltonian

$$H = \frac{1}{2}p^2 + \frac{1}{2}\omega^2 q^2. \tag{16.1}$$

By introducing the "lowering" (or "annihilation") operator

$$a \equiv \sqrt{\frac{\omega}{2}}\, q + i\sqrt{\frac{1}{2\omega}}\, p \tag{16.2}$$

we can rewrite H as

$$H = \omega\left(a^\dagger a + \frac{1}{2}I\right) \tag{16.3}$$

where $a^\dagger$ is referred to as the "raising" (or "creation") operator, and we have the commutation relations

$$[a, a^\dagger] = I, \quad [H, a] = -\omega a. \tag{16.4}$$

It then follows that in the Heisenberg representation, the position operator, q_H, is given by

$$q_H = \sqrt{\frac{1}{2\omega}}\left(e^{-i\omega t}a + e^{i\omega t}a^\dagger\right) \tag{16.5}$$

so that a is seen to be the positive frequency part of the Heisenberg position operator. The ground state, $|0\rangle$, of the harmonic oscillator is determined by

$$a|0\rangle = 0. \tag{16.6}$$

All other states of the harmonic oscillator are obtained by successive applications of $a^\dagger$ to $|0\rangle$.

Consider, now, a free Klein–Gordon scalar field, ϕ, in Minkowski spacetime. Classically, ϕ satisfies the wave equation

$$\partial^a \partial_a \phi - m^2 \phi = 0. \tag{16.7}$$

To avoid technical awkwardness, it is convenient to imagine that the scalar field resides in a cubic box of side L with periodic boundary conditions. In that case, $\phi(t, \vec{x})$ can be decomposed in terms of a Fourier series in $\vec{x}$. In terms of the Fourier coefficients

$$\phi_{\vec{k}} \equiv L^{-3/2}\int e^{-i\vec{k}\cdot\vec{x}}\phi(t, \vec{x})\, d^3 x, \tag{16.8}$$

where

$$\vec{k} = \frac{2\pi}{L}(n_1, n_2, n_3).$$ (16.9)

We have

$$H = \sum_{\vec{k}} \frac{1}{2}(|\dot{\phi}_{\vec{k}}|^2 + \omega_{\vec{k}}^2|\phi_{\vec{k}}|^2),$$ (16.10)

where

$$\omega_{\vec{k}}^2 = |\vec{k}|^2 + m^2.$$ (16.11)

Thus, a free Klein–Gordon field, ϕ, is seen to be nothing more than an infinite collection of decoupled harmonic oscillators. The quantum field theory associated to ϕ can therefore be obtained by quantizing each of these oscillators. It follows immediately that the Heisenberg field operator $\phi(t, \vec{x})$ should be given by the formula

$$\phi(t, \vec{x}) = L^{-3/2} \sum_{\vec{k}} \frac{1}{2\omega_{\vec{k}}}(e^{i\vec{k}\cdot\vec{x} - i\omega_{\vec{k}}t}a_{\vec{k}} + e^{-i\vec{k}\cdot\vec{x} + i\omega_{\vec{k}}t}a_{\vec{k}}^\dagger).$$ (16.12)

However, the sum on the right side of this equation does not converge in any sense that would allow one to define the operator ϕ at the point $(t, \vec{x})$. Roughly speaking, the infinite number of arbitrarily high frequency oscillators fluctuate too much to allow $\phi(t, \vec{x})$ to be defined. However, this difficulty can be overcome by "smearing" ϕ with an arbitrary "test function," f (i.e., f is a smooth function of compact support), so as to define

$$\phi(f) = \int f(t, \vec{x})\phi(t, \vec{x})d^4x$$ (16.13)

rather than $\phi(t, \vec{x})$. The resulting formula for $\phi(f)$ can be shown to make rigorous mathematical sense, thus defining ϕ as an "operator-valued distribution."

The ground state, $|0\rangle$, of ϕ is simply the simultaneous ground state of all of the harmonic oscillators that comprise ϕ, i.e., it is the state satisfying $a_{\vec{k}}|0\rangle = 0$ for all $\vec{k}$. In quantum field theory, this state is interpreted as representing the "vacuum." A state of the form $(a^\dagger)^n|0\rangle$ is interpreted as a state where a total of n particles are present. In an interacting theory, the state of the field may be such that the field behaves like a free field at early and late times. In that case, we would have a particle interpretation of the states of the field at early and late times. The relationship between the early and late time particle descriptions of a state—given by the S-matrix—contains a great deal of the dynamical information about the interacting theory, and, indeed, contains all of the information relevant to laboratory scattering experiments.

The particle interpretation/description of quantum field theory in flat space-time has been remarkably successful—to the extent that one might easily get the impression from the way the theory is normally described that, at a fundamental level, quantum field theory is really a theory of particles. However, the definition of particles relies on the decomposition of ϕ into annihilation and creation operators in eq. (16.12). This decomposition, in turn, relies heavily on the time translation symmetry of Minkowski spacetime, since the "annihilation part" of ϕ is its positive frequency part with respect to time translations. In a curved spacetime that does not possess a time translation symmetry, it is far from obvious how a notion of "particles" should be defined.

16.2 The Development of Quantum Field Theory in Curved Spacetime from the Mid-1960s Through the Mid-1970s

Beginning in the mid-1960s, Parker investigated effects of particle creation in an expanding universe (Parker 1966, 1969). Consider a spatially flat Friedmann–Lemaître–Robertson–Walker spacetime, with metric

$$ds^2 = -dt^2 + a^2(t)[dx^2 + dy^2 + dz^2]. \tag{16.14}$$

Consider first the (highly artificial) case where $a(t)$ is constant for $t < t_0$ and is again constant for $t > t_1$, but goes through a time-dependent phase at intermediate times, $t_0 \leq t \leq t_1$. In the "in" region $t < t_0$, spacetime is locally indistinguishable from a corresponding portion of Minkowski spacetime, so a given state of a free quantum field will have a particle interpretation in that region; i.e., it can be characterized by its "particle content." Similarly, in the "out" region, $t > t_1$, the same state will also have a particle interpretation. However, on account of the time dependence of the metric in the intermediate region, a classical solution of the Klein–Gordon equation that corresponds to a purely positive frequency solution in the "in" region will not correspond to a purely positive frequency solution in the "out" region. This means that the "in" and "out" annihilation and creation operators of the quantum field (corresponding to the decomposition of eq. (16.12) for the "in" and "out" regions) will be different. This, in turn, implies that the particle content in the "in" and "out" regions will be different. In other words, the expansion of the universe will result in spontaneous particle creation. Quite generally, the relationship between the "in" and "out" annihilation and creation operators is given by a *Bogoliubov transformation*, whose coefficients are determined by the classical scattering. It is not difficult to derive a general formula for the resulting S-matrix in terms of these coefficients and, in particular, an expression for the spontaneous particle creation from the vacuum (see, e.g., (Wald 1994) for a general derivation).

Of course, we do not believe that the universe began or will end in a phase where $a(t)$ is constant. How do we analyze and describe particle creation in a more realistic context? If the universe is expanding sufficiently slowly, an approximate notion of an "adiabatic vacuum state" can be defined (Parker 1969), and a notion of "particles" relative to this adiabatic vacuum can be introduced. Such a notion of particles is completely adequate in the present universe to describe quantum field phenomena on scales small compared with the Hubble radius; i.e., only when we consider modes of the field whose oscillation period is comparable to or larger than the Hubble time (i.e., 10^{10} years in the present universe) does the notion of "particles" become genuinely ambiguous. However, at or very near the "Big Bang" singularity, the notion of "particles" is highly ambiguous.

The next major steps in the development of quantum field theory in curved spacetime came from the application of the theory to black holes. By definition, a *black hole* in an asymptotically flat spacetime is a region of spacetime from which nothing can escape to infinity. A black hole is believed to be the endpoint of the complete gravitational collapse of a body. The "time reverse" of a black hole—i.e., a region of spacetime that is impossible to enter if one starts from infinity—is called a *white hole*. It is believed that white holes cannot occur in nature. (The asymmetry between the expected occurrence of black holes and the expected non-occurrence of white holes in nature is undoubtedly closely related to the second law of thermodynamics.) However, black holes are expected to "settle down" to a stationary final state, and if one extends the idealized stationary final state metric of a black hole backward in time, preserving the stationary symmetry, one obtains a spacetime containing a white hole region. Thus, although white holes are not expected to occur in nature, they do occur in the mathematically idealized solutions used to describe the final stationary states of black holes.

Since, by definition, nothing can escape from a black hole, it would seem that black holes would be one of the least promising places to seek any observable effects of particle creation. However, the study of effects of particle creation by black holes arose quite naturally for reasons that I shall now explain. Outside of a rotating black hole is a region, called the *ergosphere*, where the Killing field that describes time translations at infinity becomes spacelike. This means that an observer in the ergosphere cannot "stand still" relative to a stationary observer at infinity. In fact, an observer in the ergosphere must rotate relative to infinity in the same direction as the rotation of the black hole; this is an extreme example of the "dragging of inertial frames" effect in general relativity. The prime importance of the ergosphere is that, since the time translation Killing field is spacelike there, it is possible to have classical particles whose total energy (including rest mass energy) relative to infinity is negative. Consequently, as Penrose (Penrose 1969) realized, one can extract energy from a black hole by sending a body into the ergosphere and having it break up into two fragments, one of which has negative total energy. The negative energy fragment then falls into the black hole (thereby reducing its mass), but it can be arranged that the positive energy fragment emerges to infinity, carrying greater total energy than the original body.

Not long after Penrose's discovery, Misner (unpublished), Zel'dovich (1971), and Starobinski (1973) realized that there is a wave analogue of the Penrose energy extraction process. Instead of sending in a classical particle and having it break up into two fragments, one can simply have a classical wave impinge upon a rotating black hole. Part of this wave will be absorbed by the black hole and part will return to infinity. However, if the frequency and angular dependence of the wave are chosen to lie in the appropriate range, then the part of the wave that is absorbed by the black hole will carry negative energy relative to infinity. The portion of the wave which returns to infinity will thereby have greater energy and amplitude than the initial wave. This phenomenon is known as *superradiant scattering*.

Thus, when a wave of superradiant frequency and angular dependence impinges upon a rotating black hole, the black hole amplifies the wave just like a laser. Superradiant scattering thus appears to be a direct analogue of stimulated emission. However, in quantum theory, it is well known that in circumstances where stimulated emission occurs, spontaneous emission will also occur. This suggested that for a rotating black hole, "spontaneous emission"—i.e., spontaneous particle creation from the vacuum—should occur. This was noted by Starobinski (1973) and confirmed by Unruh (1974).

The fact that spontaneous particle creation occurs near rotating black holes did not cause much surprise or excitement. The effect is negligibly small for macroscopic black holes such as those that would be produced by the collapse of rotating stars, so unless tiny black holes were produced in the early universe, the effect is not of astrophysical importance. While it is an interesting phenomenon as a matter of principle, it was not surprising or unexpected in view of the ability to extract energy from a rotating black hole by classical processes. However, it led directly to a further development that caused a genuine revolution.

The calculation of particle creation by a rotating black hole was done in the idealized spacetime representing the stationary final state of the black hole. As explained above, this spacetime also contains a white hole. Consequently, in the particle creation calculation one has to impose initial conditions on the white hole horizon that express the condition that no particles are emerging from the white hole. In the calculation of Unruh (1974), a seemingly natural choice of "in" vacuum state on the white hole horizon was made. But it was not obvious that this choice was physically correct.

In 1974, Hawking (1975) realized that this difficulty could be overcome by considering the more physically relevant case of a spacetime describing gravitational collapse to a black hole rather than the idealized spacetime describing a stationary black hole (and white hole). When he carried out the calculation, he found that the results were significantly altered from the results obtained for the idealized stationary black hole using the seemingly natural choice of vacuum state on the white hole horizon. Remarkably, Hawking found that, even for a non-rotating black hole, particle creation occurs at late times and produces a steady, non-zero flux of particles to infinity. Even more remarkably, he found that, for a non-rotating black hole, the spectrum of particles emitted to infinity at late times is precisely thermal in character, at a temperature $T = \kappa/2\pi$, where κ denotes the surface gravity of the black hole.

The implications of Hawking's result were enormous. It established that black holes are perfect black bodies in the thermodynamic sense at a non-zero temperature. This tied in beautifully with the mathematical analogy that had previously been discovered between certain laws of black hole physics and the ordinary laws of thermodynamics, giving clear evidence that the similarity of these laws is much more than a mere mathematical analogy. The identification of these laws led to the identification of $A/4$ as representing the physical entropy of a black hole, where A denotes the area of the event horizon. These and other ramifications of Hawking's results have provided us with some of the deepest insights we presently have regarding the nature of quantum gravity.

However, the Hawking calculation also had other major ramifications for the development of quantum field theory in curved spacetime, and these are the ones that I wish to emphasize here. Although Hawking's results were too beautiful to be disbelieved, there was a very disturbing feature of the calculation: Using a seemingly natural notion of "particles" near the event horizon of the black hole, there appeared to be a divergent density of ultra-high-frequency particles present there. What do these "particles" mean? Does their presence destroy the black hole?

To gain insight into this issue, Unruh (1976) proceeded by taking a purely operational viewpoint regarding the notion of "particles": A "particle" is a state of the field that makes a particle detector register. Unruh then showed that, in Minkowski spacetime, when a quantum field is in its ordinary vacuum state, a particle detector carried by an accelerating observer will become excited. Indeed, he showed that a uniformly accelerating observer "sees" an exactly thermal spectrum of particles, at a temperature $T = a/2\pi$, where a denotes the acceleration of the observer. This result provided an explanation of the meaning of the divergent density of ultra-high-frequency particles present near the event horizon of a black hole. Such particles would be "seen" by a stationary observer just outside the black hole. Indeed, such an observer would have to undergo an enormous acceleration in order to remain stationary, and what this observer sees corresponds exactly to the Unruh effect in Minkowski spacetime. However, an observer who freely falls into the black hole would not "see" these particles, just as an inertial observer in Minkowski spacetime does not see the particles associated with the accelerating observer. Furthermore, there are no significant stress-energy effects associated with the quantum field near the horizon of the black hole, so the presence of these "particles" as would be seen by a hypothetical stationary observer outside the black hole does not have a significant back-reaction effect and, in particular, does not destroy the black hole.

The clear lesson from Unruh's work is that one cannot view the notion of "particles" as fundamental in quantum field theory. As its name suggests, quantum field theory is truly the quantum theory of *fields*, not particles. If one views the local fields as the fundamental objects in the theory, the Unruh effect is seen to be a simple consequence of how these fields interact with other quantum mechanical systems (i.e., "particle detectors"). If one attempts to view "particles" as the fundamental entities in the theory, the Unruh effect becomes incomprehensible.

Furthermore, with the exception of stationary spacetimes (and certain other spacetimes with very special properties), there is no preferred notion of a "vacuum state" in quantum field theory in curved spacetime and, correspondingly, there is no preferred notion of "particles." The difficulty is not that there is no notion of a vacuum state but rather that there are many, and, in a general spacetime, none can be uniquely singled out as having distinguished properties. Thus, for this reason alone, it clearly would be preferable to have a formulation of quantum field theory in curved spacetime that does not require one to specify a vacuum state or a notion of "particles" at the outset.

The usual way of constructing the theory of a free quantum field would be to choose a vacuum state and then take the Hilbert space of states to be the Fock space based upon this choice of vacuum state. The field operator can then be defined (as an operator-valued distribution) by the analogue of eq. (16.12). If different choices of vacuum state corresponded to a mere relabeling of the states in terms of their particle content, then it would make sense to construct the theory in this manner despite the lack of a preferred vacuum state. However, in general, it turns out that different choices of vacuum state will give rise to unitarily inequivalent theories, so the choice does matter. Since there does not appear to be a preferred construction, how does one formulate quantum field theory in a general curved spacetime?

By the mid-1980s, it was well understood—via the efforts of Ashtekar and Magnon (1975), Sewell (1982), Kay (1978, 1985), and others—that the theory of a free quantum field in curved spacetime could be formulated in an entirely satisfactory manner via the algebraic approach. I shall now describe this formulation.

16.3 The Algebraic Formulation of Free Quantum Field Theory in Curved Spacetime

In the algebraic formulation of quantum field theory in an arbitrary globally-hyperbolic, curved spacetime, (M, g_{ab}), one begins by specifying an algebra of field observables. For a free Klein–Gordon field, a suitable algebra can be defined as follows. Start with the free *-algebra, $\mathcal{A}_0$, generated by a unit element I and expressions of the form "$\phi(f)$," where f is a test function on M. In other words, $\mathcal{A}_0$ consists of all formal finite linear combinations of finite products of ϕ's and ϕ^*'s; e.g., an example of an element of $\mathcal{A}_0$ is $c_1\phi(f_1)\phi(f_2) + c_2\phi^*(f_3)\phi(f_4)\phi^*(f_5)$. Now impose the following relations on $\mathcal{A}_0$: (i) linearity of $\phi(f)$ in f; (ii) reality of ϕ: $\phi^*(f) = \phi(\overline{f})$, where $\overline{f}$ denotes the complex conjugate of f; (iii) the Klein–Gordon equation: $\phi([\nabla^a\nabla_a - m^2]f) = 0$; (iv) the canonical commutation relations:

$$[\phi(f), \phi(g)] = -i\Delta(f, g)I, \qquad (16.15)$$

where Δ denotes the advanced minus retarded Green's function. The desired *-algebra, $\mathcal{A}$, is simply $\mathcal{A}_0$ factored by these relations. Note that the observables in $\mathcal{A}$ correspond to the correlation functions of the quantum field ϕ.

In the algebraic approach, a *state*, ω, is simply a linear map $\omega : \mathcal{A} \to \mathbf{C}$ that satisfies the positivity condition, $\omega(A^*A) \geq 0$, for all $A \in \mathcal{A}$. The quantity $\omega(A)$ is interpreted as the expectation value of the observable A in the state ω.

States in the usual Hilbert space sense give rise to algebraic states as follows: Suppose $\mathcal{H}$ is a Hilbert space which carries a representation, π, of $\mathcal{A}$, i.e., for each $A \in \mathcal{A}$, the quantity $\pi(A)$ is an operator on $\mathcal{H}$, and the association $A \to \pi(A)$ preserves the algebraic relations of $\mathcal{A}$. Let $\Psi \in \mathcal{H}$ be such that it lies in the common domain of all operators $\pi(A)$. Then the map $\omega : \mathcal{A} \to \mathbf{C}$ given by

$$\omega(A) = \langle \Psi | \pi(A) | \Psi \rangle \qquad (16.16)$$

defines a state on $\mathcal{A}$.

Conversely, given a state, ω, on $\mathcal{A}$, we can use it to define a (pre-)inner product on $\mathcal{A}$ by

$$(A_1, A_2) = \omega(A_1^* A_2). \qquad (16.17)$$

This may fail to define an inner product on $\mathcal{A}$ because, although $(A, A) \geq 0$ for all $A \in \mathcal{A}$, there may exist non-zero elements for which $(A, A) = 0$. However, if this happens, we may factor the space by such zero-norm vectors. We may then complete the resulting space to get a Hilbert space $\mathcal{H}$ which carries a natural representation, π, of $\mathcal{A}$. The vector $\Psi \in \mathcal{H}$ corresponding to $I \in \mathcal{A}$ then satisfies $\omega(A) = \langle \Psi | \pi(A) | \Psi \rangle$ for all $A \in \mathcal{A}$. The usual quantum-mechanical probability rules for determining values of the operator $\pi(A)$ in the state Ψ can then be taken over to define probability rules for the observable A in the state ω, provided that $\pi(A)$ defines a self-adjoint operator on $\mathcal{H}$. We thereby obtain a complete specification of the quantum field theory of a Klein–Gordon field on an arbitrary globally-hyperbolic curved spacetime insofar as the local field observables appearing in $\mathcal{A}$ are concerned; i.e., in any state we can provide the probabilities for measuring the possible values of all observables in $\mathcal{A}$. No preferred notion of "vacuum state" or "particles" need be introduced, although, of course, one is free to introduce such notions in particular spacetimes if one wishes.

As has just been seen, every state in the algebraic sense corresponds to a state in the usual Hilbert space sense. What, then, is the advantage of formulating the theory via the algebraic approach? The main advantage is that one is not forced to make a particular choice of representation at the outset; i.e., one may simultaneously consider all states arising in all Hilbert space constructions of the theory. As a result, one may define the theory without first having to make a choice of "vacuum state" or introduce any other problematical notions. In addition, it is worth noting that the algebraic notion of states dispenses with the unphysical states in the Hilbert space that do not lie in the domain of the observables of the theory; vectors in a Hilbert space representation of the theory that do not lie in the domain of all $\pi(A)$ do not define states in the algebraic sense.

The above provides a completely satisfactory construction of a free Klein–Gordon field in curved spacetime insofar as observables in $\mathcal{A}$ are concerned. Similar constructions can be done for all other free (i.e., non-self-interacting) quantum fields. However, the overall situation is still quite incomplete and unsatisfactory for at least the following two reasons: First, even if we were only interested in the theory of a free Klein–Gordon field, there are many observables of interest that are not represented in $\mathcal{A}$. Indeed, the observables in $\mathcal{A}$ are merely the n-point functions of the linear field ϕ; they do not even include polynomial functions of ϕ and its derivatives ("Wick polynomials"). A prime example of an observable of great physical importance that is not represented in $\mathcal{A}$ is the stress-energy tensor, T_{ab}, which would be needed to estimate back-reaction effects of the quantum field on the spacetime metric. Therefore, we would like to enlarge the algebra $\mathcal{A}$ so that it includes at least the Wick polynomials of ϕ. Second, we do not believe that the quantum fields occurring in nature are described by free fields, so we would like to extend the theory to non-linear fields. Even in Minkowski spacetime, one understands how to do this only at a perturbative level, but one would like to at least generalize these perturbative rules to curved spacetime. These perturbative rules require that one be able to define Wick polynomials of the free field as well as time-ordered products of polynomial expressions in the field. Again, we need to enlarge the algebra $\mathcal{A}$ so that it includes such quantities.

If a quantum field were well defined at a (sharp) spacetime event p, it would be straightforward to define polynomial quantities in ϕ as well as time-ordered products. However, we have already noted below eq. (16.12) that a quantum field makes sense only as a distribution on spacetime. Consequently, *a priori*, a naive attempt to define, say, $[\phi(p)]^2$ is not likely to make any more sense than an attempt to define the square of a Dirac delta function. In particular, it would be natural to attempt to define the smeared Wick power $\phi^2(f)$ by a formula like

$$\phi^2(f) = \lim_{n \to \infty} \int \phi(x)\phi(y)f(x)F_n(x,y)d^4x d^4y \qquad (16.18)$$

where $F_n(x,y)$ is a sequence of smooth functions that approaches the Dirac delta function $\delta(x,y)$. However, the right-hand side diverges in the limit, so some sort of "regularization" of this expression must first be done in order to make the limit well behaved.

Once the Wick powers $\phi^k(f)$ have been defined, it would be an easy matter to define the time-ordered product $T(\phi^{k_1}(f_1) \ldots \phi^{k_n}(f_n))$ by a straightforward "time ordering" of the factors in the case where the supports of $f_1, \ldots, f_n$ have suitable causal properties so that they can be put in a well-defined time order. Indeed, using induction in the number, n, of factors, it is straightforward to define $T(\phi^{k_1}(f_1) \ldots \phi^{k_n}(f_n))$ whenever the intersection of the supports of $f_1, \ldots, f_n$ vanishes. However, it is not straightforward to extend this distribution to the "total diagonal," i.e., to the case where the supports of $f_1, \ldots, f_n$ have non-vanishing mutual intersections.

From the way I have described the regularization issues above, it might seem that the most difficult problem would be to define Wick polynomials and that the definition of time-ordered products would be a minor addendum to this problem. In fact, however, in Minkowski spacetime it is well known that Wick polynomials can be defined by a "normal ordering" prescription (Wightman and Garding 1964), which can be interpreted as subtracting off the vacuum expectation value of the field quantities before taking the kind of limit appearing on the right side of eq. (16.18). On the other hand, the problem of extending time-ordered products to the total diagonal corresponds to the problem of renormalizing all Feynman diagrams—an extremely difficult and complex problem.

There are major issues of principle that must be overcome in order to extend the Minkowski spacetime regularization and renormalization prescriptions to curved spacetime. The normal ordering prescription for defining Wick polynomials in Minkowski spacetime relies on the existence of a preferred vacuum state, with respect to which the "normal ordering" is carried out. However, we have already seen that, in a general curved spacetime, there does not appear to exist any notion of a preferred vacuum state. Furthermore, the renormalization prescriptions used to define time-ordered products in Minkowski spacetime make use of "momentum space methods" (i.e., global Fourier transforms of quantities) and/or "Euclidean methods" (i.e., analytic continuation of expressions defined on Euclidean space rather than Minkowski spacetime). These methods, in turn, require Poincaré symmetry, the existence of a preferred, Poincaré-invariant vacuum state, and/or the ability to "Euclideanize" Minkowski spacetime by the transformation $t \rightarrow it$. All of these features are absent in a general, curved spacetime.

It was already understood by the late 1970s that it should be possible to define the stress-energy tensor, T_{ab}, of a quantum field ϕ only on a restricted class of states, namely, the so-called Hadamard states, ω_H, whose two-point distribution $\omega_H(\phi(x), \phi(y))$ has a short distance singularity structure as $y \rightarrow x$ of a particular form (see, e.g., (Wald 1994)). For Hadamard states, a prescription for defining the expectation value, $\omega_H(T_{ab})$, can be given that involves the subtraction from $\omega_H(\phi(x), \phi(y))$ of a locally and covariantly constructed Hadamard parametrix rather than a vacuum expectation value (Wald 1994). The resulting prescription defines $\omega_H(T_{ab})$ in an entirely satisfactory manner that does not require a choice of vacuum state. Indeed, this prescription is local and covariant in the sense that the value of $\omega_H(T_{ab})$ at a point p depends only on the spacetime geometry and the behavior of ω_H in an arbitrarily small neighborhood of p. It is not difficult to show that, even if one could make a unique choice of vacuum state in all spacetimes, normal ordering would not provide a local and covariant definition of $\omega_H(T_{ab})$.

However, although the above prescription provides a satisfactory definition of the expectation value of the stress-energy tensor in Hadamard states and can be generalized to define the expectation value of higher Wick powers, it does not define T_{ab} or other Wick powers as elements of an enlarged algebra. Indeed, in a Hilbert space representation of the theory, the above prescription would merely define T_{ab} as a quadratic form on Hadamard states rather than as an operator-valued distribution, so no probability rules for measuring the possible values of T_{ab} would be available. Furthermore, it is worth mentioning that the characterization of "Hadamard states" in terms of their short-distance singularity structure is extremely cumbersome to work with. By the mid-1990s it was still very far from clear how to perform the much more difficult and complex renormalizations that would be needed to define time-ordered products in curved spacetime.

16.4 Progress Since the Mid-1990s

During the past decade, the algebra of observables for a free quantum field has been extended to include all Wick polynomials and time-ordered products, so that, in particular, the perturbative renormalization of interacting quantum fields in curved spacetime is now rigorously well defined. Much of this progress has resulted from the importation of methods of "microlocal analysis" (Hormander 1985) into the theory. In essence, microlocal analysis provides a refined characterization of the singularities of a distribution. If one has a distribution, α, defined in a neighborhood of point p on a manifold M, one can multiply α by a smooth function f with support in an arbitrarily small neighborhood of p, such that $f(p) \neq 0$. One can then examine the decay properties of the Fourier transform of $f\alpha$ (where the Fourier transform can be defined by choosing an arbitrary embedding of a neighborhood of the support of $f\alpha$ into Euclidean space). If α were smooth in a neighborhood of p, then, for f with support in this neighborhood, the Fourier transform of $f\alpha$ would decay rapidly in all directions in Fourier transform space, k, as $|k| \to \infty$. Thus, the failure of the Fourier transform of $f\alpha$ to decay rapidly characterizes the singular behavior of α at p. If, for all choices of f, the Fourier transform of $f\alpha$ does not decay rapidly in a neighborhood of the direction k, then one says that the pair (p, k) lies in the *wavefront set*, $\mathrm{WF}(\alpha)$, of α. One can naturally identify $\mathrm{WF}(\alpha)$ with a subset of the cotangent bundle of the manifold, M. The wavefront set thereby provides a characterization of not only the *points* in M at which α is singular, but also the *directions* (in the cotangent space) at which it is singular. This refined characterization of the singularities of distributions can enable one to define operations that normally are ill defined. For example, if α and β are distributions, then it normally will not make mathematical sense to take their product. However, if it is the case that whenever $(p, k) \in \mathrm{WF}(\alpha)$, we have that $(p, -k) \notin \mathrm{WF}(\beta)$, then the product $\alpha\beta$ can be defined in a natural way via the Fourier convolution formula.

By providing rules for, e.g., when products of distributions are well defined as distributions, microlocal analysis provides an extremely useful calculus for determining whether proposed regularization/renormalization schemes are well defined. Since the analysis is completely local in nature, it provides an ideal tool for analyzing the behavior of local field observables.

The first significant application of microlocal analysis to quantum field theory in curved spacetime occurred in the Ph.D. thesis of Radzikowski (Radzikowski 1992, 1996), a student of Wightman. Radzikowski was concerned with proving a conjecture, due to Kay, which stated that if a quantum state had a two-point function whose short-distance singularities are of the Hadamard form, then it could not have any additional singularities at large spacelike separations ("local Hadamard form implies global Hadamard form"). Radzikowski employed the tools of microlocal analysis to prove this conjecture. In particular, in the course of his analysis, he proved that the (quite cumbersome) characterization of Hadamard states in terms of the detailed local singularity structure of $\omega_H(\phi(x), \phi(y))$ is equivalent to a very simple condition on the wavefront set of this distribution, namely, that $\mathrm{WF}[\omega_H(\phi(x), \phi(y))]$ is the subset of the cotangent bundle of $M \times M$ consisting of all points $(x, y; k, l)$ such that x and y are connected by a null geodesic γ with future-directed tangent $k^a = g^{ab}k_b$ at x and with l_a being minus the parallel transport along γ of k_a to y.

It is worth mentioning that there was an interesting historical interplay between microlocal analysis and quantum field theory in curved spacetime. In the late 1960s Hormander visited the Institute for Advanced Study in Princeton and interacted with Wightman. Wightman explained to Hormander what the "Feynman propagator" is in Minkowski spacetime, and a characterization of "Feynman parametrices" in a general curved spacetime in terms of wavefront set properties can be found in the classic paper of Duistermaat and Hormander (Duistermaat and Hormander 1972). Conversely, Wightman realized that the methods of microlocal analysis could potentially be useful in the formulation of quantum field theory in curved spacetime. For example, in de Sitter spacetime, there is no globally timelike Killing field and therefore no global notion of energy that is positive. Therefore, it does not appear that one could impose a global spectral condition on a quantum field analogous to the requirement of positivity of energy in the Minkowski case. However, one might be able to impose a "microlocal spectral condition" on the local quantum field observables. Shortly after his interactions with Hormander, Wightman had a student, Fulling, who was interested in quantum field theory in curved spacetime, and he suggested to Fulling that he investigate the possible application of microlocal analysis to quantum field theory in curved spacetime. However, after spending some effort in studying microlocal analysis, Fulling decided that his efforts would be better spent on other projects. Among the other projects that Fulling then investigated in his thesis was the inequivalence of different quantization schemes. In particular, he showed that quantization in the "Rindler wedge" of Minkowski spacetime using a Lorentz boost Killing field to define a notion of "time translations" gave rise to a different notion of "vacuum state" than the restriction of the usual Minkowski vacuum to this region. This work provided the mathematical basis for Unruh's subsequent analysis discussed above (Unruh 1976). However, Wightman had to wait another twenty years before he had another student interested in quantum field theory in curved spacetime. When Radzikowski began to apply the methods of microlocal analysis to analyze Kay's conjecture, Wightman was well prepared to provide plenty of encouragement.

After Radzikowski's work, it became clear to Fredenhagen and collaborators that microlocal analysis should provide the needed tools for analyzing the divergences occurring in quantum field theory in curved spacetime. Brunetti, Fredenhagen, and Kohler (1996) showed that, if one considers a Fock representation associated with an arbitrary Hadamard vacuum state ω_0, then normal ordering can be used to define Wick polynomials as operator-valued distributions on this Hilbert space. Indeed, a larger algebra, W, of field observables—large enough to include all time-ordered products—can be defined in this manner. Brunetti and Fredenhagen (2000) also formulated a microlocal spectral condition that should be imposed on time-ordered products. However, as previously mentioned, a normal ordering prescription cannot yield a local and covariant definition of Wick polynomials. Furthermore, the construction of W given in (Brunetti, Fredenhagen, and Kohler 1996) invokes an arbitrary choice of Hadamard vacuum state ω_0. Nevertheless, it can be shown that, as an abstract algebra, W does not depend on the choice of ω_0, so it is a legitimate candidate for the desired enlarged algebra of observables. Thus, the key remaining issue was to determine which elements of W properly represent the "true" Wick polynomials and time-ordered products.

A key condition to be imposed on the definition of Wick polynomials and time-ordered products is that they be local and covariant fields. As mentioned in the previous section, this condition had been imposed on the definition of the expectation value of the stress-energy. However, the formulation of this notion given in (Wald 1994) was not adequate for the present purpose, and a more general formulation had to be given (Hollands and Wald 2001; Brunetti, Fredenhagen, and Verch 2003).

With these key ideas and constructions in place, it was possible to prove the following results (Hollands and Wald 2001, 2002, 2003, 2005). (1) There exists a well-defined prescription for defining all Wick polynomials that is local and covariant and satisfies a list of additional reasonable properties, including appropriate scaling behavior and continuous/analytic variation under continuous/analytic changes in the metric (Hollands and Wald 2001). This prescription is unique up to certain "local curvature ambiguities." For example, for a Klein–Gordon field, ϕ, the prescription for ϕ^2 is unique up to

$$\phi^2 \to \phi^2 + (c_1 R + c_2 m^2)I \tag{16.19}$$

where c_1, c_2 are arbitrary constants, R denotes the scalar curvature, and I denotes the identity element of $\mathcal{W}$. For a massless field in Minkowski spacetime, all of the ambiguities disappear, and the prescription agrees with normal ordering with respect to the usual Minkowski vacuum state. However, on a general curved spacetime, the prescription for defining ϕ^2 and other Wick polynomials does not agree with normal ordering with respect to any choice of vacuum state. (2) There exists a prescription for defining all time-ordered products that is local and covariant, that satisfies the microlocal spectral conditions (Brunetti and Fredenhagen 2000), and that satisfies a list of additional reasonable properties (Hollands and Wald 2002). This prescription is unique up to "renormalization ambiguities" of the type expected from Minkowski spacetime analyses, but with additional local curvature ambiguities. (3) Theories that are renormalizable in Minkowski spacetime remain renormalizable in curved spacetime. For renormalizable theories, renormalization group flow can be defined in terms of the behavior of the quantum field theory under scaling of the spacetime metric, $g_{ab} \to \lambda^2 g_{ab}$ (Hollands and Wald 2003). (4) Additional renormalization conditions on time-ordered products can be imposed so that, order by order in perturbation theory, for an arbitrary (not necessarily renormalizable) interaction, (i) the interacting field satisfies the classical interacting equation of motion and (ii) the stress-energy tensor of the interacting field is conserved (Hollands and Wald 2005). All of the above results have been obtained without any appeal to a notion of "vacuum" or "particles."

These and other results of the past decade have demonstrated that quantum field theory in curved spacetime has a mathematical structure that is comparable in depth to such theories as classical general relativity. In particular, it is highly non-trivial that quantum field theory in curved spacetime appears to be mathematically consistent. Although quantum field theory in curved spacetime cannot be a fundamental description of nature since gravity itself is treated classically, it seems hard to believe that it is not capturing some fundamental properties of nature.

The above results suffice to define interacting quantum field theory in curved spacetime at a perturbative level. However, it remains very much an open issue as to how to provide a non-perturbative formulation of interacting quantum field theory in curved spacetime. It is my hope that significant progress will be made on this issue in the coming years.

Acknowledgements

This research was supported in part by NSF grant PHY-0456619 to the University of Chicago.

References

Ashtekar, A. and Magnon, A. 1975. *Proc. Roy. Soc. London* A346: 375.
Brunetti, R. and Fredenhagen, K. 2000. *Commun. Math. Phys.* 208: 623.
Brunetti, R., Fredenhagen, K. and Kohler, M. 1996. *Commun. Math. Phys.* 180: 633.
Brunetti, R., Fredenhagen, K. and Verch, R. 2003. *Commun. Math. Phys.* 237: 31.
Duistermaat, J. J. and Hormander, L. 1972. *Acta Math.* 128: 183.
Hawking, S. W. 1975. *Commun. Math. Phys.* 43: 199.
Hollands, S. and Wald, R. M. 2001. *Commun. Math. Phys.* 223: 289.
——. 2002. *Commun. Math. Phys.* 231: 309.
——. 2003. *Commun. Math. Phys.* 237: 123.
——. 2005. *Rev. Math. Phys.* 17: 227.
Hormander, L. 1985. *The Analysis of Linear Partial Differential Operators I*. Berlin: Springer-Verlag.
Kay, B. S. 1985. *Commun. Math. Phys.* 62: 55
——. 1978. *Commun. Math. Phys.* 100: 57.
Parker, L. 1966. Ph.D. thesis, Harvard University (unpublished).
——. 1969. *Phys. Rev.* 183: 1057.
Penrose, R. 1969. *Rev. Nuovo Cimento* 1: 252.
Radzikowski, M. J. 1992. Ph.D. thesis, Princeton University (unpublished).
——. 1996. *Commun. Math. Phys.* 179: 529.
Sewell, G. L. 1982. *Ann. Phys.* 141: 201.
Starobinski, A. A. 1973. *JETP* 37: 28.
Unruh, W. G. 1974. *Phys. Rev.* D10: 3194.
——. 1976. *Phys. Rev.* D14: 870.
Wald, R. M. 1984. *General Relativity*. Chicago: University of Chicago Press.
——. 1994. *Quantum Field Theory in Curved Spacetime and Black Hole Thermodynamics*. Chicago: University of Chicago Press.
Wightman, A. S. and Garding, L. 1964. *Ark. Fys.* 28: 129.
Zel'dovich, Ya. B. 1971. *JETP Lett.* 14: 180.

17

The Border Between Relativity and Quantum Theory

Tevian Dray

Oregon State University, USA

17.1 Introduction

Many efforts have been made to fulfill Einstein's dream of unifying general relativity and quantum theory, including the study of quantum field theory in curved space, supergravity, string theory, twistors, and loop quantum gravity. While all of these approaches have had notable successes, unification has not yet been achieved. After a brief tour of the progress which has been made, we focus on the role played by spinors in several of these approaches, suggesting that spinors may be the key to combining classical relativity with quantum physics. We conclude by outlining one possible generalization of traditional spinor language, involving the octonions, and speculate on its relevance to quantum gravity.

Each of the articles Einstein published during his *annus mirabilis* was at a border between major fields of physics. His work on the light quantum (Einstein 1905a) combined thermodynamics and electrodynamics into quantum mechanics (QM), while his work on kinetic theory (Einstein 1905b) blended thermodynamics with mechanics into statistical mechanics (SM). And his remaining publications during this miracle year (Einstein 1905c, 1905d) merged electrodynamics with mechanics, resulting in special relativity (SR). The relationships between these overlapping branches of physics can be displayed in a Venn diagram (Renn 2007), as shown in Figure 17.1.

That Einstein was able to contribute to all three of these border regions is indeed remarkable. But Figure 17.1 shows that there is a further piece of the puzzle, which combines all three areas of electrodynamics, mechanics, and thermodynamics, or equivalently quantum theory and relativity. Parts of this last border region, the realm of quantum field theories including quantum gravity (QG), remain uncharted territory 100 years later.

Another representation of these borders in physics is given by the Bronstein cube (Bronstein 1933; Stachel 1999), which is shown in Figure 17.2. Here, we begin with Galilean physics, which can be viewed as the limiting case of Newtonian physics as G goes to zero, as the limiting case of quantum mechanics as $\hbar$ goes to zero, or as

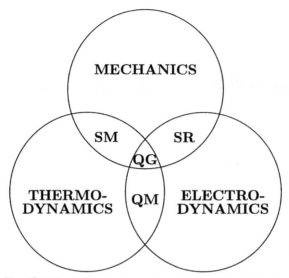

Fig. 17.1. The borders between three major fields of physics.

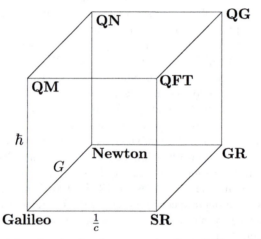

Fig. 17.2. The Bronstein cube, showing the progression from Galilean physics as one incorporates the fundamental constants c, G, and $\hbar$.

the limiting case of special relativity as c goes to infinity. Running these limits backwards, we can regard each of these last three theories as generalizations of Galilean physics, in each case introducing one of the fundamental physical constants G, $\hbar$, or c.

The borders between these regions again lead to new descriptions of nature: Combining quantum mechanics with special relativity leads to quantum field theory

(QFT), combining gravity with special relativity leads to general relativity (GR), and combining gravity with quantum mechanics leads to a theory which we refer to as "quantum Newtonian" (QN). Again, there is a final piece, the combination of all three generalizations at once, leading to quantum gravity.

In what follows, we describe some of the progress which has been made in the search for a theory of quantum gravity. Along the way, we describe the use of spinors in several of these approaches, and speculate that spinor-like objects will play a fundamental role in any theory which combines classical relativity theory with quantum physics. Finally, we outline the properties of a particular way to generalize this framework, involving the octonions.

17.2 Attempts at Unification

As the preceding diagrams suggest, there are several directions from which a theory of quantum gravity can be sought. Any such theory must reconcile the apparently conflicting roles played by time in quantum theory (preferred evolution parameter) and relativity (coordinate like any other).

We very briefly summarize four of the most popular approaches, two of which have roots in quantum theory (quantum field theory in curved space, string theory), and two of which grew out of classical relativity (loop quantum gravity, twistor theory).

This short paper cannot do justice to any of these theories; the reader in search of a more extensive treatment is encouraged to start with Rovelli's encyclopedic book (Rovelli 2004). We would nonetheless be remiss if we failed to mention several other noteworthy approaches to quantum gravity, including Sorkin's work on causal sets (Bombelli *et al.* 1987), the use of noncommutative geometry (Chamseddine and Connes 1996), the sum-over-histories approach to the wave function of the universe (Hartle and Hawking 1983), and the possibility of gravity-induced quantum state reduction (Penrose 1986).

17.2.1 Quantum Field Theory in Curved Space

One way to combine quantum theory with relativity is to investigate the behavior of quantum fields on a fixed space-time background. This simple idea provides a deep insight into some of the problems of quantum gravity—and turns out to be anything but simple.[1]

It was realized quite early (Fulling 1973; Davies 1976; Unruh 1976) that Rindler space leads to a quantum vacuum state in flat space which differs from the standard, Minkowski vacuum. With this realization came another, namely, that the very notion of what a particle is turns out to be observer dependent (Ashtekar and Magnon 1975; Unruh 2005); there is in general no preferred vacuum state.

---

[1] On a personal note, this intriguing combination of quantum theory and relativity led the author to an early collaboration with two other conference participants (Dray, Renn, and Salisbury 1983), reunited here for the first time in 20 years.

These unexpected consequences of applying ideas from standard quantum field theory to curved space led to a better understanding of the theoretical underpinnings of quantum field theory. As Wald has pointed out (Wald 2012), the essential ideas of quantum field theory are better understood in curved space than Minkowski space! This has led to extensive development of the algebraic approach to quantum field theory, and its application to curved space (Wald 1994).

But the crowning achievement of this approach was its application to black hole thermodynamics. Hawking showed (Hawking 1975) that the vacuum states of observers at early and late times in a Schwarzschild black hole space-time not only differ, but that the difference corresponds to black body radiation at a temperature related to the gravitational field of the black hole. This remarkable synthesis of quantum physics, relativity, and thermodynamics would turn out to be quite robust, being rederived later in other leading approaches to quantum gravity.

17.2.2 Supergravity and String Theory

Despite the remarkable results obtained by Hawking, a quantum field theory on a fixed space-time background is not a theory of quantum gravity. Adopting an approach modeled on particle physics, the search began for a covariant quantum field theory of fluctuations in the metric itself. Early work (DeWitt 1964; Feynman 1963) focused on constructing Feynman rules for general relativity. But 't Hooft and others soon showed that such a theory was non-renormalizable ('t Hooft 1973). The focus then shifted to supergravity, an 11-dimensional Kaluza–Klein theory for gravity, whose lower order expansion was finite. It was therefore hoped that supergravity would be finite, but this was later shown not to be the case.

This was the setting for the renaissance of string theory, with the discovery that supersymmetric versions of the theory were anomaly free (Green and Schwarz 1984). Superstring theory is a natural successor to supergravity, in which the fundamental objects are strings rather than pointlike objects. The field equations then describe evolution on the worldsheet of the string, and involve both bosonic (Π^μ) and fermionic (θ) degrees of freedom. Classical supersymmetry is only possible in certain space-time dimensions (3, 4, 6, and 10); 4-dimensional physics is obtained via compactification on Calabi–Yau manifolds starting from the 10-dimensional case.

A major success of string theory was to provide a statistical foundation for black hole thermodynamics, yielding independent descriptions of black hole entropy and Hawking radiation. And string theory is said to "contain" general relativity at low energies, by which is meant that the Einstein–Hilbert action appears in the Green–Schwarz action. Yet many unanswered questions remain. At least as originally formulated, string theory is not background independent. Furthermore, the requisite supersymmetric partners of fundamental particles have not been observed in nature. Another issue is that many of the physical predictions of string theory, such as the particle spectrum, are tied to a choice of Calabi–Yau space, of which there are many.

17.2.3 Canonical Quantum Gravity and Loop Quantum Gravity

Canonical quantum gravity has its roots in the Hamiltonian theory of constrained systems, whose early development was spearheaded by Bergmann and Dirac (Bergmann 1949; Dirac 1950). Milestones along the way include the ADM formalism (Arnowitt, Deser, and Misner 1962) for splitting space-time into space and time, which introduced new variables (h_{ab}, π^{ab}), and the later introduction of complex "new variables" $(A_a^i, \tilde{e}_i^a)$, now called Ashtekar variables, satisfying polynomial constraints (Ashtekar 1986). Another ingredient was the Wheeler–DeWitt equation, a Schrödinger equation for gravity (DeWitt 1967).

An important step forward was triggered by the discovery (Jacobson and Smolin 1988) of loop-like solutions of the Wheeler–DeWitt equation, written in terms of Ashtekar variables. This opened the door to a loop representation of quantum gravity, which has been an active area of research ever since (Rovelli 2004, Ashtekar 2012). Loop quantum gravity is background independent, but views space as naturally discrete. This results in there being no ultraviolet divergences, and provides a possible mechanism to control singularities such as the Big Bang. While fundamental issues such as Lorentz invariance and the semiclassical limit have not yet been resolved, a major success of loop quantum gravity has been the independent computation of the Bekenstein–Hawking formula for black hole entropy.

17.2.4 Twistor Theory

Another approach to quantum gravity which starts from relativity, rather than quantum theory, is Penrose's twistor theory (Penrose 1967), which exploits the combinatorics of null angular momentum. More recently, Sparling has proposed a spinor description of space-time which shares some features with twistor theory (Sparling 2006).

The fundamental object in twistor theory is a pair of spinors $Z^\alpha = (\omega^A, \pi_{A'})$ satisfying appropriate commutation relations and certain other constraints. Twistors describe massless objects, with a natural notion of helicity. Not only massive objects, but also the space-time in which they are perceived, are not fundamental in twistor theory; they must be derived. Further distinguishing features of twistor theory are that it describes complex Minkowski space, and that it is nonlocal, with local information being encoded in global fields.

However, while twistor theory is clearly beautiful mathematics, it is less obvious that it contains any physics. Wilczek has commented that, "... at present twistor ideas appear more as the desire for a physical theory than the embodiment of one" (Wilczek 2005), although Witten has pointed out a connection between twistor theory and string theory (Witten 2004).

17.3 The Present: Spinors

Are there any observable consequences to a quantum theory of gravity? Perhaps. There are some constraints, such as the sensitivity of astrophysical particle production rates due to possible changes in dispersion relations at the Planck scale. And some authors argue that there is no crisis here; we may not know the details of quantum gravity, but for all practical purposes we understand how gravity works.

Leaving aside for the moment the physics of quantum gravity, let us focus briefly on the mathematics.

As has been repeatedly emphasized by Baylis, classical relativistic physics has a spinorial formulation in terms of Clifford algebras that is closely related to the language of quantum mechanics, thus providing "an illuminating probe of the quantum/classical interface" (Baylis 1999).

With this in mind, it is perhaps not surprising that spinors play a key role in several approaches to quantum gravity, including twistor theory, string theory, and loop quantum gravity. Twistor theory is written in terms of pairs of complex spinors Z, string theory is written in terms of a (division algebra) spinor θ, and loop quantum gravity is written in terms of the covariant spinor derivative operator, namely, the spin connection A_a^i.

But what are spinors? Consider the following repackaging of an ordinary space-time vector x into a 2×2 complex matrix X:

$$
x = \begin{pmatrix} t \\ x \\ y \\ z \end{pmatrix} \longleftrightarrow X = \begin{pmatrix} t+z & x-iy \\ x+iy & t-z \end{pmatrix} \tag{17.1}
$$

Note that X is Hermitian

$$
X^\dagger = \overline{X}^T = X \tag{17.2}
$$

and that its determinant is (minus) the Lorentzian inner product

$$
-\det(X) = -t^2 + x^2 + y^2 + z^2 \tag{17.3}
$$

Furthermore, X can be written in terms of the Pauli matrices σ_m as

$$
X = t\,I + x\,\sigma_x + y\,\sigma_y + z\,\sigma_z \tag{17.4}
$$

which can alternatively be taken as the definition of the Pauli matrices themselves.

This repackaging reflects the fact that $SO(3,1)$ is locally isomorphic to $SL(2,\mathbb{C})$. The 2-component complex "vectors" in this formalism, such as

$$
v = \begin{pmatrix} \bar{b} \\ c \end{pmatrix} \tag{17.5}
$$

are (Penrose) spinors. Note that

$$
vv^\dagger = \begin{pmatrix} |b|^2 & \bar{b}\bar{c} \\ cb & |c|^2 \end{pmatrix} \tag{17.6}
$$

with

$$
\det(vv^\dagger) = 0 \tag{17.7}
$$

that is, the square of a spinor is a null vector. Spinors thus succeed in bridging the gap between quantum theory, where they describe spin-$\frac{1}{2}$ particles, and relativity, where they are naturally associated with lightlike objects. In fact, spinors have long been associated with classical relativity, with applications ranging from the Newman–Penrose formalism (Newman and Penrose 1962) to the classification of exact solutions (Karlhede 1980).

17.4 The Future: Octonions?

Each of the leading contenders for a viable theory of quantum gravity follows a natural path along the Bronstein cube. String theory has its roots in quantum mechanics and quantum field theory, while loop quantum gravity has its roots more in special and general relativity. This is a nontrivial difference; gravity really is different from the other fundamental forces. The essence of relativity is the lightcone structure, which is nicely incorporated by the use of spinors in both theories. But we still need a revolutionary idea. We speculate here on one possibility, based on generalizing the local isometry between $SO(3, 1)$ and $SL(2, \mathbb{C})$ to the other division algebras.

There are precisely four division algebras. Two of them are old friends, namely, the real numbers $\mathbb{R}$, and the complex numbers

$$\mathbb{C} = \mathbb{R} + \mathbb{R}i \tag{17.8}$$

which are obtained from the reals by including an imaginary number i satisfying $i^2 = -1$. This process can be continued, yielding Hamilton's quaternions

$$\mathbb{H} = \mathbb{C} + \mathbb{C}j \tag{17.9}$$

whose multiplication table is given in Figure 17.3, and Cayley's octonions

$$\mathbb{O} = \mathbb{H} + \mathbb{H}\ell \tag{17.10}$$

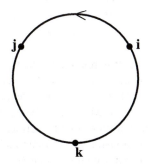

Fig. 17.3. The cyclic nature of the quaternionic multiplication table.

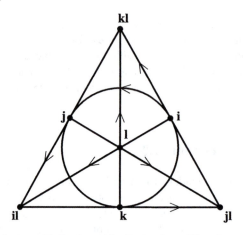

Fig. 17.4. The octonionic multiplication table. Each oriented line gives a quaternionic triple.

whose multiplication table is given in Figure 17.4. Elements of all of these algebras satisfy

$$|bc| = |b||c| \qquad (17.11)$$

which permits inverses (and division), although the quaternions fail to be commutative, and the octonions fail to be associative. It is well known that this Cayley–Dickson "doubling" process stops with the octonions; proceeding further would lead to zero divisors, in violation of (17.11).

Each of the four division algebras $\mathbb{K} = \mathbb{R}, \mathbb{C}, \mathbb{H}, \mathbb{O}$ can be used in the construction outlined in Equations (17.1)–(17.7), where the right-hand side of (17.1) is replaced by

$$\boldsymbol{X} = \begin{pmatrix} p & \bar{a} \\ a & m \end{pmatrix} \qquad (p, m \in \mathbb{R}; a \in \mathbb{K}) \qquad (17.12)$$

reflecting the local isometries

$$
\begin{aligned}
SO(2,1) &\approx SL(2,\mathbb{R}) \\
SO(3,1) &\approx SL(2,\mathbb{C}) \\
SO(5,1) &\approx SL(2,\mathbb{H}) \\
SO(9,1) &\approx SL(2,\mathbb{O})
\end{aligned}
\qquad (17.13)
$$

generalizing the complex case. It is remarkable that these four cases correspond to space-times of dimension $\dim \mathbb{K} + 2 = 3, 4, 6, 10$, precisely the dimensions in which supersymmetry is consistent. This is not a coincidence; Fairlie and Manogue used this language to show (Fairlie and Manogue 1986, 1987) that consistency follows automatically from the norm condition (17.11). Further work (Manogue and Sudbery 1989) showed that this use of "division-algebra spinors" automatically solves at least the first string equation of motion, which says that Π is null, since the trace-reversed square $\theta\theta^\dagger - \theta^\dagger\theta$ (whose derivative is Π) is also null.

More recent work using the octonions (Manogue and Dray 1999; Dray and Manogue 1999, 2000) proposes taking these local isometries themselves as the starting point. A key idea (Manogue and Dray 1999) is that the projection of a higher-dimensional null vector can be either null or timelike. Thus, starting with a theory of massless objects in 10 space-time dimensions, parameterized by the octonions, one obtains a 4-dimensional theory of both massless and massive objects simply by choosing a complex subalgebra of the octonions.

Furthermore, since 2-component spinors over the quaternions $\mathbb{H}$ are equivalent to 4-component complex spinors, a theory expressed in terms of quaternionic Penrose spinors includes ordinary Dirac spinors as a special case; the standard complex description of both massless and massive fermions can be recovered from the massless quaternionic Dirac equation. Over the octonions, there are naturally three such quaternionic descriptions, which can be interpreted as generations, having the correct number of spin/helicity states to describe leptons and their antiparticles (Manogue and Dray 1999; Dray and Manogue 2000).

The theory presented here can be elegantly rewritten in terms of Jordan matrices, as was previously done for the superparticle (Schray 1996). Jordan matrices are 3×3 octonionic Hermitian matrices

$$\mathcal{X} = \begin{pmatrix} p & a & \bar{b} \\ \bar{a} & m & c \\ b & \bar{c} & n \end{pmatrix} \qquad (p, m, n \in \mathbb{R}; a, b, c \in \mathbb{O}) \qquad (17.14)$$

which, under the Jordan product

$$\mathcal{X} \circ \mathcal{Y} = \frac{1}{2} (\mathcal{X}\mathcal{Y} + \mathcal{Y}\mathcal{X}) \qquad (17.15)$$

form the exceptional Jordan algebra (also known as the Albert algebra). This algebra satisfies the identity

$$(\mathcal{X} \circ \mathcal{Y}) \circ \mathcal{X}^2 = \mathcal{X} \circ (\mathcal{Y} \circ \mathcal{X}^2) \qquad (17.16)$$

which leads to an alternative formulation of quantum mechanics, and in fact was originally found as the unique example of exceptional quantum mechanics (Jordan *et al.* 1934).

We proposed (Dray and Manogue 1999, 2000) reinterpreting this algebra as containing both bosonic and fermionic degrees of freedom, which can be represented in the form

$$\mathcal{X} = \begin{pmatrix} \boldsymbol{\Pi} & \theta \\ \theta^\dagger & \phi \end{pmatrix} \qquad (17.17)$$

where $\boldsymbol{\Pi}$ is a vector, and θ a scalar, under Lorentz transformations. Solutions of the massless 10-dimensional Dirac equation take a particularly nice form in this language, as they correspond to the Cayley–Moufang plane

$$\mathbb{OP}^2 = \{\mathcal{X}^2 = \mathcal{X}, \operatorname{tr} \mathcal{X} = 1\} \qquad (17.18)$$
$$= \{\mathcal{X} = \psi\psi^\dagger, \psi^\dagger\psi = 1, \psi \in \mathbb{H}^3\}$$

where ψ is a 3-component octonionic "Cayley spinor." This description makes clear that the theory is invariant under a much bigger group than the Lorentz group, namely, the exceptional group E_6 defined by

$$\det(\mathcal{M}\mathcal{X}\mathcal{M}^\dagger) = \det \mathcal{X} \implies \mathcal{M} \in E_6 \qquad (17.19)$$

which suggests that it may be possible to extend the theory so as to include quarks and color. This suggestion is particularly attractive given that $SO(3,1) \times U(1) \times SU(2) \times SU(3)$ can be viewed as a subgroup of E_6.

Where would gravity fit in? Perhaps by turning global symmetries involving the octonions into local symmetries. It has also been suggested that a "Jordan" spin foam network might have special properties (Smolin 2004). In any case, the theory described here should be viewed as speculative, although we are not the first to wonder at the apparently close connection between the octonions and the symmetries of nature.

17.5 Conclusion

One might feel that the existence of so many approaches to quantum gravity indicates how little we know about the subject. The opposite conclusion is also possible, namely, that *all roads lead to quantum gravity!* From the latter point of view, the current theories are on the right track. The similarities in some of their results, such as the explanations of Hawking radiation by both string theory and loop quantum gravity, support this position.

On the other hand, it may be difficult to tell the forest from the trees. For example, as already noted, early incarnations of string theory left much of the details to the choice of a Calabi–Yau manifold, without providing any guidance on which one to choose. So all of these approaches to quantum gravity may ultimately be shown to be converging to their goal, a viable theory of quantum gravity. But do they know it?

Acknowledgements

This paper is dedicated to the memory of Peter G. Bergmann (1915–2002), father of quantum gravity, and beloved mentor.

References

Arnowitt, R., Deser, S. and Misner, C. W. (1962). The Dynamics of General Relativity. In *Gravitation: An Introduction to Current Research*. L. Witten, ed. New York: Wiley, 227–265.

Ashtekar, Abhay (1986). New Variables for Classical and Quantum Gravity. *Phys. Rev. Lett.* **57**, 2244–2247.

—— (2012). The Issue of the Beginning in Quantum Gravity. *Einstein and the Changing Worldviews of Physics.* New York: Birkhäuser Science (Einstein Studies, Volume 12).

Ashtekar, A. and Magnon, Anne (1975). Quantum Field Theory in Curved Space-Times. *Proc. Roy. Soc.* **A346**, 375–394.

Baylis, William E. (1999). *Electrodynamics: A Modern Geometric Approach.* Boston: Birkhäuser.

Bergmann, Peter G. (1949). Non-Linear Field Theories. *Phys. Rev.* **75**, 680–685.

Bombelli, L., Lee, J., Meyer, D. and Sorkin, R. (1987). Spacetime as a Causal Set. *Phys. Rev.* **59**, 521–524.

Bronstein, Matvey Petrovich (1933). K Voprosy o vozmozhnoy teorii mira kak tselogo. (On the Question of a Possible Theory of the World as a Whole.) *Usp. Astron. Nauk Sb.* **3**, 3–30.

Chamseddine, Ali H. and Connes, Alain (1996). Universal Formula for Noncommutative Geometry Actions: Unification of Gravity and the Standard Model. *Phys. Rev. Lett.* **77**, 4868–4871.

Davies, P. C. W. (1976). Scalar Particle Production in Schwarzschild and Rindler Metrics. *J. Phys.* **A8**, 609–616.

DeWitt, Bryce S. (1964). Theory of Radiative Corrections for Non-Abelian Gauge Fields. *Phys. Rev. Lett.* **12**, 742–746.

—— (1967). Quantum Theory of Gravity. I. The Canonical Theory. *Phys. Rev.* **160**, 1113–1148.

Dirac, P. A. M. (1950). Generalized Hamiltonian Dynamics. *Can. J. Math. Phys.* **2**, 129–148.

Dray, Tevian and Manogue, Corinne A. (1999). The Exceptional Jordan Eigenvalue Problem. *Int. J. Theor. Phys.* **38**, 2901–2916.

—— (2000). Quaternionic Spin. In *Clifford Algebras and their Applications in Mathematical Physics.* Rafał Abłamowicz and Bertfried Fauser, eds. Boston: Birkhäuser, 29–46.

Dray, Tevian, Renn, Jürgen and Salisbury, Donald C. (1983). Particle Creation with Finite Energy Density. *Lett. Math. Phys.* **7**, 145–153.

Einstein, Albert (1905a). Über einen die Erzeugung und Verwandlung des Lichtes betreffenden heuristischen Gesichtspunkt. *Ann. Phys.* **17**, 132–148.

—— (1905b). Über die von der molekularkinetischen Theorie der Wärme geforderte Bewegung von in ruhenden Flüssigkeiten suspendierten Teilchen. *Ann. Phys.* **17**, 549–560.

—— (1905c). Zur Elektrodynamik bewegter Körper. *Ann. Phys.* **17**, 891–921.

—— (1905d). Ist die Trägheit eines Körpers von seinem Energiegehalt abhängig? *Ann. Phys.* **18**, 639–641.

Fairlie, David B. and Manogue, Corinne A. (1986). Lorentz Invariance and the Composite String. *Phys. Rev.* **D34**, 1832–1834.

—— (1987). A Parameterization of the Covariant Superstring. *Phys. Rev.* **D36**, 475–479.

Feynman, R. (1963). Quantum Theory of Gravitation. *Acta Phys. Pol.* **24**, 697.

Fulling, S. A. (1973). Nonuniqueness of Canonical Field Quantization in Riemannian Space-Time. *Phys. Rev.* **D7**, 2850–2862.

Green, M. B. and Schwarz, J. H. (1984). Anomaly Cancellation in Supersymmetric $D = 10$ Gauge Theory Requires SO(32). *Phys. Lett.* **149B**, 117–122.

Hartle, J. B. and Hawking, S. W. (1983). Wave Function of the Universe. *Phys. Rev.* **D28**, 2960–2975.

Hawking, S. W. (1975). Particle Creation by Black Holes. *Commun. Math. Phys.* **43**, 199–220.

't Hooft, G. (1973). An Algorithm for the Poles at Dimension Four in the Dimensional Regularization Procedure. *Nucl. Phys.* **B62**, 444–460.

Jacobson, T. and Smolin, L. (1988) Nonperturbative Quantum Geometries. *Nucl. Phys.* **B299**, 295–345.

Jordan, P., von Neumann, J. and Wigner, E. (1934). On an Algebraic Generalization of the Quantum Mechanical Formalism. *Ann. Math.* **35**, 29–64.

Karlhede, A. (1980). A Review of the Equivalence Problem. *Gen. Rel. Grav.* **12**, 693–707.

Manogue, Corinne A. and Dray, Tevian (1999). Dimensional Reduction. *Mod. Phys. Lett.* **A14**, 93–97.

Manogue, Corinne A. and Sudbery, Anthony (1989). General Solutions of Covariant Superstring Equations of Motion. *Phys. Rev.* **D40**, 4073–4077.

Newman, Ezra T. and Penrose, Roger (1962). An Approach to Gravitational Radiation by a Method of Spin Coefficients. *J. Math. Phys.* **3**, 566–768.

Penrose, R. (1967). Twistor Algebra. *J. Math. Phys.* **8**, 345–366.

—— (1986). Gravity and State Vector Reduction. In *Quantum Concepts in Space and Time*. R. Penrose and C. J. Isham, eds. Oxford: Clarendon Press, 129–146.

Renn, J. (2007). Classical Physics in Disarray: The Emergence of the Riddle of Gravitation. In *The Genesis of General Relativity, vol. 1: Einsteins's Zurich Notebook: Introduction and Source*. Jürgen Renn, ed. Dordrecht: Springer, 31.

Rovelli, Carlo (2004). *Quantum Gravity*. Cambridge: Cambridge University Press.

Schray, Jörg (1996). The General Classical Solution of the Superparticle. *Class. Quant. Grav.* **13**, 27–38.

Smolin, Lee (2004). Private communication.

Sparling, George A. J. (2006). Spacetime is spinorial; new dimensions are timelike. University of Pittsburgh preprint. `http://xxx.lanl.gov/abstract/gr-qc/0610068`.

Stachel, John (1999). Space-Time Structures: What's the Point. *Proceedings of the Minnowbrook Symposium on the Structure of Space-Time* (Blue Mountain Lake, NY; 5/28–31/99). `http://physics.syr.edu/research/hetheory/minnowbrook/stachel.html`.

Unruh, W. G. (1976). Notes on Black Hole Evaporation. *Phys. Rev.* **D14**, 870–892.

—— (2005). What is a Particle? Quantum Field Theory Meets General Relativity. Talk given at the 7th International Conference on the History of General Relativity, Tenerife.

Wald, Robert M. (1994). *Quantum Field Theory in Curved Spacetimes and Black Hole Thermodynamics*. Chicago: University of Chicago Press.

—— (2012). The History and Present Status of Quantum Field Theory in Curved Spacetime. *Einstein and the Changing Worldviews of Physics*. New York: Birkhäuser Science (Einstein Studies, Volume 12).

Wilczek, Frank (2005). Physics: Treks of Imagination. *Science* **307** (11 February 2005), 852–853.

Witten, Edward (2004). Perturbative Gauge Theory as a String Theory in Twistor Space. *Commun. Math. Phys.* **252**, 189–258.

18

The Issue of the Beginning in Quantum Gravity

Abhay Ashtekar

Institute for Gravitational Physics and Geometry, Penn State University, USA
Institute for Theoretical Physics, University of Utrecht, The Netherlands

18.1 Introduction

Treatises on Time, the Beginning and the End date back at least twenty-five centuries. Does the flow of time have an objective, universal meaning beyond human perception? Or, is it fundamentally only a convenient, and perhaps merely psychological, notion? Are its properties tied to the specifics of observers such as their location and state of motion? Did the physical universe have a finite beginning, or has it been evolving eternally? Leading thinkers across cultures meditated on these issues and arrived at definite but strikingly different answers. For example, in the sixth century B.C.E., Gautama Buddha taught that 'a period of time' is a purely conventional notion; time and space exist only in relation to our experience, and the universe is eternal. In the Christian thought, however, the universe had a finite beginning, and there was debate as to whether time represents 'movement' of bodies or whether it flows only in the soul. In the fourth century C.E., St. Augustine held that time itself started with the world.

Founding fathers of modern science from Galileo to Newton accepted that God created the universe, but Newton posited an absolute time which is to run uniformly from the infinite past to the infinite future. This paradigm became a dogma over centuries. Some philosophers used it to argue that the universe itself *had* to be eternal. For, otherwise, one could ask "what was there before?" General relativity toppled this paradigm in one fell swoop. Since the gravitational field is now encoded in space-time geometry, geometry itself now becomes dynamical. The universe could have had a finite beginning—the Big Bang—at which not only matter but space-time itself is 'born'. General relativity takes us back to St. Augustine's paradigm but in a detailed, specific, and mathematically precise form. In semi-popular articles and radio shows, relativists now like to emphasize that the question "what was there before?" is rendered meaningless because the notions of 'before' and 'after' refer to a space-time geometry. We now have a new paradigm, a new dogma: In the Beginning there was the Big Bang.

But general relativity is incomplete, for it ignores quantum effects entirely. Over the last century, we have learned that these effects become important in the physics of the small and should in fact dominate in parts of the universe where matter densities become enormous. So, there is no reason to trust the predictions of general relativity near space-time singularities. The classical physics of general relativity does come to a halt at the Big Bang. But applying general relativity near a singularity is an extrapolation which has no justification whatsoever. We need a theory that incorporates not only the dynamical nature of geometry but also the ramifications of quantum physics. Does the 'correct' or 'true' physics also stop at the Big Bang in quantum gravity? Or is there yet another paradigm shift waiting in the wings?

The goal of this article is to present an up-to-date summary of the status of these age-old issues within loop quantum gravity (LQG). Detailed calculations in simple cosmological models have shown that the quantum nature of geometry does dominate physics near the big bang, altering dynamics and drastically changing the paradigm provided by general relativity. In particular, the quantum space-time may be much larger than what general relativity has led us to believe, whence the Big Bang may not, after all, be the Beginning. Some of the mathematics underlying the main results are subtle. However, I have made an attempt to also include a descriptive summary of the viewpoint, ideas, constructions, and physical ramifications of the results.

18.2 Loop Quantum Gravity

18.2.1 Conceptual Issues

Remarkably, the necessity of a quantum theory of gravity was already pointed out by Einstein in 1916. In a paper in the *Preussische Akademie Sitzungsberichte* he wrote:

> Nevertheless, due to the inneratomic movement of electrons, atoms would have to radiate not only electromagnetic but also gravitational energy, if only in tiny amounts. As this is hardly true in Nature, it appears that quantum theory would have to modify not only Maxwellian electrodynamics but also the new theory of gravitation.

Ninety years later, our understanding of the physical world is vastly richer, but a fully satisfactory unification of general relativity with quantum physics still eludes us. Indeed, the problem has now moved to the center stage of fundamental physics.[1]

A key reason why the issue is still open is the lack of experimental data with direct bearing on quantum gravity. As a result, research is necessarily driven by theoretical insights into what the key issues are and what will 'take care of itself' once this core is understood. As a consequence, there are distinct starting points which seem natural. Such diversity is not unique to this problem. However, for other fundamental forces we have had clear-cut experiments to weed out ideas which, in spite of their theoretical appeal, fail to be realized in Nature. We do not have this luxury in quantum gravity. But then, in the absence of strong experimental constraints, one would expect a rich variety of internally consistent theories. Why is

---

[1] For a brief historical account of the evolution of ideas, see, e.g., (Ashtekar 2005).

it then that we do not have a single one? The reason, I believe, lies in the deep conceptual difference between the description of gravity in general relativity and that of non-gravitational forces in other fundamental theories. In those theories, space-time is given *a priori*, serving as an inert background, a stage on which the drama of evolution unfolds. General relativity, on the other hand, is not only a theory of gravity, it is also a theory of space-time structure. Indeed, as I remarked in Section 18.1, in general relativity gravity is encoded in the very geometry of space-time. Therefore, a quantum theory of gravity has to simultaneously bring together *gravity, geometry, and the quantum*. This is a brand-new adventure for which our past experience with other forces cannot serve as a reliable guide.

LQG is an approach that attempts to face this challenge squarely.[2] Recall that Riemannian geometry provides the appropriate mathematical language to formulate the physical, kinematical notions as well as the final dynamical equations of any classical theory of relativistic gravity. This role is now assumed by *quantum* Riemannian geometry. Thus, in LQG both matter and geometry are quantum mechanical 'from birth'.

In the classical domain, general relativity stands out as the best available theory of gravity. Therefore, it is natural to ask: *Does quantum general relativity, coupled to suitable matter* (or supergravity, its supersymmetric generalization), *exist as a consistent theory non-perturbatively?* In particle physics circles the answer is often assumed to be in the negative, not because there is concrete evidence which rules out this possibility, but because of the analogy to the theory of weak interactions. There, one first had a 4-point interaction model due to Fermi which works quite well at low energies, but which fails to be renormalizable. Progress occurred, not by looking for non-perturbative formulations of the Fermi model, but by replacing the model by the Glashow–Salam–Weinberg renormalizable theory of electro-weak interactions, in which the 4-point interaction is replaced by $W^{\pm}$ and Z propagators. It is often assumed that the perturbative non-renormalizability of quantum general relativity points in a similar direction. However, this argument overlooks a crucial and qualitatively new element of general relativity. Perturbative treatments pre-suppose that space-time is a smooth continuum *at all scales* of interest to physics under consideration. This assumption is safe for weak interactions. In the gravitational case, on the other hand, the scale of interest is *the Planck length*, and there is no physical basis to pre-suppose that the continuum approximation should be valid down to that scale. The failure of the standard perturbative treatments may largely be due to this grossly incorrect assumption, and a non-perturbative treatment which correctly incorporates the physical micro-structure of geometry may well be free of these inconsistencies.

Are there any situations outside LQG where such physical expectations are borne out by detailed mathematics? The answer is in the affirmative. There exist quantum field theories (such as the Gross–Neveu model in three dimensions), in which the standard perturbation expansion is not renormalizable although the theory is *exactly soluble!* Failure of the standard perturbation expansion can occur because one insists on perturbing around the trivial, Gaussian point rather than the more physical

---

[2] For details, see, e.g., (Ashtekar and Lewandowski 2004; Rovelli 2004; Thiemann 2007).

non-trivial fixed point of the renormalization group flow. Interestingly, thanks to the recent work by Lauscher, Reuter, Percacci, Perini, and others, there is now growing evidence that this situation may be similar to that in general relativity (see (Lauscher and Reuter 2005) and references therein). Impressive calculations have shown that pure Einstein theory may also admit a non-trivial fixed point. Furthermore, the requirement that the fixed point should continue to exist in the presence of matter constrains the couplings in physically interesting ways (Percacci and Perini 2003).

Let me conclude this discussion with an important caveat. Suppose one manages to establish that non-perturbative quantum general relativity (or, supergravity) does exist as a mathematically consistent theory. Still, there is no a priori reason to assume that the result would be the 'final' theory of all known physics. In particular, as is the case with classical general relativity, while requirements of background independence and general covariance do restrict the form of interactions between gravity and matter fields and among matter fields themselves, the theory would not have a built-in principle which *determines* these interactions. Put differently, such a theory would not be a satisfactory candidate for unification of all known forces. However, just as general relativity has had powerful implications in spite of this limitation in the classical domain, quantum general relativity should lead to qualitatively new predictions, pushing further the existing frontiers of physics. In Section 18.3 we will see an illustration of this possibility.

18.2.2 Salient Features

Detailed as well as short and semi-qualitative reviews of LQG have recently appeared in the literature.[3] Therefore, here I will only summarize the key features of the theory that are used in Section 18.3.

The starting point of LQG is a Hamiltonian formulation of general relativity based on spin connections (Ashtekar 1986). Since all other basic forces of nature are also described by theories of connections, this formulation naturally leads to a unification of all four fundamental forces at a *kinematical* level. Specifically, the phase space of general relativity is the same as that of a Yang–Mills theory. The difference lies in dynamics: Whereas in the standard Yang–Mills theory the Minkowski metric features prominently in the definition of the Hamiltonian, there are no background fields whatsoever once gravity is switched on.

Let us focus on the gravitational sector of the theory. Then, the phase space Γ_{grav} consists of canonically conjugate pairs (A_a^i, P_a^i), where A_a^i is a connection on a 3-manifold M and P_a^i a vector density of weight 1, both of which take values in the Lie algebra su(2). The connection A enables one to parallel transport chiral spinors (such as the left-handed fermions of the standard electro-weak model) along curves in M. Its curvature is directly related to the electric and magnetic parts of the space-time *Riemann tensor*. P_i^a plays a double role. Being the momentum canonically

---

[3] See, e.g., (Ashtekar and Lewandowski 2004; Rovelli 2004; Thiemann 2007) and (Ashtekar 2005), respectively.

conjugate to A, it is analogous to the Yang–Mills electric field. In addition, $E_i^a :=$ $8\pi G\gamma P_i^a$ has the interpretation of a frame or orthonormal triad (with density weight 1) on M, where γ is the 'Barbero–Immirzi parameter' representing a quantization ambiguity. Each triad E_i^a determines a positive definite 'spatial' 3-metric q_{ab}, and hence the Riemannian geometry of M. This dual role of P is a reflection of the fact that now $SU(2)$ is the (double cover of the) group of rotations of the orthonormal spatial triads on M itself rather than that of rotations in an 'internal' space associated with M.

To pass to quantum theory, one first constructs an algebra of 'elementary' functions on Γ_{grav} (analogous to the phase space functions x and p in the case of a particle) which are to have unambiguous operator analogues. In LQG, the configuration variables are the holonomies h_e built from A_a^i which enable us to parallel transport chiral spinors along edges e, and the fluxes $E_{S,f}$ of 'electric fields' or 'triads' (smeared with test fields f) across 2-surfaces S. These functions generate a certain algebra $\mathfrak{a}$ (analogous to the algebra generated by operators $\widehat{\exp i\lambda x}$ and $\hat{p}$ in quantum mechanics). The first principal task is to find representations of this algebra. In that representation, *quantum* Riemannian geometry can be probed through the triad operators $\hat{E}_{S,f}$, which stem from classical orthonormal triads. Quite surprisingly, the requirement of diffeomorphism covariance on M suffices to single out a *unique* representation of $\mathfrak{a}$ (Lewandowski et al. 2005; Fleischhack 2004)! This recent result is the quantum-geometry analogue of the seminal results by Segal and others of that characterized the Fock vacuum in Minkowskian field theories. However, while that result assumes not only Poincaré invariance of the vacuum but also specific (namely free) dynamics, it is striking that the present uniqueness theorems require no such restriction on dynamics. The requirement that there be a diffeomorphism-invariant state is surprisingly strong and makes the 'background-independent' quantum geometry framework surprisingly tight.

This unique representation was in fact already introduced in the mid-1990s (Ashtekar and Lewandowski 1994, 1995a, 1995b; Baez 1994, 1996; Marolf and Mourão 1995; Rovelli and Smolin 1995) and has been extensively used in LQG since then. The underlying Hilbert space is given by $\mathcal{H} = L^2(\overline{\mathcal{A}}, d\mu_o)$ where $\overline{\mathcal{A}}$ is a certain completion of the classical configuration space $\mathcal{A}$ consisting of smooth connections on M and μ_o is a diffeomorphism-invariant, faithful, regular Borel measure on $\overline{\mathcal{A}}$. The holonomy (or configuration) operators $\hat{h}_e$ act just by multiplication. The momentum operators $\hat{P}_{S,f}$ act as Lie derivatives. In the classical theory, by taking suitable derivatives in M of holonomies h_e along arbitrary edges e, one can recover the connection from which the holonomy is built. However, in the quantum theory, the operators $\hat{h}_e$ are discontinuous, and there is no operator $\hat{A}$ corresponding to the connection itself.

Key features of this representation which distinguish it from, say, the standard Fock representation of the quantum Maxwell field are the following: while the Fock representation of photons makes a crucial use of the background Minkowski metric, the above construction is manifestly 'background independent'. Second, as remarked above, the connection itself is not represented as an operator (valued distribution). Holonomy operators, on the other hand, are well defined. Third, the 'triads' or 'electric field' operators now have purely discrete eigenvalues. Given a surface S and a region R, one can express the area A_S and volume V_R using the triads. Although they are non-polynomial functions of triads, the operators $\hat{A}_S$ and $\hat{V}_R$ are well defined and also have discrete eigenvalues. In contrast, such functions of electric fields cannot be promoted to operators on the Fock space. Finally, and most importantly, the Hilbert space $\mathcal{H}$ and the associated holonomy and (smeared) triad operators only provide a *kinematical* framework—the quantum analogue of the full phase space. Thus, while elements of the Fock space represent physical states of photons, elements of $\mathcal{H}$ are *not* the physical states of LQG. Rather, like the classical phase space, the kinematic setup provides a home for *formulating* quantum dynamics.

In the Hamiltonian framework, the dynamical content of any background-independent theory is contained in its constraints. In quantum theory, the Hilbert space $\mathcal{H}$ and the holonomy and (smeared) triad operators thereon provide the necessary tools to write down quantum constraint operators. Physical states are solutions of these quantum constraints. Thus, to complete the program, one has to: i) obtain the expressions of the quantum constraints; ii) solve the constraint equations; iii) construct the physical Hilbert space from the solutions (e.g., by the group averaging procedure); and iv) extract physics from this physical sector (e.g., by analyzing the expectation values, fluctuations of, and correlations between Dirac observables). While strategies have been developed—particularly through Thiemann's 'Master constraint program' (Thiemann 2003)—to complete these steps, important open issues remain in the full theory. However, as Section 18.3 illustrates, the program has been completed in mini and midi superspace models, leading to surprising insights and answers to some longstanding questions.

18.3 Application: Homogeneous Isotropic Cosmology

There is a long list of questions about the quantum nature of the Big Bang. For example:

- How close to the Big Bang does a smooth space-time of general relativity make sense? In particular, can one show from first principles that this approximation is valid at the onset of inflation?
- Is the Big Bang singularity naturally resolved by quantum gravity? Or is some external input such as a new principle or a boundary condition at the Big Bang essential?
- Is the quantum evolution across the 'singularity' deterministic? Since one needs a fully non-perturbative framework to answer this question in the affirmative, in the pre-Big Bang (Gasperini and Veneziano 2003) and ekpyrotic/cyclic (Khoury et al. 2001, 2002) scenarios, for example, so far the answer is in the negative.
- If the singularity is resolved, what is on the 'other side'? Is there just a 'quantum foam', far removed from any classical space-time, or is there another large, classical universe?

For many years, these and related issues had been generally relegated to the 'wish list' of what one would like the future, satisfactory quantum gravity theory to eventually address. It seems likely that these issues can be met head on only in a background-independent, non-perturbative approach. One such candidate is LQG. Indeed, starting with the seminal work of Bojowald some ten years ago (Bojowald 2001, 2002), notable progress has been made in the context of symmetry-reduced, mini superspaces. Earlier papers focused only on singularity resolution. However, to describe physics in detail, it is essential to construct the physical Hilbert space and introduce interesting observables and semi-classical states by completing the program outlined at the end of the last section. These steps have been completed recently. In this section, I will summarize the state of the art, emphasizing these recent developments. (For a comprehensive review of the older work see, e.g., (Bojowald 2005).)

Consider the spatially homogeneous, isotropic, $k = 0$ cosmologies with a massless scalar field. It is instructive to focus on this model because *each* of its classical solutions has a singularity. There are two possibilities: In one the universe starts out at the Big Bang and expands, and in the other it contracts into a "Big Crunch." The question is whether this unavoidable classical singularity is naturally tamed by quantum effects. This issue can be analyzed in the geometrodynamical framework used in older quantum cosmology. Unfortunately, the answer turns out to be in the negative. For example, if one begins with a semi-classical state representing an expanding classical universe at late times and evolves it back via the Wheeler–DeWitt equation, one finds that it just follows the classical trajectory into the Big Bang singularity (Ashtekar et al. 2006a, Ashtekar et al. 2006b).

In loop quantum cosmology (LQC), the situation is very different (Ashtekar et al. 2006a, 2006b, 2006c). This may seem surprising at first, for the system has only a finite number of degrees of freedom and von Neumann's theorem assures us that, under appropriate assumptions, the resulting quantum mechanics is unique. The only remaining freedom is factor-ordering, and this is generally insufficient to lead to qualitatively different predictions. However, for reasons I will now explain, LQC does turn out to be qualitatively different from the Wheeler–DeWitt theory (Ashtekar et al. 2003).

Because of spatial homogeneity and isotropy, one can fix a fiducial (flat) triad $^0e_i^a$ and its dual co-triad $^0\omega_a^i$. The $SU(2)$ gravitational spin connection A_a^i used in LQG has only one component c which furthermore depends only on time; $A_a^i = c \, ^0\omega_a^i$. Similarly, the triad E_i^a (of density weight 1) has a single component p; $E_i^a = p \, (\det{}^0\omega) \, ^0e_i^a$. p is related to the scale factor a via $a^2 = |p|$. However, p is not restricted to be positive; under $p \to -p$ the metric remains unchanged, but the spatial triad flips the orientation. The pair (c, p) is 'canonically conjugate' in the sense that the only non-zero Poisson bracket is given by

$$\{c, \, p\} = \frac{8\pi G\gamma}{3} \, , \tag{18.1}$$

where as before γ is the Barbero–Immirzi parameter.

Since a precise quantum-mechanical framework was not available for full geo-metrodynamics, in the Wheeler–DeWitt quantum cosmology one focused just on the reduced model, without the benefit of guidance from the full theory. A major difference in LQC is that, although the symmetry-reduced theory has only a finite number of degrees of freedom, quantization is carried out by closely mimicking the procedure used in *full* LQG, outlined in Section 18.2. Key differences between LQC and the older Wheeler–DeWitt theory can be traced back to this fact.

Recall that in full LQG diffeomorphism invariance leads one to a specific kine-matical framework in which there are operators $\hat{h}_e$ representing holonomies and $\hat{P}_{S,f}$ representing (smeared) momenta, but there is no operator(-valued distribu-tion) representing the connection A itself (Lewandowski et al. 2005; Fleishchack 2004). In the cosmological model now under consideration, it is sufficient to evalu-ate holonomies along segments $\mu \, {}^o e_i^a$ of straight lines determined by the fiducial triad ${}^o e_i^a$. These holonomies turn out almost periodic functions of c, i.e., they are of the form $N_{(\mu)}(c) := \exp i\mu(c/2)$, where the word 'almost' refers to the fact that μ can be any real number. These functions were studied exhaustively by the mathematician Harold Bohr, Niels' brother. In quantum geometry, the $N_{(\mu)}$ are the LQC analogues of the spin-network functions of full LQG.

In quantum theory, then, we are led to a representation in which operators $\hat{N}_{(\mu)}$ and $\hat{p}$ are well defined, but there is *no* operator corresponding to the connection com-ponent c. This seems surprising because our experience with quantum mechanics suggests that one should be able to obtain the operator analogue of c by differentiat-ing $\hat{N}_{(\mu)}$ with respect to the parameter μ. However, in the representation of the basic quantum algebra that descends to LQC from full LQG, although the $\hat{N}_{(\mu)}$ provide a 1-parameter group of unitary transformations, it fails to be weakly continuous in μ. Therefore, one cannot differentiate and obtain the operator analogue of c. In quan-tum mechanics, this would be analogous to having well-defined (Weyl) operators corresponding to the classical functions $\exp i\mu x$ but no operator $\hat{x}$ corresponding to x itself. This violates one of the assumptions of the von Neumann uniqueness theorem. New representations then become available which are *inequivalent* to the standard Schrödinger one. In quantum mechanics, these representations are not of direct physical interest because we need the operator $\hat{x}$. In LQC, on the other hand, full LQG naturally leads us to a new representation, i.e., to *new quantum mechanics*. This theory is inequivalent to the Wheeler–DeWitt-type theory already at a kine-matical level. In the Wheeler–Dewitt theory, the gravitational Hilbert space would be $L^2(\mathbb{R}, dc)$, operators $\hat{c}$ would act by multiplication, and $\hat{p}$ would be represented by $-i\hbar d/dc$. In LQC the 'quantum configuration space' is different from the clas-sical configuration space: Just as we had to complete the space $\mathcal{A}$ of smooth con-nections to the space $\overline{\mathcal{A}}$ of generalized connections in LQG, we are now led to consider a completion—called the Bohr compactification $\overline{\mathbb{R}}_{\text{Bohr}}$—of the '$c$-axis'. The gravitational Hilbert space is now $L^2(\overline{\mathbb{R}}_{\text{Bohr}}, d\mu_{\text{Bohr}})$ (Ashtekar et al. 2003) where $d\mu_{\text{Bohr}}$ is the LQC analogue of the measure $d\mu_o$ selected by the uniqueness results (Lewandowski et al. 2005; Fleishchack 2004) in full LQG. The operators $\hat{N}_{(\mu)}$ act by multiplication and $\hat{p}$ by differentiation. However, there is no operator $\hat{c}$.

In spite of these differences, in the semi-classical regime LQC is well approximated by the Wheeler–DeWitt theory. However, important differences manifest themselves at the Planck scale. These are the hallmarks of quantum geometry (Ashtekar and Lewandowski 2004; Bojowald 2005).

The new representation also leads to a qualitative difference in the structure of the Hamiltonian constraint operator: the gravitational part of the constraint is a *difference* operator, rather than a differential operator as in the Wheeler–DeWitt theory. The derivation (Ashtekar et al. 2003; Ashtekar et al. 2006b, 2006c) can be summarized briefly as follows. In the classical theory, the gravitational part of the constraint is given by $\int d^3x\, \epsilon^{ijk} e^{-1} E_i^a E_j^b F_{abk}$ where $e = |\det E|^{1/2}$ and F_{ab}^k the curvature of the connection A_a^i. The part $\epsilon^{ijk} e^{-1} E_i^a E_j^b$ of this operator involving triads can be quantized (Bojowald 2001; Ashtekar et al. 2003) using a standard procedure introduced by Thiemann in the full theory (Thiemann 2007). However, since there is no operator corresponding to the connection itself, one has to express F_{ab}^k as a limit of the holonomy around a loop divided by the area enclosed by the loop, as the area shrinks to zero. Now, quantum geometry tells us that the area operator has a minimum non-zero eigenvalue, Δ, and in the quantum theory it is natural to shrink the loop only until it attains this minimum. There are two ways to implement this idea in detail (see Ashtekar et al. 2003; Ashtekar et al. 2006b, 2006c). In both cases, it is the existence of the 'area gap' Δ that leads one to a difference equation. So far, most of the LQC literature has used the first method (Ashtekar et al. 2003; Ashtekar et al. 2006b). In the resulting theory, the classical big bang is replaced with a quantum bounce with a number of desirable features. However, it also has one serious drawback: at the bounce, matter density can be low even for physically reasonable choices of quantum states. Thus, that theory predicts certain departures from classical general relativity even in the low-curvature regime (for details, see (Ashtekar et al. 2006b, 2006c)). The second and more recently discovered method (Ashtekar et al. 2006c) cures this problem while retaining the physically appealing features of the first and, furthermore, has a more direct motivation. For brevity, therefore, I will confine myself only to the second method.

Let us represent states as functions $\Psi(v, \phi)$, where ϕ is the scalar field and the dimensionless real number v represents geometry. Specifically, $|v|$ is the eigenvalue of the operator $\hat{V}$ representing volume[4] (essentially the cube of the scale factor):

$$\hat{V}|v\rangle = K \left(\frac{8\pi\gamma}{6}\right)^{\frac{3}{2}} |v|\, \ell_{\mathrm{Pl}}^3 |v\rangle \quad \text{where} \quad K = \frac{3\sqrt{3\sqrt{3}}}{2\sqrt{2}}. \tag{18.2}$$

---

[4] In non-compact spatially homogeneous models, integrals of physical interest over the full spatial manifold diverge. Therefore, to obtain a consistent Hamiltonian description, one has to introduce an elementary cell $\mathcal{V}$ and restrict all integrals to $\mathcal{V}$ already in the classical theory. This is also necessary in geometrodynamics. $\hat{V}$ is the volume operator associated with $\mathcal{V}$.

Then, the LQC Hamiltonian constraint assumes the form

$$\partial_\phi^2 \Psi(v,\phi) = [B(v)]^{-1} \left(C^+(v)\,\Psi(v+4,\phi) + C^o(v)\,\Psi(v,\phi) + C^-(v)\,\Psi(v-4,\phi) \right)$$
$$=: -\Theta\,\Psi(v,\phi) \qquad (18.3)$$

where the coefficients $C^\pm(v)$, $C^o(V)$, and $B(v)$ are given by

$$C^+(v) = \frac{3\pi K G}{8}\,|v+2|\,\big||v+1| - |v+3|\big|$$

$$C^-(v) = C^+(v-4) \quad \text{and} \quad C^o(v) = -C^+(v) - C^-(v)$$

$$B(v) = \left(\frac{3}{2}\right)^3 K\,|v|\,\big||v+1|^{1/3} - |v-1|^{1/3}\big|^3\,. \qquad (18.4)$$

Now, in each classical solution, ϕ is a globally monotonic function of time and can therefore be taken as the dynamical variable representing an *internal* clock. In quantum theory there is no space-time metric, even on-shell. But since the quantum constraint (18.3) dictates how $\Psi(v,\phi)$ 'evolves' as ϕ changes, it is convenient to regard the argument ϕ in $\Psi(v,\phi)$ as *emergent time* and v as the physical degree of freedom. A complete set of Dirac observables is then provided by the constant of motion $\hat{p}_\phi$ and operators $\hat{v}|_{\phi_o}$ determining the value of v at the 'instant' $\phi = \phi_o$.

Physical states are the (suitably regular) solutions to Eq. (18.3). The map $\hat{\Pi}$ defined by $\hat{\Pi}\,\Psi(v,\phi) = \Psi(-v,\phi)$ corresponds just to the flip of orientation of the spatial triad (under which geometry remains unchanged); $\hat{\Pi}$ is thus a large gauge transformation on the space of solutions to Eq. (18.3). One is therefore led to divide physical states into sectors, each providing an irreducible, unitary representation of this gauge symmetry. Physical considerations (Ashtekar et al. 2006b, 2006c) imply that we should consider the symmetric sector, with eigenvalue +1 of $\hat{\Pi}$.

To endow this space with the structure of a Hilbert space, one can proceed along one of two paths. In the first, one defines the action of the Dirac observables on the space of suitably regular solutions to the constraints and selects the inner product by demanding that these operators be self-adjoint (Ashtekar 1991). A more systematic procedure is the 'group averaging method' (Marolf 1995a). The technical implementation (Ashtekar et al. 2006b, 2006c) of both these procedures is greatly simplified by the fact that the difference operator Θ on the right side of (18.3) is independent of ϕ and can be shown to be self-adjoint and positive definite on the Hilbert space $L^2(\overline{\mathbb{R}}_{\mathrm{Bohr}}, B(v)\mathrm{d}\mu_{\mathrm{Bohr}})$.

The final result can be summarized as follows. Since Θ is a difference operator, the physical Hilbert space $\mathcal{H}_{\mathrm{phy}}$ has sectors $\mathcal{H}_\epsilon$ which are superselected; $\mathcal{H}_{\mathrm{phy}} = \oplus_\epsilon \mathcal{H}_\epsilon$ with $\epsilon \in (0,2)$. The overall predictions are insensitive to the choice of a specific sector (for details, see (Ashtekar et al. 2006b, 2006c)). States $\Psi(v,\phi)$ in $\mathcal{H}_\epsilon$ are symmetric under the orientation inversion $\hat{\Pi}$ and have support on points $v = |\epsilon| + 4n$ where n is an integer. Wave functions $\Psi(v,\phi)$ in a generic sector solve (18.3) and are of positive frequency with respect to the 'internal time' ϕ: they satisfy the 'positive frequency' square root

$$-i\partial_\phi \Psi = \sqrt{\Theta}\, \Psi \tag{18.5}$$

of Eq (18.3). (The square root is a well-defined (positive self-adjoint) operator because Θ is positive and self-adjoint.) The physical inner product is given by

$$\langle \Psi_1 \mid \Psi_2 \rangle = \sum_{v \in \{|\epsilon|+4n\}} B(v)\,\overline{\Psi}_1(v, \phi_o)\Psi_2(v, \phi_o) \tag{18.6}$$

and is 'conserved', i.e., is independent of the 'instant' ϕ_o chosen in its evaluation. On these states, the Dirac observables act in the expected fashion:

$$\hat{p}_\phi \Psi = -i\hbar\partial_\phi\, \Psi$$
$$\hat{v}|_{\phi_o}\, \Psi(v, \phi) = e^{i\sqrt{\Theta}(\phi-\phi_o)}\, v\, \Psi(v, \phi_o). \tag{18.7}$$

What is the relation of this LQC description to the Wheeler–DeWitt theory? It is straightforward to show (Ashtekar et al. 2006c) that, for $v \gg 1$, there is a precise sense in which the difference operator Θ approaches the Wheeler–DeWitt differential operator $\underline{\Theta}$, given by

$$\underline{\Theta}\Psi(v, \phi) = 12\pi G\, v\partial_v\bigl(v\partial_v\Psi(v, \phi)\bigr). \tag{18.8}$$

Thus, if one ignores the quantum geometry effects, Eq. (18.3) reduces to the Wheeler–DeWitt equation

$$\partial_\phi^2\Psi = -\underline{\Theta}\, \Psi. \tag{18.9}$$

Note that the operator $\underline{\Theta}$ is positive definite and self-adjoint on the Hilbert space $L_s^2(\mathbb{R}, \underline{B}(v)dv)$, where the subscript s denotes the restriction to the symmetric eigenspace of Π and $\underline{B}(v) := Kv^{-1}$ is the limiting form of $B(v)$ for large v. Its eigenfunctions $\underline{e}_k$ with eigenvalue $\omega^2(\geq 0)$ are 2-fold degenerate on this Hilbert space. Therefore, they can be labelled by a real number k:

$$\underline{e}_k(v) := \frac{1}{\sqrt{2\pi}}\, e^{ik \ln |v|} \tag{18.10}$$

where k is related to ω via $\omega = \sqrt{12\pi G}|k|$. They form an orthonormal basis on $L_s^2(\mathbb{R}, \underline{B}(v)dv)$. A 'general' positive frequency solution to (18.9) can be written as

$$\Psi(v, \phi) = \int_{-\infty}^{\infty} dk\, \tilde{\Psi}(k)\, \underline{e}_k(v)e^{i\omega\phi} \tag{18.11}$$

for suitably regular $\tilde{\Psi}(k)$. This expression will enable us to show explicitly that the singularity is not resolved in the Wheeler–DeWitt theory.

With the physical Hilbert space and a complete set of Dirac observables at hand, we can now construct states which are semi-classical at late times—e.g., now—and evolve them 'backward in time' numerically. There are three natural constructions to implement this idea in detail, reflecting the freedom in the notion of semi-classical states. In all cases, the main results are the same (Ashtekar et al. 2006b, 2006c). Here

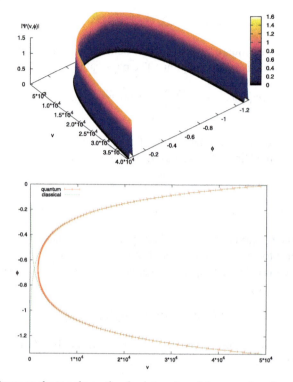

Fig. 18.1. The figure on the top shows the absolute value of the wave function Ψ as a function of ϕ and v. Being a physical state, Ψ is symmetric under $v \to -v$. The figure on the bottom shows the expectation values of Dirac observables $\hat{v}|_\phi$ and their dispersions. They exhibit a quantum bounce which joins the contracting and expanding classical trajectories marked by fainter lines. In this simulation, the parameters in the initial data are: $v^\star = 5 \times 10^4, p_\phi^\star = 5 \times 10^3 \sqrt{G\hbar}$, and $\Delta p_\phi/p_\phi = 0.0025$.

I will report on the results obtained using the strategy that brings out the contrast with the Wheeler–DeWitt theory most sharply.

As noted before, p_ϕ is a constant of motion. For the semi-classical analysis, we are led to choose a large value $p_\phi^\star$ ($\gg \sqrt{G\hbar}$). In the closed model, for example, this condition is necessary to ensure that the universe can expand out to a macroscopic size. Fix a point $(v^\star, \phi_o)$ on the corresponding classical trajectory which starts out at the Big Bang and then expands, choosing $v^\star \gg 1$. We want to construct a state which is peaked at $(v^\star, p_\phi^\star)$ at a 'late initial time' $\phi = \phi_o$ and follow its 'evolution' backward. At 'time' $\phi = \phi_o$, consider then the function

$$\Psi(v, \phi_o) = \int_{-\infty}^{\infty} dk \, \tilde{\Psi}(k) \, \underline{e}_k(v) \, e^{i\omega(\phi_o - \phi^\star)}, \quad \text{with } \tilde{\Psi}(k) = e^{-\frac{(k - k^\star)^2}{2\sigma^2}} \quad (18.12)$$

where $k^\star = -p_\phi^\star/\sqrt{12\pi G\hbar^2}$ and $\phi^\star = -\sqrt{1/12\pi G} \ln(v^\star) + \phi_o$. One can easily evaluate the integral in the approximation $|k^*| \gg 1$ and calculate mean values of the Dirac observables and their fluctuations. One finds that, as required, the state is sharply peaked at values $v^\star, p_\phi^\star$. The above construction is closely related to that of coherent states in non-relativistic quantum mechanics. The main difference is that the observables of interest are not v and its conjugate momentum but rather v and p_ϕ—the momentum conjugate to 'time', i.e., the analogue of the Hamiltonian in non-relativistic quantum mechanics. Now, one can evolve this state backwards using the Wheeler–DeWitt equation (18.9). It follows immediately from the form (18.11) of the general solution to (18.9) and the fact that p_ϕ is large that this state would remain sharply peaked at the chosen classical trajectory and simply follow it into the Big Bang singularity.

In LQC, we can use the restriction of (18.12) to points $v = |\epsilon| + 4n$ as the initial data and evolve it backwards numerically. Now the evolution is qualitatively different (see Figure 18.1). The state does remains sharply peaked at the classical trajectory till the matter density reaches a critical value:

$$\rho_{\text{crit}} = \frac{\sqrt{3}}{16\pi^2\gamma^3 G^2\hbar} , \qquad (18.13)$$

which is about 0.82 times the Planck density. However, *then it bounces*. Rather than following the classical trajectory into the singularity as in the Wheeler–DeWitt theory, the state 'turns around'. What is perhaps most surprising is that it again becomes semi-classical and follows the 'past' portion of a classical trajectory, again with $p_\phi = p_\phi^\star$, which was headed towards the big crunch. Let us summarize the forward evolution of the full quantum state. In the distant past, the state is peaked on a classical, contracting pre-Big Bang branch which closely follows the evolution dictated by Friedmann equations. But when the matter density reaches the Planck regime, quantum geometry effects become significant. Interestingly, they make gravity *repulsive*, not only halting the collapse but turning it around; the quantum state is again peaked on the classical solution now representing the post-Big Bang, expanding universe. Since this behavior is so surprising, a very large number of numerical simulations were performed to ensure that the results are robust and not an artifact of the special choices of initial data or of the numerical methods used to obtain the solution (Ashtekar et al. 2006b, 2006c).

For states which are semi-classical at late times, the numerical evolution in exact LQC can be well modelled by an effective, modified Friedmann equation:

$$\frac{\dot{a}^2}{a^2} = \frac{8\pi G}{3}\,\rho\left[1 - \frac{\rho}{\rho_{\text{crit}}}\right] \qquad (18.14)$$

where, as usual, a is the scale factor. In the limit $\hbar \to 0$, ρ_{crit} diverges and we recover the standard Friedmann equation. Thus, the second term is a genuine quantum correction. Equation (18.14) can also be obtained analytically from (18.3) by a systematic procedure (Ashtekar et al. 2004; Willis 2004). But the approximations involved are only valid well outside the Planck domain. It is therefore surprising that the bounce predicted by the exact quantum equation (18.3) is well approximated by a naive extrapolation of (18.14) across the Planck domain. While there is some understanding of this seemingly 'unreasonable success' of the effective equation (18.14), further work is needed to fully understand this issue.

Finally, let us return to the questions posed in the beginning of this section. In the model, LQC has been able to answer all of them. One can deduce from first principles that classical general relativity is an excellent approximation until very early times, including the onset of inflation in standard scenarios. Yet quantum geometry effects have a profound, global effect on evolution. In particular, the singularity is naturally resolved without any external input and there is also a classical space-time in the pre-Big Bang branch. LQC provides a deterministic evolution which joins the two branches.

18.4 Discussion

Even though there are several open issues in the formulation of full quantum dynamics in LQG, detailed calculations in simple models have provided hints about the general structure. It appears that the most important non-perturbative effects arise from the replacement of the local curvature term F_{ab}^i by non-local holonomies. This non-locality is likely to be a central feature of the full LQG dynamics. In the cosmological model considered in Section 18.3, it is this replacement of curvature by holonomies that is responsible for the subtle but crucial differences between LQC and the Wheeler–DeWitt theory.[5]

By now a number of mini superspaces and a few midi superspaces have been studied in varying degrees of detail. In all cases, the classical, space-like singularities are resolved by quantum geometry *provided one treats the problem non-perturbatively*. For example, in anisotropic mini superspaces, there is a qualitative difference between perturbative and non-perturbative treatments: If anisotropies are treated as perturbations of a background isotropic model, the Big Bang singularity is not resolved, while if one treats the whole problem non-perturbatively, it is (Bojowald et al. 2001).

A qualitative picture that emerges is that the non-perturbative quantum geometry corrections are '*repulsive*'. While they are negligible under normal conditions, they dominate when curvature approaches the Planck scale and halt the collapse that would classically have led to a singularity. In this respect, there is a curious similarity with the situation in the stellar collapse where a new repulsive force comes into play when the core approaches a critical density, halting further collapse and leading to stable white dwarfs and neutron stars. This force, with its origin in the Fermi–Dirac statistics, is *associated with the quantum nature of matter*. However, if

---

[5] Because early presentations emphasized the difference between $B(v)$ of LQC and $\underline{B}(v) = Kv^{-1}$ of the Wheeler–DeWitt theory, there is a misconception in some circles that the difference in quantum dynamics is primarily due to the non-trivial 'inverse volume' operator of LQC. This is not correct. Indeed, in the model considered here, qualitative features of quantum dynamics, including the bounce, remain unaffected if one replaces by hand $B(v)$ with $\underline{B}(v)$ in the LQC evolution equation (18.3).

the total mass of the star is larger than, say five solar masses, classical gravity overwhelms this force. The suggestion from LQC is that a new repulsive force *associated with the quantum nature of geometry* may come into play near Planck density, strong enough to prevent the formation of singularities irrespective of how large the mass is. Since this force is negligible until one enters the Planck regime, predictions of classical relativity on the formation of trapped surfaces and dynamical and isolated horizons would still hold. But assumptions of the standard singularity theorems would be violated. There would be no singularities, no abrupt end to space-time where physics stops. Non-perturbative, background-independent quantum physics would continue.

Returning to the issue of the Beginning, the Big Bang in particular appears to be an artifact of the assumption that the continuum, classical space-time of general relativity should hold at all scales. LQC strongly suggests that this approximation breaks down when the matter reaches Planck density. One might have thought at first that, since this is a tiny portion of space-time, whatever quantum effects there may be would have negligible effect on global properties of space-time and hence almost no bearing on the issue of the Beginning. However, detailed LQC calculations have shown that this intuition may be too naive. The 'tiny portion' may actually be a bridge to another large universe. The physical, quantum space-time could be significantly larger than what general relativity led us to believe. The outstanding open issue is whether this scenario persists when inhomogeneities are adequately incorporated into the analysis.

Note Added in Proof: Since this article was written, LQC has advanced in several directions. The closed, $k = 1$ models as well as anisotropic models were analyzed in detail; the cosmological constant and inflationary potentials have been incorporated; ramifications of effective equations were worked out to show that singularities other than the Big Bang and the Big Crunch are also resolved in isotropic models with perfect fluids; the relation to the path integral approach and spin foams has been investigated; and analysis of quantum fields on cosmological quantum space-times was initiated. For a detailed review of these developments, see (Ashtekar and Singh 2011).

Acknowledgements

I would like to thank Martin Bojowald, Jerzy Lewandowski, and especially Tomasz Pawlowski and Parampreet Singh for collaboration and numerous discussions. This work was supported in part by the NSF grants PHY-0354932 and PHY-0456913, the Alexander von Humboldt Foundation, the Kramers Chair program of the University of Utrecht, and the Eberly research funds of Penn State.

References

Ashtekar, A. 1986. New variables for classical and quantum gravity. *Phys. Rev. Lett.* 57: 2244–2247.

——. 1987. New Hamiltonian formulation of general relativity. *Phys. Rev.* D36: 1587–1602.

——. 1991. *Lectures on non-perturbative canonical gravity.* Notes prepared in collaboration with R. S. Tate, chap. 10. Singapore: World Scientific.

——. 2005. Gravity and the quantum. *New J. Phys.* 7: 198; arXiv:gr-qc/0410054.

Ashtekar, A., Bojowald, M. and Lewandowski, J. 2003. Mathematical structure of loop quantum cosmology. *Adv. Theor. Math. Phys.* 7: 233–268; gr-qc/0304074.

Ashtekar, A., Bojowald, M. and Willis, J. 2004. Corrections to Friedmann equations induced by quantum geometry, IGPG preprint.

Ashtekar, A. and Lewandowski, J. 1994. Representation theory of analytic holonomy algebras, in *Knots and Quantum Gravity*, ed J. Baez. Oxford: Oxford University Press.

——. 1995a. Differential geometry on the space of connections using projective techniques. *J. Geom. Phys.* 17: 191–230.

——. 1995b. Projective techniques and functional integration. *J. Math. Phys.* 36: 2170–2191.

——. 2004. Background independent quantum gravity: A status report. *Class. Quant. Grav.* 21: R53–R152; arXiv:gr-qc/0404018.

Ashtekar, A., Pawlowski, T. and Singh, P. 2006a. Quantum nature of the big bang. *Phys. Rev. Lett.* 96: 141301; arXiv:gr-qc/0602086.

——. 2006b. Quantum nature of the big bang: An analytical and numerical investigation I; arXiv:gr-qc/0604013.

——. 2006c. Quantum nature of the big bang: Improved dynamics; arXiv: gr-qc/0607039.

Ashtekar, A. and Singh, P. 2011. Loop quantum cosmology: A Status Report. *Class. Quantum Grav.* arXiv:1108.0893 (in preparation).

Baez, J. C. 1994. Generalized measures in gauge theory. *Lett. Math. Phys.* 31: 213–223.

——. 1996. Spin networks in non-perturbative quantum gravity, in *The Interface of Knots and Physics*, ed. Kauffman L. Providence: American Mathematical Society, pp. 167–203.

Bojowald, M. 2001. Absence of singularity in loop quantum cosmology. *Phys. Rev. Lett.* 86: 5227–5230; arXiv:gr-qc/0102069.

——. 2002. Isotropic loop quantum cosmology. *Class. Quant. Grav.* 19: 2717–2741; arXiv:gr-qc/0202077.

——. 2005. Loop quantum cosmology. *Liv. Rev. Rel.* 8: 11; arXiv:gr-qc/0601085.

Bojowald, M., Hernandez, H. H. and Morales-Tecotl, H. A. 2001. Perturbative degrees of freedom in loop quantum gravity: Anisotropies. *Class. Quant. Grav.* 18: L117–L127; arXiv:gr-qc/0511058.

Fleishchack, C. 2004. Representations of the Weyl algebra in quantum geometry; arXiv:math-ph/0407006.

Gasperini, M. and Veneziano, G. 2003. The pre-big bang scenario in string cosmology. *Phys. Rep.* 373: 1; arXiv:hep-th/0207130.

Khoury, J., Ovrut, B. A., Steinhardt, P. J. and Turok, N. 2001. The ekpyrotic universe: Colliding branes and the origin of the hot big bang. *Phys. Rev.* D64, 123522, hep-th/0103239.

Khoury, J., Ovrut, B., Seiberg, N., Steinhardt, P. J. and Turok, N. 2002. From big crunch to big bang. *Phys. Rev.* D65, 086007, hep-th/0108187.

Lauscher, O. and Reuter, M. 2005. Asymptotic safety in quantum Einstein gravity: nonperturbative renormalizability and fractal spacetime structure; arXiv: hep-th/0511260.

Lewandowski, J., Okolow, A., Sahlmann, H. and Thiemann, T. 2005. Uniqueness of diffeomorphism invariant states on holonomy flux algebras; arXiv: gr-qc/0504147.

Marolf, D. 1995a. Refined algebraic quantization: Systems with a single constraint; arXives:gr-qc/9508015.

———. 1995b. Quantum observables and recollapsing dynamics. *Class. Quant. Grav.* 12: 1199–1220.

Marolf, D. and Mourão, J. 1995. On the support of the Ashtekar-Lewandowski measure. *Commun. Math. Phys.* 170: 583–606.

Percacci, R. and Perini, D. 2003. Asymptotic safety of gravity coupled to matter. *Phys. Rev.* D68: 044018.

Rovelli, C. 2004. *Quantum Gravity*. Cambridge: Cambridge University Press.

Rovelli, C. and Smolin, L. 1995. Spin networks and quantum gravity. *Phys. Rev.* D52: 5743–5759.

Thiemann, T. 2003. The Phoenix project: Master constraint program for loop quantum gravity; arXiv:gr-qc/0305080.

———. 2007. *Introduction to Modern Canonical Quantum General Relativity.* Cambridge: Cambridge University Press.

Willis, J. 2004. On the low energy ramifications and a mathematical extension of loop quantum gravity. Ph.D. dissertation, The Pennsylvania State University, University Park, PA.